Molecular and Applied Aspects of Oxidative Drug Metabolizing Enzymes

NATO ASI Series

Advanced Science Institutes Series

A series presenting the results of activities sponsored by the NATO Science Committee, which aims at the dissemination of advanced scientific and technological knowledge, with a view to strengthening links between scientific communities.

The series is published by an international board of publishers in conjunction with the NATO Scientific Affairs Division

A	Life Sciences	Kluwer Academic / Plenum Publishers
B	Physics	New York, Boston, Dordrecht, London, Moscow
C	Mathematical and Physical Sciences	Kluwer Academic Publishers Dordrecht, Boston, and London
D	Behavioral and Social Sciences	
E	Applied Sciences	
F	Computer and Systems Sciences	Springer-Verlag
G	Ecological Sciences	Berlin, Heidelberg, New York, London,
H	Cell Biology	Paris, Tokyo, Hong Kong, and Barcelona
I	Global Environmental Change	

PARTNERSHIP SUB-SERIES

1. Disarmament Technologies	Kluwer Academic Publishers
2. Environment	Springer-Verlag
3. High Technology	Kluwer Academic Publishers
4. Science and Technology Policy	Kluwer Academic Publishers
5. Computer Networking	Kluwer Academic Publishers

The Partnership Sub-Series incorporates activities undertaken in collaboration with NATO's Cooperation Partners, the countries of the CIS and Central and Eastern Europe, in Priority Areas of concern to those countries.

Recent Volumes in this Series:

Series A: Life Sciences

Molecular and Applied Aspects of Oxidative Drug Metabolizing Enzymes

Edited by

Emel Arinç
Middle East Technical University
Ankara, Turkey

John B. Schenkman
University of Connecticut Health Center
Farmington, Connecticut

and

Ernest Hodgson
North Carolina State University
Raleigh, North Carolina

Kluwer Academic / Plenum Publishers
New York, Boston, Dordrecht, London, Moscow
Published in cooperation with NATO Scientific Affairs Division

Proceedings of a NATO Advanced Study Institute on
Molecular and Applied Aspects of Oxidative Drug Metabolizing Enzymes,
held August 31 – September 11, 1997,
in Antalya, Turkey

NATO-PCO-DATA BASE

The electronic index to the NATO ASI Series provides full bibliographical references (with keywords and/or abstracts) to about 50,000 contributions from international scientists published in all sections of the NATO ASI Series. Access to the NATO-PCO-DATA BASE is possible via a CD-ROM "NATO Science and Technology Disk" with user-friendly retrieval software in English, French, and German (©WTV GmbH and DATAWARE Technologies, Inc. 1989). The CD-ROM contains the AGARD Aerospace Database.

The CD-ROM can be ordered through any member of the Board of Publishers or through NATO-PCO, Overijse, Belgium.

Library of Congress Cataloging-in-Publication Data

Molecular and applied aspects of oxidative drug metabolizing enzymes /
    edited by Emel Arınç, John B. Schenkman, and Ernest Hodgson.
        p.   cm. -- (NATO ASI series. Series A, Life sciences ; v.
    303)
    Includes bibliographical references and index.
    ISBN 0-306-46048-3
    1. Xenobiotics--Metabolic detoxication--Congresses. 2. Drugs-
-Metabolism--Congresses. 3. Cytochromes--Congresses.
4. Monooxygenases--Congresses. I. Arınç, Emel. II. Schenkman,
John B. III. Hodgson, Ernest, 1932-   . IV. Series.
QP529.M64 1998
572'.791--dc21   98-44331
                                                          CIP

ISBN 0-306-46048-3

© 1999 Kluwer Academic / Plenum Publishers, New York
233 Spring Street, New York, N.Y. 10013

10 9 8 7 6 5 4 3 2 1

A C.I.P. record for this book is available from the Library of Congress.

Printed in the United States of America

PREFACE

The NATO Advanced Study Institute of "Molecular and Applied Aspects of Oxidative Drug Metabolizing Enzymes" was held in Tekirova, Antalya, Turkey, from August 31 to September 11, 1997. This Institute was the third of a series of the NATO ASIs on a similar topic relating to the enzymes of oxidative metabolism of xenobiotics. The first NATO ASI in this series, entitled "Molecular Aspects of Monooxygenases and Bioactivation of Toxic Compounds" (NATO ASI Series A: Life Sciences, Vol. 202), was held in Çesme, Izmir, Turkey, in 1989. The Institute dealt with the potential dangers of drugs, pesticides, pollutants, and carcinogens in the environment resulting from the enzymes of xenobiotic metabolism. The second NATO ASI was entitled "Molecular Aspects of Oxidative Drug Metabolizing Enzymes: Their Significance in Environmental Toxicology, Chemical Carcinogenesis, and Health" (NATO ASI Series H: Cell Biology, Vol. 90). This Institute was held in Kusadasi, Aydin, Turkey, from June 20 to July 2, 1993, and updated and extended the coverage of the first Institute to aquatic species, and delved deeper into the subject of genotoxicity and carcinogenicity. In this third Institute, greater emphasis was put on the human enzymes of oxidative xenobiotic biotransformation, including the flavin-containing monooxygenases and particularly those of the cytochrome P450 family. Considerable effort was made to make the participants aware of the potential dangers of polymorphisms in the enzymes of xenobiotic metabolism, and several of the lecturers addressed this issue.

Today, we are at a crossroads on the information highway. The area of molecular biology has expanded exponentially, with many investigators using new tools to develop methods and techniques for delving into the genetic material of the cell. Others are turning to molecular biology as a means to approach older questions from a different angle: at the level of transcription and translation. Another group of investigators has begun to make use of these methods to study the enzymes of xenobiotic metabolism using the newer tools for the generation of and perturbation of the enzymes. In order to understand newer developments in the field of the oxidative xenobiotic metabolizing enzymes, it is necessary to be well grounded in the methods and procedures currently being used in the studies. In this Institute, efforts were made to put the subject matter into perspective by giving the participants an overview of the various routes that a compound foreign to the body may take after gaining entrance until it is eliminated as a metabolite. Our goal was to provide a ready source of up-to-date information for the student and expert in the field alike, containing sufficient background information for the former to follow the subject matter and the latest information on the subject to satisfy the latter. Hopefully, you, the reader, will find that we have reached this goal.

Emel Arinç
John B. Schenkman
Ernest Hodgson

CONTENTS

THE FATE OF XENOBIOTICS IN THE BODY

Enzymes of Metabolism

John B. Schenkman

Department of Pharmacology
University of Connecticut Health Center
MC-1505
Farmington, Connecticut 06030

1. ROLE OF BIOTRANSFORMATION IN ELIMINATION OF XENOBIOTICS

Xenobiotics are chemicals that are taken up into the body from exogenous sources, either ingested with the food we eat or imbibed with the water we drink, but may also be contained in the air we breathe and absorbed through our skin and lungs. Most of these chemicals are lipophilic, *i.e.*, they have a greater solubility in lipid than in aqueous media. As a result they pose a potential problem in elimination, because the peritubular cells lining the renal tubules all have cell membranes composed of phospholipid and proteins. As the glomerular filtrate passes through the renal tubules, solutes contained in it are either selectively removed along with the water or become concentrated. Compounds with a sufficiently high lipid/water partition coefficient are absorbed down the concentration gradient, dissolving into the cell membranes. From there they pass through the cells back into the body. In this simplistic model the only elimination of the compound would be that present in the excreted water, the urine and this would be at the concentration present in the plasma and glomerular filtrate. Since the fraction of filtered body water eliminated daily is about 0.8% of the total ($k_e = 0.008$ day^{-1}), the half-life of such a compound, *e.g.*, hexobarbital (see Table 1), would be in excess of 87 days,

$$T_{\frac{1}{2}} = \frac{0.693}{k_e}$$

and the agent would retain its pharmacological or toxicological properties. For example, a lipophilic hypnotic like hexobarbital would have a duration in the body of about 1.6 years. Actually, the half-life of hexobarbital is a short 3.7 hrs, the result of active metabolism. Table 1 shows the relationship between the partition coefficients of a series of barbiturates

Molecular and Applied Aspects of Oxidative Drug Metabolizing Enzymes,
edited by Arınç *et al.* Kluwer Academic / Plenum Publishers, New York, 1999.

Table 1. Parameters influencing duration of barbiturate action

Drug	Class	Partition coef. τ(o/w)	Metabolism (nmol/min)	% unchanged in urine
Hexobarbital	ultrashort	7.62	23.4	0
Amobarbital	intermediate	4.85	13.1	0
Phenobarbital	long	1.03	7.4	25-30
Barbital	long	0.15	0.15	70-90

Modified from Jansson, et al. (1972) Arch. Biochem. Biophys., *151*,391.

and their rates of metabolism. Those agents with a high partition coefficient have a shorter duration of action and a smaller proportion of the drug is found unchanged in the urine. Although there appears to be a faster rate of metabolism with increased lipophilicity, no direct correlation exists. Enzymes of xenobiotic metabolism are found throughout the cell, in the cytosol and in the particulate fractions of homogenates of most tissues, with highest levels generally residing in the liver.

2. ROUTES OF XENOBIOTIC METABOLISM

2.1. Phase I and Phase II Metabolism

Perhaps the most important concept to grasp is that pathways of xenobiotic metabolism are similar to the other pathways of intermediary metabolism in living organisms, lipid metabolism, carbohydrate metabolism, protein metabolism and nucleic acid metabolism, in that there are two legs to the pathway. These two legs are a synthetic branch and a degradative branch. Like other synthetic routes of intermediary metabolism, *e.g.,* glycogen formation or nucleic acid formation, that of xenobiotic conjugation is energy consuming. Unlike the degradative pathways of lipid, carbohydrate and protein metabolism, but like nucleic acid degradation, catabolic pathways of xenobiotic metabolism are generally not energy yielding. The two branches have been called Phase I and Phase II, indicating a relationship between them (Figure 1). Lipophilic drugs and chemicals enter the body and are biotransformed into inactive, more polar, readily excretable metabolites (Phase I) by the process of unmasking existing functional groups or by the creation of new functional groups. These may also be conjugated with more polar biochemicals of endogenous origin (Phase II), generally rendering the metabolites inactive and more readily excreted. In many instances inactive agents, called prodrugs, may be administered to individuals; Such chemicals may be eliminated directly, or may be conjugated by Phase II enzymes and eliminated, or may be activated by Phase I drug metabolizing enzymes to the pharmacologically active agent. However, in many instances inert chemicals may also be activated to metabolites that have toxic, carcinogenic, mutagenic, or teratogenic potential. Table 2 outlines some of the most frequently utilized pathways of xenobiotic metabolism, both those of the Phase I and the

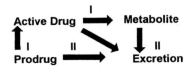

Figure 1. Relationship between Phase I and Phase II metabolic pathways.

Table 2. Outline of pathways of xenobiotic biotransformation

A. Phase I Reactions
  1. Hydrolytic Enzymes
    a) amidases
    b) esterases
    c) epoxide hydrolases
  2. Reductive Enzymes
    a) azo reductases
    b) disulfide reductase
    c) aldo-keto reductases
    d) nitro reductases
    e) reductive dehalogenation
  3. Oxidative Enzymes
    a) flavin-containing monooxygenases
    b) amine oxidases
    c) alcohol and aldehyde dehydrogenases
    d) cytochrome P450 monooxygenases
B. Phase II Enzymes
  1. UDP-glucuronyl transferases
  2. Glutathione transferases
  3. Glycine N-acetyl transferases
  4. Sulfotransferases
  5. Acetyl CoA transferases

Phase II routes. In the first two Phase I pathways, hydrolytic and reductive enzymes function to unmask existing functional groups on xenobiotics. The third group of enzymes function to form new functional groups on xenobiotic molecules. The Phase II enzymes use components from carbohydrate metabolism, protein metabolism fat metabolism and nucleic acid metabolism for complexation of xenobiotic molecules.

2.2. Phase I Reactions: Unmasking or Generating New Functional Groups

2.2.1. Hydrolytic Reactions. The hydrolytic reactions all involve the nucleophilic attack of water at an electrophilic carbon atom with the cleavage of an oxygen or nitrogen bond to the carbonyl carbon and its replacement by a water hydroxyl group. The reaction can be described by the equation:

$$\tag{1}$$

An ester, such as acetylcholine or aspirin, will be converted to a carboxylic acid (R^1) plus an alcohol (R^2). In contrast, an amide, like procainamide, will be converted to its R^1-carboxylic acid and its R^2-amine. There exists a very large number of amidases and esterases, and they are found in just about every tissue, cellular compartment and also in the plasma. An early review of these enzymes describes many of the properties.[1] The kinetic properties of the enzymes will depend upon the length of both the acyl residue and the alkyl residue.[2] Amidases generally will also hydrolyze esters and esterases will usually also hydrolyze some amides, and both often will also hydrolyze thioesters.[3] In addition, some esterases exist (A-esterases which can hydrolyze phosphate esters.[4] Hepatic endoplasmic reticulum itself has at least eight specifically different amidase/esterases.[3]

The epoxide hydrolases, which are discussed in a later chapter, differ, since they catalyze an SN2 attack by the nucleophilic water at a side opposite from a strained epoxide ring. This results in stereospecific formation of a vicinal diol in the trans configuration. Steric hindrance will influence the site of attack by the epoxide hydrolases, *i.e,* at R^1 or at R^2. The epoxide hydrolase is located mainly in the endoplasmic reticulum of cells, but cytosolic and mitochondrial forms have also been described.[5] The endoplasmic reticulum enzyme is inducible, responding to challenge by xenobiotics.[6,7]

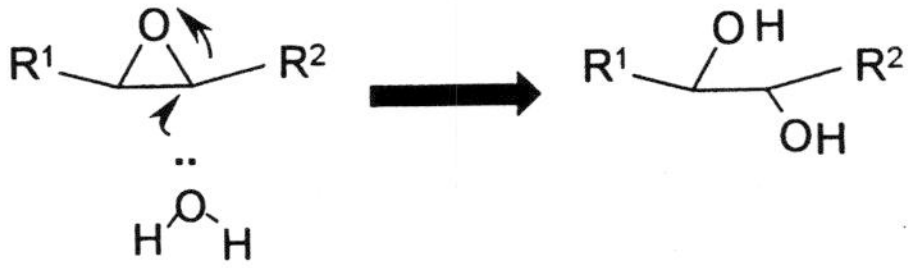

$$(2)$$

2.2.2. Reductive Reactions. <u>2.2.2.1. Azo Reductase.</u> Reductive reactions include two that are reductive cleavages and three that are direct reductions. Azo group reductions to amino compounds and disulfide reductions to thiols are similar in that they involve reductive cleavage (Figure 2), thereby unmasking amino and thiol groups. The azo group reductive cleavage mainly involves microsomal enzymes, NADPH-cytochrome P450 reductase and cytochrome P450.[8] However, large number of enzymes appear to be capable of catalyzing and contributing to the *in* vivo azo reduction of compounds, from microsomal NADPH-cytochrome P450 reductase and cytochrome P450, to cytosolic proteins like xanthine oxidase and DT-diaphorase. The reactions probably take place in oxygen-poor regions of cells and tissue, as demonstration of the microsomal activity requires anaerobic conditions.[9] Amaranth azo reductase activity is elevated by inducers of cytochrome P450, and carbon monoxide inhibits this P450-mediated azo reductase activity.[10] However, activity can be stimulated by addition of FMN or FAD, and such enhanced azoreductase activity is not CO sensitive.[9] Suggestions

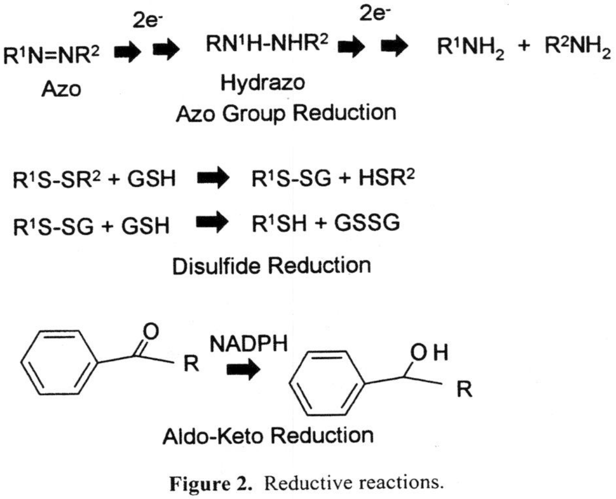

Figure 2. Reductive reactions.

have been made that an azo anion radical is an intermediate in the reduction and this is the oxygen sensitive species. Hydrazo metabolites (2 electron reduced) are also produced, and may be the result of disproportionations between azo anions.

2.2.2.2. Thioltransferase. Reductive cleavage of disulfides and thiosulfate esters appears to be the work of thioltransferases (glutathione:disulfide oxidoreductase) that use glutathione (GSH) as a co-reactant, reducing a disulfide to a thiol (R^2SH) and mixed disulfide with GSH (Figure 2). The resultant mixed disulfide then reacts with a second GSH to form oxidized glutathione and the thiol (R^1SH) of the second half of the disulfide.[11] In this manner drugs like disulfiram are inactivated. The reaction proceeds in two steps and results in formation of oxidized glutathione, the regeneration of which requires NADPH. Mixed small molecule disulfides, protein disulfides and mixed protein small molecule disulfides are all substrates of the different thioltransferases, also called glutaredoxins. Low molecular weight, cytosolic thioltransferases have been purified,[12] and at least one membranous thiol-disulfide exchange enzyme has been isolated which has also been called an insulin thioltransferase. These latter enzymes are thought to function in the repair of protein thiols to their natured state after oxidation by thiols and thiosulfates.

2.2.2.3. Aldo-Keto Reductases. Aldehyde- and ketone-containing chemicals are widely distributed in nature and many have diverse effects in the body. They are metabolized *in vivo* by a family of functional group-specific enzymes that recognize both endogenous and exogenous aldehydes and ketones.[13] Although the physiological roles of these enzymes remains to be established, many are thought to have developed as a means to detoxify reactive aldehydes. Substrates include many aromatic aldehydes and ketones (Figure 2), which are reduced by NADPH-dependent enzymes. Examples of xenobiotic substrates include the anticancer anthracyclines daunorubicin and doxyrubicin. The aldehyde sidechains of these are reduced by aldo-keto reductases. Similarly, the anti-clotting agent warfarin is a substrate for aldo-keto reductases. The enzymes are found in the cytosol of many tissue cells in the body. There is a superfamily of aldo-keto reductases (AKR), and a new nomenclature has just been devised for them.[14] The enzymes are monomeric proteins of 30–40 kDa that bind NADPH and have an extremely wide range of aliphatic and aromatic substrates. At least 39 different genes have been identified in rat, and these may be divided into seven different families, based upon amino acid sequence.[14] At least seven different forms have been identified in man and a similar number have been found in rat. The highest levels are found in liver and in kidney cortex, but enzymes are found in just about every tissue.[15] The enzymes have an interesting mechanism involving first binding of NADPH, followed by an isomerization of the enzyme. The enzyme is then able to accept a substrate and reduce it. Release of the reduction product is followed by isomerization of the enzyme again, and release of the oxidized pyridine nucleotide.[15]

2.2.2.4. Nitro Reductases. Among the reductive reactions that occur in the body are nitro group reductions. These are carried out by a very diverse group of enzymes in the body, including xanthine oxidase, several reductases, including NADPH-cytochrome P450 reductase, cytosolic dicumarol-sensitive NAD(P)H:quinone reductase, and cytochrome P450. The reaction can be viewed as a series of 1-electron transfers to the nitro group, yielding a nitro anion radical (Figure 3), which has been detected by EPR spectroscopy, then another 1 electron reduction to the nitroso intermediate. A third and a 4th -electron reduce the metabolite to a hydronitroxide anion and then to hydroxylamine, respectively. A final two electrons form the 6-electron reduced amino compound.[16] In 1968, Gillette's

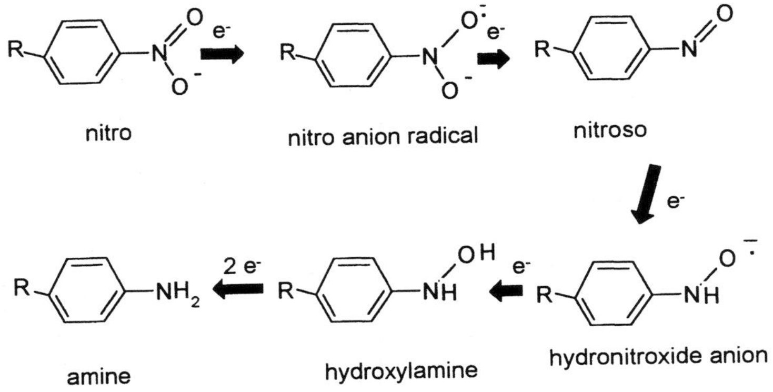

Figure 3. Nitro group reduction.

group isolated the nitroso and the hydroxylamine intermediates of nitrobenzene during formation of aminobenzoate from nitrobenzenzoate. A number of nitro compounds are drugs, or agricultural reagents such as parathion. Recently two additional cytosolic hepatic nitroreductases have been identified, one an NADH-dependent, dicumarol-insensitive enzyme, and the other a pyridine-nucleotide non-specific, dicumarol-sensitive enzyme (quinone reductase; DT-diaphorase).[17] In fact, it was postulated that oxygen and the nitro group compete for binding to cytochrome P450. Early studies showed that with some compounds (not all) carbon monoxide was actually inhibitory of nitroreductase activity. As with azo reductase activity, it is only possible to carry out nitro group reduction under conditions of strict anaerobiosis, and it is suggested that the rate-limiting, oxygen sensitive step is formation of the nitro anion radical, and that other steps may not all be oxygen sensitive. This would also mean that, since reduced metabolites are found, they are probably produced in oxygen poor regions in the body tissues. Presently, reductive activation of nitrogen compounds to metabolites that bind to cellular DNA is being studied. Indications are that many such reactive intermediates are produced in the gut by nitro reductases of gut bacteria. Studies are also in progress[18] which identify oxygen poor regions of tissue by demonstration of nitro reductase activities (binding to tissue of labeled nitroimidizoles). To the extent that a nitro chemical may be metabolized by a form of cytochrome P450, nitroreductase activity will be elevated by compounds that induce that form of cytochrome P450. Drugs such as chloramphenicol and clonazepam are substrates of nitro reductases, based upon metabolites excreted in the urine. Intestinal microorganism also contain nitroreductases, and these organisms may play a role both in detoxifying xenobiotics, and in susceptibility of their hosts to toxic and carcinogenic effects of these agents.

2.2.2.5. _Reductive Dehalogenation._ It has been known since the 1940's that the mammalian system has the ability to cleave the carbon halogen bond, releasing halides. Halogenated anesthetics such as chloroform and halothane were once thought to be inert in the body. However, subsequent studies revealed a small proportion of these underwent dehalogenation. For example, the gaseous anesthetic halothane is dehalogenated, with some trifluoroacetic acid appearing in the urine. The reaction is catalyzed by cytochrome P450. The reaction (Figure 4) proceeds stepwise initially releasing bromide.[19] Two one-electron transfers to the halothane results in removal of bromide and chloride. In the absence of sufficient levels of oxygen reductive metabolites react with macromolecules such

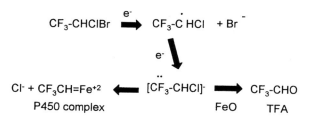

Figure 4. Reductive dehalogenation of Halothane.

as membrane phospholipids, or with cytochrome P450 (Figure 4), in the latter case forming a stable, reduced, inactive, spectrophotometrically visible complex.[19,20] Not all isozymes of cytochrome P450 form the spectrophotometrically observable complex with halothane metabolites, indicating non-reactive forms either do not metabolize this halocarbon or produce different metabolites. In the presence of oxygen the reaction is inhibited due to competition of oxygen for electrons. Carbon monoxide can also inhibit the reaction in microsomes, by stabilizing the reduced hemoprotein. At low oxygen tensions, trifluoracetaldehyde (TFA) can form after elimination of chloride (Figure 4) which then is oxidized by other enzymes in the body to the trifluoroacetic acid found in the urine. Further, studies using cytochrome $P450_{cam}$ have shown a 2-electron reduction, a 1:1 stoichiometry of NADH consumption and substrate consumption, yielding $ClCH_2CF_3$. With haloethanes the two electron transfer yielded haloalkenes, due to concomitant reduction and elimination of two vicinal halogens.[21] With $CFCl_3$ elimination of a chloride and formation of a proposed carbene marked the 2-electron transfer. This intermediate then releases a second chloride and undergoes hydrolysis to carbon monoxide, HF and HCl. Studies on the reductive dehalogenations of haloalkanes have indicated that the order of decreasing activity is I>Br>Cl>F. The reverse order appears to hold for halogenated anilines.[22] Formation of p-aminophenol from the 4-haloanilines had greatest V_{max} values with fluorine and lowest values with iodine.[22] This dehalogenation was suggested to be "oxidative dehalogenation", i.e., the result of attack by active oxygen. A likely mechanism might involve ionization of the halogen after formation of the imine, followed by binding of the active oxygen atom to the C4 carbon, yielding the quinoneimine. This would yield the aminophenol after subsequent reduction.

2.2.3. Oxidative Enzymes. Oxidative enzymes catalyze reactions that result in the appearance of new functional groups containing oxygen. A very wide number of diverse oxidative enzyme families are involved in the metabolism of drugs and other xenobiotics.

<u>2.2.3.1. Flavin-Containing Monooxygenase.</u> Flavin-containing monooxygenase (FMO) is a multigene family of enzymes that metabolize drugs and chemicals with nucleophilic heteroatom groups such as sulfur, nitrogen, phosphorus, or selenium[23] to their respective oxides. Attack does not occur on carbon atoms. Primary, secondary and tertiary amines are substrates of the flavin-containing monooxygenases (Table 3) and products of nitrogen attacks are amine oxides, hydroxylamines and oximes (Figure 5). With sulfur-containing substrates sulfoxides and sulfones are produced. Reaction with phosphines also occurs, as with phonophos or triphenylphosphine.[24] The gene family has almost as wide a range of substrates as the cytochrome P450 gene family. As with the cytochrome P450 enzymes, there appears to be a gender-specific expression of FMOs in rodents, and develop-

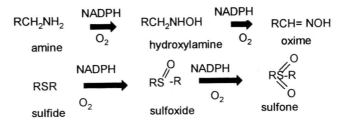

Figure 5. Substrate oxidation by flavin-containing monooxygenase.

mental changes in expression.[25] Human fetal liver expresses FMO1, while in adult liver FMO3 is the predominant form and a small amount of FMO4 is detected. FMO1 is, however, produced in the adult human kidney.[23] Unlike the cytochrome P450 enzymes, however, the enzymes are constitutively expressed and do not appear to be inducible by xenobiotics. The members of this family are membrane proteins, located in the endoplasmic reticulum of tissue, particularly in the liver and kidney and are markedly thermally labile. Five gene families have been identified in rabbit (FMO1-5) with amino acid sequences 50–58% identical between families.[25] Mammalian orthologous forms in another species are assigned based upon sequence identity of greater than 80% with the orthologous rabbit gene product.[23] It appears that each gene family has a single gene. Some forms are tissue specific. The enzymes exist in cells as a stable activated complex, formed by interaction with NADPH and oxygen.

Flavin monooxygenases have as a prosthetic a single FAD molecule. The enzyme binds NADPH (Figure 6) which is oxidized by the FAD prosthetic group. The $FADH_2$ then reacts with molecular oxygen at the 4α-position of the flavin. The activated oxygen form, the hydroperoxy flavin enzyme, is a stable entity in the absence of substrate, but which, in its presence, reacts with the substrate at a nucleophilic center releasing water, the oxidized substrate and the NADP. As noted by Beaty and Ballou,[26] the system is tightly coupled.

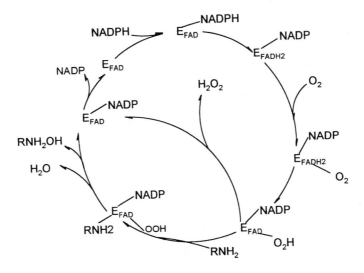

Figure 6. The flavin-containing monooxygenase cycle.

Table 3. Substrates of flavin-containing monooxygenases

Nitrogen containing	Examples	Sulfur containing	Examples	Phosphorus containing	Examples
1° amines	N-octylamine	thiols,	dithiothreitol, thioridazine	Phosphines	Diethylphenyl-phosphine
acyclic 3°-amines	N-methylaniline chlorpromazine, tamoxifen	Mercaptopurines Thiocarbamides	6-thioguanine thiourea, phenylthiourea	Phosphonothioate	Fonofos
cyclic amines hydroxylamines	nicotine N-hydroxyamino-azobenzene	Thioamides sulfides, disulfides	thioacetamide dimethylsulfide, phorate, butyl disulfide		

Only minute amounts of H_2O_2 leak away from the enzyme under normal condition.[25] The complex will interact with a substrate (Figure 6), inserting oxygen at the appropriate site. A wide variety of compounds, from monoamine oxidase inhibitors to antihistamines, to pesticides, are substrates for this family of enzymes (Table 3). All appear to be soft nucleophiles. Several forms of cytochrome P450 produce the same products with some drugs as flavin monooxygenase. Some of these, such as mercaptopyrimidines and thiocarbamides, may generate more reactive (toxic) intermediates. FMOs are also important in metabolism of endogenous substrates and compounds generated by the gut bacteria. Recent studies on "fish odor" have indicated a genetic deficiency in FMO prevents the N-oxidation of trimethylamine, which, when secreted in sweat and urine produces a fishy aroma.[25]

2.2.3.2.Amine Oxidases. Monoamine oxidase (MAO), is present in the outer membrane of mitochondria of presynaptic nerve terminals and neurons as well as being widely distributed in the body tissues. Highest levels are found in liver, kidney and intestine. It functions in the deamination of a number of biogenic amines, such as epinephrine and norepinephrine that are not reutilized. In the intestines monoamines taken up from the gut also serve as substrates. A large number of aromatic amines are substrates for the monoamine oxidase. All substrates have the amino group attached to an unsubstituted methylene group, e.g., benzylamine. Aniline is not a substrate. The enzyme is a flavoprotein, containing FAD as the prosthetic group. The flavin oxidizes the amino group to an imine, and this subsequently hydrolyzes to an aldehyde and ammonia. Oxygen then regenerates the oxidized enzyme, producing hydrogen peroxide. At present two forms of MAO are known, A and B forms, based upon the differential action of different inhibitors.[27]

$$RCH_2NH_2 + E_{FAD} \longrightarrow [RCH=NH + E_{FADH2}] \longrightarrow RCH_2CHO + NH_3$$

$$E_{FADH2} + O_2 \longrightarrow E_{FAD} + H_2O_2 \tag{3}$$

Diamine oxidases (DAO) are similar to monoamine oxidase in ability to oxidize several naturally occurring diamines like histamine and putricine, as well as polyamines. They are cytosolic enzymes found in plasma and in most tissue, especially the intestine. DAOs are copper-containing enzymes and at least some have recently been reported to utilize a quinone cofactor, oxidized 2,4,5-trihydroxyphenylalanine (topaquinone) as a prosthetic group. The C5 carbonyl of the quinone is believed to be the site of binding of the substrate amine.[28] The indication is that the copper of the enzyme serves to stabilize the topaquinone by interacting with one of its carbonyl groups and does not undergo redox changes. In the absence of molecular oxygen aldehyde is released, but ammonia is not.

$$E_Q Cu^{++} + NH_2CH_2CH_2NH_2 \longrightarrow E_{QH2}HN^+=CHCH_2NH_2$$

$$E_{QH2}HN^+=CHCH_2NH_2 + H_2O \longrightarrow E_{QH2}N^+H_3 + O=CHCH_2NH_2$$

$$E_{QH2}N^+H_3 + O_2 \longrightarrow E_Q Cu^{++} + H_2O_2 + NH_3 \qquad (4)$$

2.2.3.3. Alcohol and Aldehyde Dehydrogenases. Two families of alcohol dehydrogenases exist in nature, the medium-chain and the short chain dehydrogenase/reductase family. These are dimeric proteins. The short chain family members have shorter subunits. Hundreds of enzymes constitute the two families, each of which can be divided into two sublines on the basis of reductase activity or dehydrogenase activity. The individual enzymes differ with respect to substrates. There are seven mammalian medium chain alcohol dehydrogenases subdivided into six classes, of which the Class I is the typical hepatic enzyme. All of the medium-chain alcohol dehydrogenases are dimeric zinc-containing enzymes. Three genes, α (gene 1), β (gene 2), and γ (gene 3), are members of Class I, explaining the different isozymes of hepatic alcohol dehydrogenase.[29] Genetic polymorphisms have been seen in human and horse alcohol dehydrogenases as well as in human aldehyde dehydrogenase, accounting for the great variability between individuals in the handling of alcohol. More than a dozen and a half forms of alcohol dehydrogenase are known. The human hepatic dimeric Class I enzyme, ADH_2, may be composed, for example, of one of three polymorphic alleles, β_1, β_2, or β_3. The β_2 allele has much higher activity than either of the other allelic forms; In addition, alleles γ_1 and γ_2 exist. Ethanol is metabolized in the body fairly specifically by the Class I alcohol dehydrogenases, NAD-utilizing enzymes that also metabolizes methanol and a number of short-chain, branched-chain and aromatic alcohols. The product of the reaction with ethanol is acetaldehyde. Aldehyde dehydrogenase oxidizes the product of ethanol oxidation to acetic acid, and of methanol to the toxic formic acid. Metabolism is at a constant rate, *i.e.,* proceeds by zero order reaction due to saturation of the enzyme. The availability of the oxidized pyridine nucleotide, NAD, is the limiting factor for this enzyme. The Class IV enzyme, ADH6 gene product, has even higher ethanol oxidizing activity than the Class I enzyme. The gene has hormone responsive elements and is expressed primarily in epithelial cells, especially in the stomach where it receives first contact with imbibed ethanol.[30] It also has high retinol dehydrogenase activity, aiding in retinoic acid formation., and the ability to oxidize products of lipid peroxidation.[29] ADH6 reportedly is responsible for a significant amount of the orally consumed ethanol in men but not women, exhibiting a gender-selective extent of expression that probably contributes to the lower alcohol tolerance of females.[30]

Cytochrome P450 forms, too, will oxidize alcohols, consuming NADPH in the process, yielding the same aldehyde product. One form, CYP2E1, is more active in this regard than other forms of P450, and this enzyme is also induced by chronic ethanol use. However, even when maximally induced its extent of ethanol metabolism is a small fraction of that of the alcohol dehydrogenase. Several aldehyde dehydrogenase isozymes have been identified. Two of these, the cytosolic $ALDH_1$ and the mitochondrial $ALDH_2$ are considered to be the major contributors to acetaldehyde metabolism. Polymorphisms are also known for aldehyde dehydrogenase. About half of all Orientals lack $ALDH_2$ in their livers due to point mutations.[31] The variant is negligibly low in Caucasians, and no variants of $ALDH_1$ have been reported. Another aldehyde dehydrogenase, $ALDH_3$ is expressed in stomach and hepatoma cells, but is not found in liver.[31]

2.2.3.4. Cytochrome P450 Monooxygenase. The cytochrome P450 monooxygenases are a very large family of enzymes present in most tissues in the body. This enzyme system is perhaps one of the most important of the xenobiotic metabolizing enzymes in that it has perhaps the greatest spectrum of substrates, overlapping that of most of the other metabolizing enzymes. As an indication of its importance cytochrome P450 forms are found in just about every phylum in which it has been sought. They are heme containing enzymes that utilize reducing equivalents from a reductase, NADPH-cytochrome P450 reductase, to sequentially reduce dioxygen to a reactive species that attacks lipophilic substrates. The active center of the enzyme contains a heme prosthetic group with coordinated iron in its center. The iron undergoes reduction and oxidation in the course of the catalytic cycle. The name of the enzyme is derived from the absorption peak of the reduced, carbon monoxide complexed enzyme.[32] Figure 7 shows the cytochrome P450 monooxygenase cycle. Substrate binds to the hemoprotein and shifts the enzyme to the high spin configuration, which is more receptive to reduction (see review [33]). The enzyme then accepts an electron from NADPH via NADPH-cytochrome P450 reductase. Oxygen binds to the ferrous hemoprotein forming the oxycytochrome P450, which then accepts a second electron from the reductase, or from ferrous cytochrome b_5. This raises the oxygen to the redox level of peroxide, and the heme iron-bound oxygen then disproportionates, releasing a molecule of water. The remaining atom of oxygen is thought to be bound to the iron is what has been called a perferryl complex, a very powerful oxidant. This is suggested to abstract a hydrogen atom from the substrate held in juxtaposition by the enzyme, yielding a substrate radical and a hydroxyl radical, which react to form the oxidized substrate, which is subsequently released from the enzyme.[34]

Cytochrome P450 is probably one of the more versatile of the xenobiotic metabolizing enzymes in the body. Some isozymes are present in the mitochondria, functioning in steroidogenesis. Other forms are located in the endoplasmic reticulum of most tissue, some functioning in steroidogenesis, others functioning in nitric oxide synthetase, and still others in formation of active messenger production from arachidonate.[35] Over 30 forms are known in rat, and it is estimated that a similar number is present in tissues of humans. However, most forms are involved in the catabolic metabolism of drugs and other xenobi-

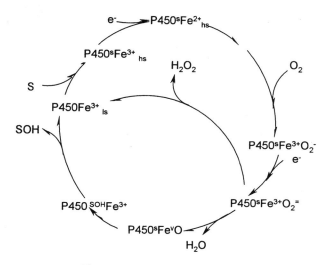

Figure 7. Cytochrome P450 cycle.

otics.[36] What makes these enzymes so versatile is their ability to accept as substrate molecules as small as ethanol and as large as the polycyclic hydrocarbon benzo(a)pyrene or the antibiotic erythromycin. There does not seem to be any structural requirements for substrates, the only similarity is that they appear to be somewhat lipophilic.

Different forms of cytochrome P450 are present in the newborn animal and these change as the animal proceeds from the neonate state through puberty toward adulthood.[37] Studies over the past few years have indicated that these enzymes are under hormonal[38,39] control as well as respond to various external stimuli, such as nutritional status and challenge by external chemicals. Levels of individual forms may be changed by a factor of 20 or higher, and forms not normally present, such as CYP1A1, may appear when challenged by aromatic hydrocarbons or halogenated hydrocarbons.[40]

Cytochrome P450 can catalyze a very wide number of oxidations on the very large number of xenobiotics that serve as substrates (Table 4). In most of the reactions the site of attack is on a carbon atom,[41] even in the case of the oxidative deamination reaction.[42] Attack on carbon atoms will produce primary, secondary and tertiary alcohols in a reaction that may be depicted as abstraction of a hydrogen atom by iron-bound oxygen atom:

$$Fe^{3+}O_2^= + 2H^+ \longrightarrow Fe^VO + HOH$$
$$Fe^VO + RH \longrightarrow Fe^{3+}OH^\cdot + R^\cdot \longrightarrow ROH$$

(5)

Table 4. Types of cytochrome P450-mediated biotransformation reactions

Reaction Type	Reaction	Examples
Aliphatic hydroxylation	R-H → R-OH	pentobarbital, phenylbutazone
Aromatic hydroxylation!	Ar-H → Ar-OH	coumarins, phenytoin
Epoxidation (RH, ArH)	R-CH=CH-R' → R-CH - CH-R' (with O bridging, epoxide)	styrene, aflatoxins, benzo(a)pyrene
N-dealkylation	R-NH-CH₂R' → R-NH₂ + R'CHO	diazepam, methadone, erythromycin
O-dealkylation	R-O-CH₂R' → R-OH + R'CHO	codeine, ethylmorphine
S-dealkylation	R-S-CH₂R' → R-SH + R'CHO	6-methylthiopurine, azathioprine
Oxidative deamination	R-CH-CH₂R' (NH₂) → R-C-CH₂R' (O)	amphetamine, diazepam
Oxidative desulfuration	R-C-CH₂R' (S) → R-C-CH₂R' (O)	chlorpromazine, thiopental, parathione
N-hydroxylation	R-N-R' (H) → R-N-R' (OH)	amphetamine, aniline
Sulfoxidation	R-S-R' → R-S-R' (O)	chlorpromazine, 6-thiopurine, thioguanine
Dual hydrogen abstraction	R-CH-R' (OH) → R-C-R' (O)	testosterone
Reductive dehalogenation	R-CH-R' (X) → R-C·H-R' + X⁻	halothane, carbon tetrachloride

Table 5. Energy utilization in phase II reactions

Reaction type	Reaction
Glucuronidation	glucose 1P + UTP → UDP-glucose (UDPG) + pyrophosphate (PPi)
	UDPG + 2 NAD$^+$ + H$_2$O → UDP-glucuronic acid (UDPGA)
Amidation (glycine; glutamine)	ATP + R-COOH + CoA → R-CO~SCoA + AMP + PPi
Glutathione conjugation	L-cysteine + L-glutamate + ATP → γ-glutamylcysteine
	γ-glu-cys + glycine + ATP → γ-glutamylcysteinylglycine (GSH)
Sulfation	SO$_4^=$ + ATP → Adenosine 5′-phosphosulfate (APS)
	ATP + APS → 3′-phospho-APS (PAPS)
Acetylation	CH$_3$-CO-OH + CoA + ATP → CH$_3$-CO~SCoA + AMP + PPi
Methylation	ATP + methionine → S-adenosylmethionine (SAM) + PPi + Pi

Equation 5 depicts the way hydrogen abstraction may be visualized. The heme iron-bound oxygen atom would be a perferryl complex, the oxygen stabilized by electrons of the iron. The oxygen is shown as pulling a hydrogen atom from the substrate, forming a hydroxyl radical and a substrate radical which combine as the alcohol product enzyme.[34] In the case of heteroatom oxidation, such as N-oxidation and S-oxidations (Table 4), analogous attack on these atoms may similarly be visualized, just as with flavin-containing monooxygenase. Two reactions are different, that of dual hydrogen abstraction and reductive dehalogenation, since these do not appear to involve the insertion of oxygen into the substrate. For example, in the formation of androstenedione from testosterone[43] lack of oxygen incorporation into the product was demonstrated using ^{18}O$_2$. Also, formation of 17β-hydroxy-4,6-androstadiene-3-one from testosterone was reported, and it was suggested that this, too, involved dual hydrogen abstraction, rather than oxygen insertion and dehydration.[44] In the case of the dehalogenation reactions, as indicated above, the reactions are oxygen-sensitive, and it appears that the dehalogenation proceeds directly via electron input to the molecule from the cytochrome P450.

2.3. Phase II Reactions: Conjugation of Xenobiotics

The different conjugation pathways of metabolism all require energy to activate molecules for complexation with xenobiotic drugs and chemicals (Table 5). In all instances the xenobiotic is complexed with some endogenous compound derived from the pathways of carbohydrate metabolism (UDPGA, acetyl CoA), protein metabolism (glycine, glutamine, methionine, and glutathione), and fat metabolism (acetyl CoA). In most instances the acti-

Table 6. Conjugatable functional groups

Functional group	Reaction
Hydroxyl **-OH**	glucuronidation
	sulfation
	methylation
Carboxyl **-COOH**	glucuronidation
	glycine amidation
	glutamine amidation
Amino **-NH$_2$**	acetylation
Sulfhydryl **-SH**	methylation
	glucuronidation
	sulfation
Heterocyclic amine **NH, N**	methylation

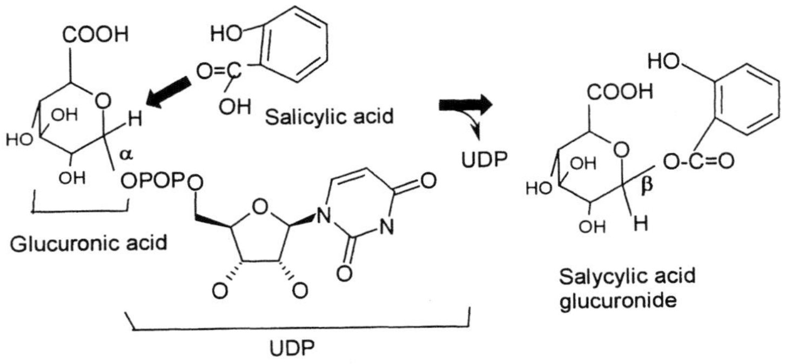

Figure 8. Mechanism of glucuronidation.

vated compound is the endogenous one. However, with the amidation reactions, hippurate and glutaminate formation, it is the xenobiotic that is activated. The types of reactions carried out on a drug or other xenobiotic depend upon the functional groups it possesses. The more functional groups on a molecule the greater the number of potential metabolites that may be produced *in vivo*. For any one functional group a number of potential reactions may occur (Table 6). The predominant reaction is usually that which proceeds at the fastest rate.

2.3.1. Glucuronidations. Perhaps the major reaction used by the body in facilitating excretion of drugs and chemicals is the glucuronidation pathway. This pathway is responsible for the conjugation of nucleophilic compounds and metabolites to more polar metabolites for excretion. The reaction uses activated glucuronic acid, UDP-glucuronic acid, and the enzyme UDP-glucuronyl transferase (UDPGT), a microsomal enzyme, and couples this with phenols, carboxylic acids, aromatic amines, and N-oxidized metabolites. Many substrates exhibit stereoselectivity in metabolite formation; For example, (+)-morphine is conjugated preferentially at the 6-hydroxyl while (-)- morphine is conjugated at the 3-hydroxyl. Although found in most tissue, the major sites of the enzymes are the liver, intestinal mucosa and the kidney.[45] The enzymes are glycoproteins located on the luminal side of the endoplasmic reticulum. Most conjugated metabolites are biologically inactive, perhaps because the pK$_a$ of the glucuronic acid moiety is so low that the molecule is ionized and doesn't get taken up into cells readily. However, some glucuronides are reactive intermediates and carcinogenic. For example, naphthylamine glucuronides are secreted in the urine, and are bladder carcinogens as a result of cleavage of the glucuronide in the acid pH. Two gene families (UDPGT1 and UDPGT2) have been identified, based upon primary structure, as well as substrates conjugated. One gene family contains the phenol and bilirubin-conjugating forms, while the other family contains the steroid-conjugating forms. Ten different UDP-glucuronyl transferases have been identified in humans, and these exhibit different, but overlapping. substrate preferences. The enzymes are named UDPGT and given a Roman number indicating family 1 or 2. Family 1 members, which contain phenol and bilirubin-metaboizing forms, are given a * and Roman number indicating gene number. For example, the planar phenol UDPGT is UDPGT1*6. This enzyme conjugates acetaminophen, while the complex phenol UDPGT, UDPGT1*02, conjugates propranolol, ethinylestradiol and other complex phenols, and UDPGT1*1 metabolizes ethinylestradiol and bilirubin. The UDPGT1 is a gene complex that contains 6 different exon 1 components that are alternatively spliced to common exons 2–5.[46] Family 2 members are the steroid-metabolizing forms and these are given the letter B and a

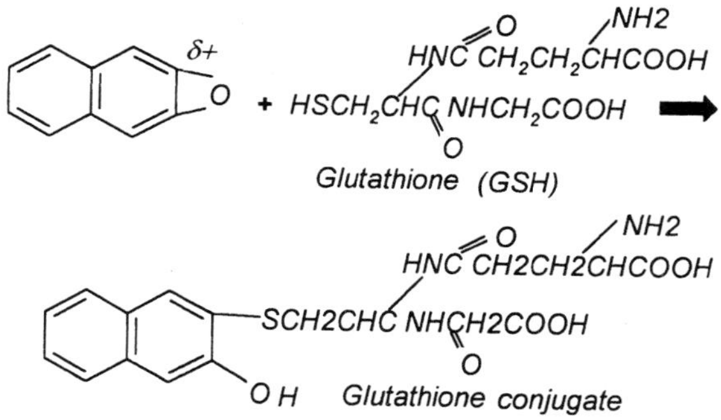

Figure 9. Glutathionine conjugation of naphthalene oxide.

number. For example, UDPGT2B8 conjugates estradiol and 1-naphthylamine, and UDPGT2B4 conjugates estriol, and some hyodeoxycholic acid, and some conjugate drugs like clofibrate and valproate.[47] UDPGTs are inducible enzymes, with different inducers that differentially elevate the cytochrome P450 enzymes also elevating different forms of UDPGT. Reactions catalyzed by the UDPGTs include O-glucuronide, N-glucuronide and S-glucuronide formation (Table 5). Substrates of the different forms range from the endogenous substrates bilirubin, 5-hydroxytryptamine, and estrone, to drugs like morphine, chloramphenicol and acetaminophen, to chemicals like benzo(a)pyrene quinol and 4-hydroxybiphenyl. The reaction catalyzed by the enzyme is an SN2 type of nucleophilic attack by the substrate on the UDP-glucuronic acid (Figure 8). Genetic deficiencies in bilirubin glucuronidation is seen occasionally in infants (Crigler-Najjar Syndrome), and can result in severe kernicterus [47] Another disease, Gilbert Syndrome, also involves the conjugation of bilirubin to the diglucuronide. In the fetus and in newborns the levels of many of the drug metabolizing enzymes are low, including the forms of UDPGT that metabolizes chloramphenicol. Low levels of this enzyme put newborns and infants at risk of chloramphenicol toxicity, since they depend on this enzyme as the main mechanism for elimination of the antibiotic. Gray baby syndrome, a manifestation of chloramphenicol toxicity is the direct result of inability to metabolize the drug.

2.3.2. Glutathione Transferases. The *in vivo* metabolism of a number of different drugs and chemicals results in the appearance in the urine of mercapturic acids, N-acetyl cysteine conjugates of metabolites of these xenobiotics. Production of these metabolites involves several enzymes working in series, the first of which is one or another member of a large gene family of cytosolic proteins, the glutathione S-transferases. The glutathione transferases (GST) catalyze the nucleophilic addition of glutathione to substrates that have electrophilic functional groups or atoms, chemicals that could act as alkylating agents in cells. Four different gene classes of GST exist, designated alpha, mu, pi and theta (α, μ, π, and θ).[48] The functional enzymes exist as dimers. In rats, homodimers and intraclass heterodimers are formed. Interclass heterodimers have not been reported. Structural identification of eight of the human GST forms indicates they are all homodimers. Developmental changes in levels of the enzymes are seen. They are very low at birth and increase rapidly to adult levels by puberty in experimental animals. However, relatively high levels are found in human fetal liver and in placenta. The enzymes are present in most tissues but are highest in content in liver and kidney.[49] The activated component in the reaction is glutathione, a tripeptide (Table 5). It is the nucleophilic thiol group of the cysteine component that reacts with the wide variety of com-

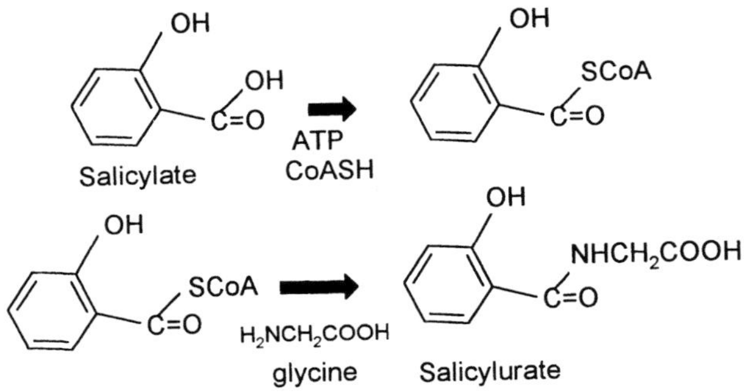

Figure 10. Formation of a hippurate (glycine conjugation of salicylate).

pounds with electrophilic centers, *i.e.*, carbon, oxygen, nitrogen, and sulfur (Figure 9). The mechanism of action involves the binding of GSH to the enzyme; GSH is present in healthy cells at 1–5 mM levels, and the dissociation constant for binding to the enzyme is considerably lower (10–30 μM), so it probably binds to the enzyme first. A proton is removed forming the thiolate anion (GS⁻), a reactive nucleophile, in the active site of the enzyme. It reacts with electrophilic groups and atoms.[50] Glutathione transferases have also been called ligandins, as they have a high affinity for many lipophilic, toxic compounds such as bilirubin, as well as a number of other lipophilic compounds as steroids, cephalosporins and dyes. The types of reactions catalyzed by GSH-transferase are extremely varied. It includes thioether formation, steroid isomerase, thiolysis (removal of a sulfate), dehalogenation, organic nitrate ester reductase (removal of a nitrate, as from nitroglycerin) and epoxide and hydrocarbon addition. Substrates range from nitroglycerin, to ethacrynic acid to a number of alkyl thiocyanates to a wide range of structurally diverse compounds containing electrophilic atoms.[49]

2.3.3. N-Acyltransferases. A number of amino acids are conjugated with activated xenobiotic carboxylic acids in different species. In man, primarily glycine and glutamine are used. The conjugates, are termed hippurates, after the species in which these conjugates were first studied, the horse. The carboxylic acid-containing xenobiotics are activated by mitochondrial Coenzyme A ligases (mainly medium chain acyl CoA ligase) in the liver and kidneys [51] and then conjugated with the amino acids by N-aminoacid transferases in the mitochondrial matrix (Figure 10). The CoA-generating ligases are the limiting step in hippurate formation. In most species, including man, renal glycine N-acyltransferase activities are generally the major source of hippurates, with liver carrying an appreciable portion of the remainder of the activity. In man, phenylacetic acids and indoleacetic acids are preferentially coupled to glutamine by N-arylacyltransferase.[51] At low exposure levels, xenobiotic carboxylic acids are almost quantitatively converted to the corresponding hippurates and glutaminates. As the body load of the xenobiotic rises the proportion of glucuronide metabolites increase. The transferases are not generally inducible by the classical cytochrome P450 inducers, but aspirin appears to increase the rate of its own conjugation with glycine, suggesting a fairly specific induction.[51]

2.3.4. Sulfotransferases. Sulfonation is a reaction that is carried out in both plants and animals. With the development of molecular biology it has become apparent that the enzymes of sulfonation comprise a superfamily of enzymes. In mammalian systems enzymes of sulfonation are used both for formation of ground substance, the sulfonation of polysaccharides, and the sulfonation of tyrosyl residues in proteins, as well as the sulfonation of endogenous hormones and exogenous xenobiotics. Sulfonation of these compounds makes them more polar and thereby more readily excretable. The sulfotransferase superfamily subdivides into three families, the phenol sulfotransferases, the hydroxysteroid sulfotransferases an the flavenol sulfotransferase of plants.[52] There are five sulfotransferes that have been identified in man. These are all cytosolic and include an estrogen sulfotransferase, three phenolsulfotransferases, and a hydroxysteroidsulfotransferase. There is a broad, overlapping substrate specificity of these enzymes, and as a result, confusion reigned for a while as to the number of such enzymes. Substrates of the phenolsulfotransferases include compounds like 2,6-dichloro-4-nitrophenol, dopamine, dihydroepiandrosterone (DHEA), estrone and naphthylamine (anubi group sulfonated). One of the phenolsulfotransferases is thermolabile and another is thermostable. The latter is a high affinity enzyme, active with micromolar substrate levels.[53]

2.3.5. N-Acetyltransferases (NAT). Acetylation reactions occur in a wide range of tissues and of a number of reaction types. Functional groups that undergo acetylation include primary amines, hydrazines, sulfhydryls and hydroxyls. Acetylation is one of the major routes of conjugation in mammals. The reaction, like that of aminoacid conjugations, results in amide formation, when the substrate is a primary amine. Unlike the hippurate conjugation reaction, it is the compound of endogenous origin that is activated in the acetylation reactions. A number of endogenous acetylations also utilize the cofactor acetyl CoA, and include formation of choline. The main site of acetylation is the liver, although almost all tissue carry out some xenobiotic acetylation. While most acetylations inactivate the xenobiotic, in a number of instances chemicals acetylated become activated as carcinogens, *e.g.,* N-hydroxylamine becomes a bladder carcinogen.[54] N-acetyltransferases are cytosolic enzymes that have a very broad specificity. The acetyltransferase activities are inhibited by N-ethylmaleimide (NEM), iodoacetate and p-chloromercuribenzoate (PCMB). The activities do not appear to be inducible. There is a polymorphism that involves N-acetylase activity, with some individuals being slow acetylators and some rapid acetylators. Serious neurotoxicities were seen as a result of this polymorphism, for example with the drug isoniazid, an antituberculosis drug. Early studies on the enzymes from slow acetylators and rapid acetylators indicated they had similar K_m and K_i values but V_{max} values that differed by an order of magnitude for isoniazid. While slow acetylators might be at greater risk of toxicity due to excessive plasma drug concentrations, fast acetylators might be at greater risk with agents that are activated by acetylation. Two NAT genes, NAT1 and NAT2, were been identified. NAT2 gene is responsible for the observed hereditary polymorphism in the metabolism of isoniazid. The polymorphic proteins are identical, however, and the reason for the slow acetylator phenotype appears to be a lower amount of enzyme, the result of a mutation in the control of the rate of translational expression of the functional protein.[55] NAT1 and NAT2 have 80% amino acid identity and overlapping substrate selectivity. Catalytic activity of NAT2 toward some substrates like sulfamethazine is almost two orders of magnitude greater that NAT1, while for p-aminosalicylic acid NAT1 appears to have more than a 3 order of magnitude higher activity.[55] A number of allelic variants have been noted for NAT2, and allelic variants for NAT1 may also exist.

3. CONCLUSIONS

A very large number of enzymes exist in the body that function to aid in removal of lipophilic xenobiotics and waste products from the body. These enzymes make use of a very large number of chemical reactions with very broad and overlapping substrate specificities. The different enzymes are able to react with just about every potential functional group that may be presented to the body, and in the absence of a functional group on a molecule, can create one. The enzymes fit into two broad categories, those which unmask or create additional functional groups on a molecule (Phase I), and those that complex existing functional groups with endogenous compounds of intermediary metabolism (Phase II). The Phase I pathway reactions can make use of water or of electron transport reactions and molecular oxygen to carry out their reactions. The Phase II reactions all require some form of energy in their anabolic reactions. The creation of a functional group or complexation with a structure with a more polar nature generally make the chemical more polar or water-soluble and thereby more readily excreted from the body. The action of these xenobiotic metabolizing enzymes will in most instances inactivate such molecules, removing their toxicological or pharmacological activities. In some cases, however, inert compounds may be made pharmacologically active or toxicologically reactive. In a number of instances inherited diseases and disorders have pointed to the importance of the xenobiotic metabolizing enzymes. The emergence of molecular biology techniques has aided in understanding such diseases and in many cases has revealed the presence of polymorphisms in the different subfamilies of xenobiotic metabolizing enzymes. It is expected that future studies will continue to expand the understanding of the role these enzyme play in homeostasis in the body.

REFERENCES

1. Heymann E (1980) Carboxyesterases and amidases. *in Enzymztic Basis pf Detoxication*Vol. I (Jacoby W, eds.), pp. 291–323, Academic Press, NY
2. Junge W, Heymann E (1979) Characterization of the isozymes of pig liver esterase. Kinetic studies. *Eur. J. Biochem.* **95**:519–525.
3. Mentlein R, Suttorp M, Heymann E (1984) Specificity of purified monoacylglycerol lipase, palmitoyl-CoA hydrolase, palmitoyl-carnitine hydrolase, and nonspecific carboxylesterase from rat liver microsomes. *Arch. Biochem. Biophys.* **228**:230–246.
4. Reiner E (1993) Recommendations of hte IUBMB nomenclature committee: Comments concerning classification and nomenclature of esterases hydrolyzing organophosphorus compounds. *Chemico-Biological Interactions* **87**:15–16.
5. Gill S, Hammock B (1981) Epoxide hydrolase activity in the mitochondrial and submitochondrial fractions of mouse liver. *Biochem. Pharmacol.* **30**:2111–2120.
6. Oesch F (1980) Epoxide hydrolase. *in Enzymatic Basis of Detoxication.*Vol. II (Jacoby W, eds.), pp. 291–323, Academic Press, NY
7. Oesch F (1972) Mammalian epoxide hydrolases: Inducible enzymes catalysing the inactivation of carcinogenic and cytotoxic metabolites derived from aromatic and olefinic compounds. *Xenobiotica* **3**:305–340.
8. Gillette J (1971) Reductive Enzymes. *in Concepts in Pharmacology, Part 2* (Brodie B, Gillette J, eds.), pp. 349–361, Springer-Verlag, NY
9. Mallett A, King L, Walker R (1985) Solubilization, purification and reconstitution of hepatic microsomal azoreductase activity. *Biochem. Pharmacol.* **34**:337–342.
10. Zbaida S, Levine W (1990) Characteristics of two classes of azo dye reductase activity associated with rat liver microsomal cytochrome P450. *Biochem. Pharmacol.* **40**:2415–2423.
11. Mannervik B (1980) Thioltransferases. *in Enzymatic Basis of Detoxificaiton*Vol. II (Jakoby W, eds.), pp. 229–244, Academic Press, New York
12. Benard O, Balasubramanian (1996) Purification and properties of thioltransferase from monkey small intestinal mucosa: Its role in protein-s-thiolation. *Int. J. Biochem. Cell Biol.* **28**:1051–1059.
13. Bachur N (1976) Cytoplasmic aldo-keto reductases: A class of drug metabolizing enzymes. *Science* **193**:595–597.
14. Jez J, Flynn T, Penning T (1997) A nomenclature system for the aldo-keto reductase superfamily. *Advances Experimental Medicine & Biology* **414**:579–589.

15. Barski O, Gabbay K, Bohren K (1997) Aldehyde reductase: Catalytic mechanism and substrate recognition. *Advances in Experimental Biology & Medicine* **414**:443–451.

16. Peterson F, Mason R, Hovsepian J, Holtzman J (1979) Oxygen-sensitive and -insensitive nitroreduction by Escherichia coli and rat hepatic microsomes. *J. Biol. Chem.* **254**:4009–4014.

17. Benson A (1993) Conversion of 4-nitroquinoline 1-oxide (4NQO) to 4-hydroxyaminoquinoline 1-oxide by a dicumarol-resistant hepatic 4NQO nitroreductase in rats and mice. *Biochem. Pharmacol.* **46**:1217–1221.

18. Linder K, Chan Y, Cyr J, Malley M, Nowotnik D, Nunn A (1994) TcO(PnAO-1-(2-nitroimidazole)) [BMS-181321], a new technetium-containing nitroimidazole complex for imaging hypoxia: Synthesis, characterization, and xanthine oxidase-catalyzed reaction. *Journal of Medicinal Chemistry* **37**:9–17.

19. van Dyke R, Wood C (1975) In vitro studies on irreversible binding of halothane metabolite to microsomes. *Drug Metab. Dispos.* **3**:51–57.

20. Nastainczyk W, Ullrich V, Sies H (1978) Effect of oxygen concentration on the reaction of halothane with cytochrome P450 in liver microsomes and isolated perfused rat liver. *Biochem. Pharmacol.* **27**:387–392.

21. Li H, Mani C, Kupfer D (1993) Reversible and time-dependent inhibition of the hepatic cytochrome-P450 steroidal hydroxylases by the proestrogenic pesticide methoxychlor in rat and human. *Journal of Biochemical Toxicology* **8**:195–206.

22. Cnubben N, Vervoort J, Boersma M, Rietjens I (1995) The effect of varying halogen substituent patterns on the cytochrome P450 catalyzed dehalogenation of 4-halogenated anilines to 4-aminophenol metabolites. *Biochem. Pharmacol.* **49**:1235–1248.

23. Hines R, Cashman J, Philpot R, Williams D, Ziegler D (1994) The mammalian flavin-containing monooxygenases: Molecular characterization and regulation of expression. *Toxicology Applied Pharmacology* **125**:1–6.

24. Jakoby WB, D.M. (1990) The enzymes of detoxification. *The Journal of Biological Chemistry* **265**:20715–20718.

25. Cashman J (1995) Structural and catalytic properties of the mammalian flavin-containing monooxygenase. *Chemical Research in Toxicology* **8**:165–181.

26. Beaty N, Ballou D (1980) Transient kinetic study of liver microsomal FAD-containing monooxygenase. *J. Biol. Chem.* **255**:3817–3819.

27. Tipton K (1980) Monoamine oxidase. in *Enzymtic Basis of Detoxication* Vol. I (Jakoby W, eds.), pp. 355–370, Academic Press, New York

28. He Z, Zou Y, Greenbaway F (1995) Cyanide inhibition of porcine kidney diamine oxidase and bovine plasma amine oxidase: Evidence for multiple interaction sites. *Arch. Biochem. Biophys.* **319**:185–195.

29. Jörnvall H, Danielsson O, Persson B, Shafqat J (1995) The alcohol dehydrogenase system. *Advances Experimental Medicine & Biology* **372**:281–294.

30. Yoshida N, Osawa Y (1991) Purification of human placental aromatase cytochrome- P-450 with monoclonal antibody and its characterization. *Biochemistry* **30**:3003–3010.

31. Yoshida A (1994) Genetic polymorphisms of alcohol metabolizing enzymes relataed to alcohol sensitivity and alcoholic diseases. *Alcohol and Alcoholism* **29**:693–696.

32. Omura T, Sato R (1962) A new cytochrome in liver microsomes. *J. Biol. Chem.* **237**:1375–1376.

33. Schenkman J, Sligar S, Cinti D (1981) Substrate interaction with cytochrome P-450. *Pharmacol. Ther.* **12**:43–71.

34. White REa, Coon MJ (1980) Oxygen activation by cytochrome P450. *Annual Review of Biochemistry* **49**:315–356.

35. Zimniac P, Waxman D (1993) Liver cytochrome P450 metabolism of endogenous steroid hormones, bile acids and fatty acids. in *Cytochrome P450*, Handbook of Experimental Pharmacology Vol. **105** (Schenkman J, Greim H, eds.), pp. 123–144, Springer-Verlag, Berlin

36. Schenkman J, Kupfer D. Hepatic Cytochrome P450 Monooxygenase System. Sartorelli A, editor. New York: Pergamon Press; 1982. 841 p.

37. Ryan D, Levin W (1993) Age- and Gender-related expression of rat liver cytochrome P450. in *Cytochrome P450*, Handbook of Experimentl Pharmacology Vol. **105** (Schenkman J, Greim H, eds.), pp. 461–476, Springer-Verlag, Berlin

38. Waxman D, Morrissey J, LeBlanc G (1989) Hypophysectomy differentially alters P-450 protein levels and enzyme activities in rat liver: pituitary control of hepatic NADPH cytochrome P-450 reductase. *Mol. Pharmacol.* **35**:519–525.

39. Mode A, Wiersma-Larsson E, Gustafsson J-A (1989) Transcriptional and posttranscriptional regulation of sexually differentiated rat liver cytochrome P-450 by growth hormone. *Molecular Endocrinology* **3**:1142–1147.

40. Bresnick E (1993) Induction of cytochromes P450 1 and P450 2 by xenobiotics. in *Cytochrome P450* Vol. **105** (Schenkman J, Gleim H, eds.), pp. 503–524, Springer-Verlag, Berlin

41. Guengerich F (1993) Metabolic reactions: Types of reactions of cytochrome P450 enzymes. *in Cytochrome P450*, Handbook of Experimental Pharmacology Vol. **105** (Schenkman J, Greim H, eds.), pp. 89–104, Springer-Verlag, Berlin

42. Parli C, McMahon R (1973) The mechanism of microsomal deamination: Heavy isotope studies. *Drug Metab. Dispos.* **1**:337–340.

43. Cheng K-C, Schenkman J (1983) Testosterone metabolism by cytochrome P-450 isozymes RLM3 and RLM5 and by microsomes. Metabolite identification. *J. Biol. Chem.* **258**:11738–11744.

44. Nagata K, Liberato D, Gillette J, Sasame H (1986) An unusual metabolite of testosterone. 17B-hydroxy-4,6-androstadiene-3-one. *Drug Metab. Dispos.* **14**:559–565.

45. Mulder G, Coughtrie M, Burchell B (1990) Glucuronidation. *in Conjugation Reactions in Drug Metabolism: An Integrated Approach.* (Mulder G, eds.), pp. 51–105, Taylor & Francis, London

46. Ritter J, Chen F, Sheen Y, Tran, HN, Kimura S, Yeatman M (1992) A novel complex locus UGT1 encodes human bilirubin, phenol, and other UDP-glucuronosyltransferase isozymes with identical carboxyl termini. *J. Biol. Chem.* **267**:3257–3261.

47. Clarke D, Burchell B (1994) Transferases and hydrolases involved in phase II conjugation-deconjugation reactions, genetic polymorphism and regulation of expression. *in Conjugation-Deconjugation reactions in drug metabolism and toxicity.*, Handbook of Experimental Pharmacology Vol. **112** (Kauffman F, eds.), pp. 1–44, Springer-Verlag, Berlin

48. Mannervik B, Awasthi Y, Board P, Hayes J, Di Ilio C, Ketterer B, Listowski I, Morgenstern R, Muramatsu M, Pearson W and others (1992) Nomenclature for human glutathione transferases. *Biochem. J.* **282**:305–308.

49. Ketterer B, Mulder G (1990) Glutathione conjugation. *in Conjugation Reactions in Drug etabolism: An Integrated Approach.* (Mulder G, eds.), pp. 308–364, Tayor & Francis, London

50. Armstrong R (1991) Glutathione S-transferases: Reaction mechanism, structure and function. *Chemical Research Toxicology* **4**:131–140.

51. Hutt A, Caldwell J (1990) Amino acid conjugation. *in Conjugation Reactions in Drug Metabolism.* (Mulder G, eds.), pp. 274–305, Taylor & Francis, London

52. Weinshilboum R, Otterness D, Aksoy I, Wood T, Her C, Raftogianis R (1997) Sulfotransferase molecular biology: cDNAs and genes. *FASEB J.* **11**:3–14.

53. Weinshilboum R, Otterness D (1994) Sulfotransferase enzymes. *in Conjugation-Deconjugation Reactions in Drug Metabolism and Toxicity.*, Handbook of Experimental Pharmacoogy Vol. **112** (Kauffman F, eds.), pp. 45–107, Springer-Verlag, Berlin

54. Weber W, Levy G, Hein D (1990) Acetylation. *in Conjugation Reactions in Drug Metabolism.* (Mulder G, eds.), pp. 163–191, Taylor & Francis, London

55. Vatsis K, Weber W (1994) Human N-acetyltransferases. *in Conjugation-Deconjugation Reactions in Drug Metabolism and Toxicity.*, Handbook of Experimental Pharmacology Vol. **112** (Kauffman F, eds.), pp. 109–130, Springer-Verlag, Berlin.

PROTEIN-PROTEIN INTERACTIONS IN THE P450 MONOOXYGENASE SYSTEM

John B. Schenkman,[1] Ingela Jansson,[1] Gary Davis,[2] Paul P. Tamburini,[2] Zhongqing Lu,[3] Zhe Zhang,[3] and James F. Rusling[3]

[1]Department of Pharmacology
University of Connecticut Health Center
MC-1505
Farmington, Connecticut 06030
[2]Bayer Corporation
Pharmaceutical Division
West Haven, Connecticut 06516
[3]Department of Chemistry
University of Connecticut
Storrs, Connecticut 06269

1. CYTOCHROME b$_5$ INTERACTIONS

1.1. Cytochrome b$_5$

Cytochrome b$_5$ is a small acidic hemoprotein that functions as an electron transfer protein. It is a ubiquitous mammalian membrane protein, found in the endoplasmic reticulum of most tissues. Its primary structure is highly conserved with sequence identities of greater than 89% in mammals and greater than 71% between rat and chicken (Figure 1). Even in comparisons between plant (rice and tobacco) cytochrome b$_5$ with that of mammals sequence identity is greater than 35%. Twenty-five of the one hundred and thirty-three amino acids of rabbit cytochrome b$_5$ are acidic residues, glutamate or aspartate. Cytochrome b$_5$ has a number of its very highly conserved acidic residues around an exposed heme edge (Figure 2), some of which are delineated in Figure 1 (bold type). Residues around the exposed heme and the heme are shown in ball and stick model. It was shown quite early that cytochrome b$_5$ uses its acidic residues for stabilization of interactions with its electron transfer (redox) partners. Although not its normal redox partner, cytochrome c has been used to study electron transfer with cytochrome b$_5$, and, since the crystal structure of both proteins have been elucidated,[1-3] surface maps of the topologies of cytochrome b$_5$ and cytochrome c were constructed and fitted together making use of the

Molecular and Applied Aspects of Oxidative Drug Metabolizing Enzymes,
edited by Arinç *et al.* Kluwer Academic / Plenum Publishers, New York, 1999.

```
         20            30           40      *    * 50               *        70
TLEEIQKHKDSKSTWVILHHKVYDLTKFLEEHPGGEEVLREQAGGDATENFEDVGHSTDAR-RAT
TLEEIQKHNHSKSTWLILHHKVYDLTKFLEEHPGGEEVLREQAGGDATENFEDVGHSTDAR-HUM
TLEEIKKHNHSKSTWLILHHKVYDLTKFLEEHPGGEEVLREQAGGDATENFEDVGHSTDAR-RAB
TLEEIKKHNHSKSTWLILHHKVYDLTKFLEDHPGGEEVLREQAGGDATENFEDIGHSTDAR-HOR
TLEEIQKHNNSKSTWLILHHKVYDLTKFLEEHPGGEEVLREQAGGDATENFEDVGHSTDAR-PIG
RLEEVQKHNNSQSTWIIVHHRIYDITKFLDEHPGGEEVLREQAGGDATENFEDVGHSTDAR-CHICK
TLEEVAKHNSKDDCWLIIGGKVYNVSKFLEDHPGGDDVLLSSTGKDATDDFEDVGHTTTAR-RICE
TLAEVSNHNNAKDCWLIISGKVYNVTKFLEDHPGGGEVLLSATGKDATDDFEDIGHSSSAR-TOBAC
```

Figure 1. Alignment of cytochrome b₅ forms. The numbering is that used in the crystal structure. Asterisks indicate amino acids most frequently implicated in protein-protein interactions.

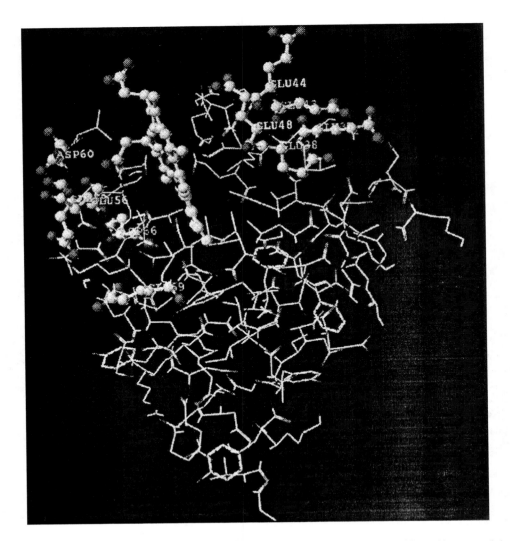

Figure 2. Relationship of acidic residues of cytochrome b₅ to the exposed heme. The acidic residues around the exposed heme and the heme are shown in ball and stick model.

invariant anionic and cationic charges respectively, ringing the exposed heme prosthetic groups as a means for determining the docking domains of the two proteins.[3] Interaction models between cytochrome b_5 and methemoglobin have likewise been constructed which make use of charge pairings.[4] Interaction between erythrocyte cytochrome b_5 and its reductase was greatly inhibited by increased ionic strength,[5] which suggested that such interaction also utilizes complementary charge pairing for efficient electron transfer. Similarly, neutralization of lysyl residues on the reductase or carboxylate residues on the cytochrome b_5 resulted in inhibition of electron transfer to microsomal cytochrome b_5 from its reductase.[6] In time, chemical modification and cross-linking of charge-pairing residues identified the cytochrome b_5 carboxylate residues E^{47}, E^{48} and E^{52} (residues E^{43}, E^{44} and E^{48} from structural analyses [2]) as well as a side-chain of the cytochrome b_5 heme prosthetic group in interactions with NADH-cytochrome b_5 reductase [7]. Subsequently the same anionic surface on cytochrome b_5 was implicated in interactions with NADPH-cytochrome P450 reductase and with stearoyl CoA desaturase.[8] Overall, modeling experiments have implicated other residues on cytochrome b_5 in interactions as well, such as E^{37}, E^{38}, and D^{60}.[3]

1.2. Cytochrome b_5 Involvement with Cytochrome P450

Early studies indicated that the NADPH-dependent microsomal monooxygenase was greatly stimulated by the co-addition of NADH.[9] This NADH synergism of the NADPH-dependent reaction was suggested as possibly involvement of the NADH electron transfer pathway,[10] and cytochrome b_5 was thought to provide the second electron needed in the monooxygenase reaction.[11] Antibodies to cytochrome b_5 were found to inhibit the NADH synergism, but not the NADPH-supported metabolism of drug substrates,[12,13] indicating a need for the cytochrome b_5 in the NADH-dependent electron transfer. In 1981 Chiang[14] reported that cytochrome b_5 interaction with cytochrome P450 caused a shift in the spin state of the latter hemoprotein. The shift seen involves a change in the equilibrium between high and low spin cytochrome P450 (Figure 3). In the figure we see that addition of cytochrome b_5 to cytochrome P450 causes a shift from the 420 nm low spin peak to the

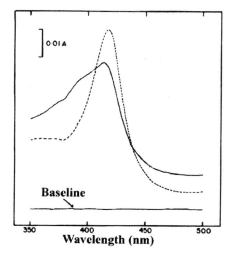

Figure 3. Spectra of interaction of cytochrome P450 2B4 with cytochrome b_5. Equimolar amounts of cytochrome b_5 were added to sample and reference cuvet to eliminate spectral overlap, and shift the absorbance of the cytochrome P450 toward the high spin form spectrum (solid line). Data from ref. 15.

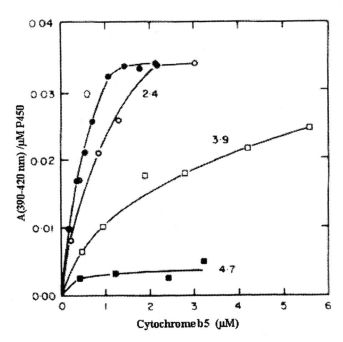

Figure 4. Titration of the spin equilibrium of CYP2C11 by cytochrome b_5 with different amounts of carboxyl residues neutralized with EDC. From ref. 15 with permission.

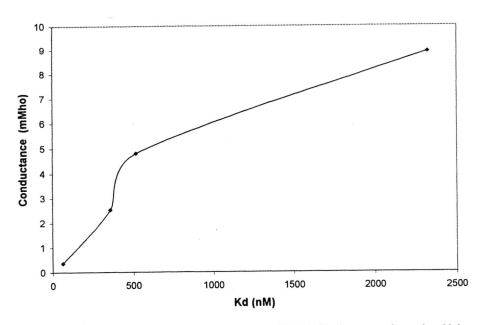

Figure 5. Change in the dissociation constant for cytochrome P450 2C11 binding to cytochrome b_5 with increasing ionic strength (conductance) of the medium. The salt concentration (sodium phosphate) was increased from 5 mM to 200 mM and the dissociation constants were measured from the spectral interactions.

395 nm high spin peak. The addition of a substrate causes further increases in the extent of the shift.[15]

1.3. Ionic Nature of the Interaction between Cytochrome b_5 and Cytochrome P450

We investigated the interaction between the two hemoproteins and were able to quantify it and demonstrate that it required the intact carboxyl residues of cytochrome b_5.[15] As shown in Figure 4, when carboxyl residues on cytochrome b_5 were neutralized by reacting with the water soluble 1-ethyl-3-(2-dimethylaminopropyl) carbodiimide HCl (EDC) and methylamine its affinity for CYP2C11 was greatly diminished. The carbodiimide activates the carboxyl residues and the charge-pairing amino group of the methylamine reacts forming an amide linkage. As shown in Figure 4, the neutralization of as few as 2.4 carboxyl residues per cytochrome b_5 resulted in a decrease in the affinity of the cytochrome P450 for cytochrome b_5. Neutralization of 2.9 carboxyl residues per cytochrome b_5 molecule caused an even greater loss in the affinity for cytochrome b_5. However, it was still possible to cause a maximal spin shift in the cytochrome P450. On the other hand, when a much larger number of carboxyl residues were neutralized the ability to bind to cytochrome P450 and perturb the spin equilibrium was lost (Figure 4). The Kd values for cytochrome b_5-CYP2B4 and for cytochrome b_5-CYP2C11 both increased to undetectable values due to decreases in the spectral changes to below measurable levels[15] when as few as 8 carboxyls/cytochrome b_5 were neutralized. A further indication that interaction between cytochrome b_5 and CYP2C11 is by complementary charge-pairing was demonstrated by the effect of increasing the ionic strength of the medium. As seen in Figure 5, there was an increase in the dissociation constant with increasing conductance of the medium. This has the effect of decreasing the ionic activity coefficient and thereby the ability of ions in solution to charge-pair. A further demonstration that interaction between these proteins required charge-pairing could be demonstrated by the ratio of the association constants of the interaction as a function of the ionic activity coefficients of the medium (Figure 6). The data fit a straight line, with the equation for the line being $\log (Ka_{app}/Ka_0) = -2mn \log(y)$, where y is the ionic activity coefficient, Ka_{app} is the association constant at any ionic strength, Ka_0 is the association constant at zero ionic strength and 'm' and 'n' are

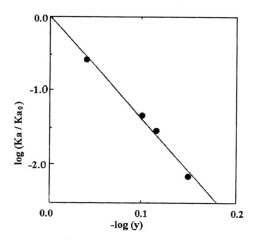

Figure 6. Dependence of the association constant of cytochrome b_5-CYP2C11 interaction on the ionic activity coefficient. Data from ref 16.

the interacting charges on the respective proteins. The equation predicts a straight line with a slope depending on the charge product.[17] The charge-product, mn, in this study was 9, suggesting that interaction between the two proteins that effected the observed spectral changes involved charge-pairing between 3 oppositely charged residues on each protein. In studies in which modification of anionic residues was carried out there was a similar indication that three specific amino acids of cytochrome b_5 were involved in complementary charge pairing with its redox partners.[7] Of interest, in studies on the interaction between cytochrome c and cytochrome b_5 the energy of interaction was decreased only about 14% by substitutions of uncharged residues for E^{44}, E^{48}, D^{60}, and the heme propionate of cytochrome b_5. This was taken as an indication that electrostatic interactions may only be a minor force in the redox complexation between these two proteins,[18] and it was suggested that perhaps hydrophobic interactions and hydrogen binding are more important processes.

2. INFLUENCE OF CYTOCHROME b_5 ON CYTOCHROME P450 MONOOXYGENATIONS

2.1. Influence of Cytochrome b_5 on Microsomal Monooxygenation

The addition of cytochrome b_5 to most forms of cytochrome P450 enhances the rate of monooxygenation of many of the substrates of the enzymes. Early studies on the microsomal metabolism of drugs and chemicals revealed that the reaction required NADPH.[19–21] While NADH could provide some activity, it could not replace NADPH with more than a small fraction of the monooxygenase activity. However, when NADH was added to the NADPH supported reaction it provided a synergy.[9,22,23] Antibodies to cytochrome b_5 were found to inhibit the NADH-supplied stimulation of the monooxygenation of substrates in phenobarbital-treated rat liver microsomes, but did not inhibit the metabolism of ethylmorphine or aniline supported by NADPH alone.[12] In a similar manner, antibodies to cytochrome b_5 were inhibitory of the NADH synergism of NADPH-supported aminopyrine dealkylation and of the reaction supported by NADH alone, but did not affect the NADPH-supported reaction in untreated rat liver microsomes.[13]

2.2. Influence of Cytochrome b_5 in the Reconstituted Monooxygenase System

The first attempt to determine the role of cytochrome b_5 on the monooxygenase in the reconstituted system revealed it to be inhibitory of the phenobarbital induced cytochrome P450 as well as the 3-methylcholanthrene induced hemoprotein (P448), although the latter was less sensitive to inhibition.[24] Similar inhibition was seen when the cytochrome b_5 was added to untreated or phenobarbital treated rat liver microsomes, suggesting the added cytochrome b_5 competes with the cytochrome P450, perhaps draining electrons needed for the monooxygenase reaction.[25,26] In contrast to these results, the effect of cytochrome b_5 on the NADPH-supported metabolism of chlorobenzene in the reconstituted system appeared to depend upon the form of cytochrome P450 utilized.[27] With the phenobarbital induced form of cytochrome P450 the rate of chlorobenzene hydroxylation was doubled by the addition of cytochrome b_5, while with the 3-methylcholanthrene induced form cytochrome b_5 was without effect with this substrate. Using highly purified forms of cytochrome P450 it was also seen that the influence of cytochrome b_5 varied, depending upon the substrate and the form of cytochrome P450 examined. It was inhibitory of benzphetamine metabolism but stimulatory of 7-ethoxycoumarin metabolism and very stimulatory of acetanilide oxidation with CYP2B4. In contrast, CYP2C3 (LM3b) metabolism of benzphetamine was increased by cytochrome b_5 while its metabolism of 7-ethoxycoumarin and acetanilide were both inhibited.[28] Our own studies were consistent with these observations. With p-nitroanisole as substrate in the reconstituted system containing CYP2B4, CYP2C11 or CYP2A2 the addition of cytochrome b_5 caused stimulations of 6.5 fold, 2.6 fold and 1.4 fold respectively. In contrast, with aminopyrine as substrate cytochrome b_5 caused stimulations of 245%, 30% and 11% respectively.[29] Further complexity was seen when reports appeared of a form of cytochrome P450 for which cytochrome b_5 was *obligatory* for the metabolism of substrates. In 1979 Yamano's group reported on the purification of a form of cytochrome P450, P450B1, from rabbit liver using a cytochrome b_5 affinity column.[30] The protein was shown to catalyze NADPH-dependent p-nitroanisole dealkylation and to *require* cytochrome b_5. The addition of nonionic detergent enhanced the p-nitroanisole demethylation activity, but the presence of NADH-cytochrome b_5 reductase only afforded a small increase in the presence of NADH. Based upon the partial amino-terminal sequence obtained for the protein [31] it was probably CYP3A6. Another member of the 3A subfamily, rat CYP3A1, is also catalytically inactive in the absence of cytochrome b_5,[32] catalyzing testosterone metabolism in the presence of cytochrome b_5 and microsomal lipid. A nonionic detergent was also necessary. Interestingly, even a human form of CYP3A protein, CYP3A4, showed the same requirement for cytochrome b_5 and detergent.[33] Clearly, the influence of cytochrome b_5 on different forms of cytochrome P450 varies with the substrate metabolized and perhaps reflects a complex role.

2.3. Cytochrome b_5 and Electron Transfer

In our studies on the role of cytochrome b_5 in cytochrome P450 monooxygenations we have shown, using manganese heme-substituted cytochrome b_5, that the ability of cytochrome b_5 to stimulate CYP2B4-dependent monooxygenations depends upon the cytochrome b_5 being able to undergo redox change.[34] This is true with CYP2C11 as well (Table 1).The manganous heme in cytochrome b_5 did not affect the spectrally determined complex formation, nor alter the dissociation constant, but it did cause a loss in the ability of cytochrome b_5 to stimulate the rate of drug oxidation. The heme was removed from the native cytochrome b_5 by acid acetone and re-added to reconstitute the cytochrome b_5, or manganous heme was added to form the poorly reducible manganous b_5. The turnover of p-nitroanisole was very slow with CYP2C11, and the addition of cytochrome b_5 stimulated turnover by one order of magnitude. While manganous b_5 did not prevent the turnover of

Table 1. Requirement for redox changes in cytochrome b_5 for stimulation of CYP2C11 oxidation of p-nitroanisole

Addition	Turnover (min^{-1})
None	0.04
1 μM native b_5	0.60
1 μM reconstituted b_5	0.58
1 μM Manganous b_5	0.05

The 2C11 and Fp_T concentrations were each 0.19 μM.

Table 2. Electron flow to cytochrome b_5 and CYP2B4 alone and together in a reconstituted system and on benzphetamine turnover

System	K_{fast} P450	K b_5 (lo I*)	K b_5 (hi I+)	Turnover
Complete assay	$0.55\ s^{-1}$	$0.012\ s^{-1}$	$0.20\ s^{-1}$	$0.32\ min^{-1}$
+ cyt b_5 or P450	$0.58\ s^{-1}$	$0.183\ s^{-1}$	$0.21\ s^{-1}$	$1.73\ min^{-1}$

*lo indicates 10 mM and +hi indicates 100 mM sodium phosphate buffer. P450 reduction and turnover are at low ionic strength where binding of cytochrome b_5 is strong.

the substrate it did cause a loss of the stimulatory action, indicating a need for electron transfer in the stimulation. In contrast, with rabbit forms of cytochrome P450 titration with manganous cytochrome b_5 resulted in increasing inhibition below the rate in the absence of cytochrome b_5, regardless of whether native cytochrome b_5 was stimulatory of the monooxygenation.[28]

In our more recent examination of the mechanism of the stimulation of the NADPH-supported monooxygenation reaction afforded by cytochrome b_5 we were able to demonstrate that the enhancement of metabolism is the result of an increased rate of cytochrome b_5 reduction[35] and is not due to an effect on the fast phase (K_{fast}) of electron flow to cytochrome P450 (Table 2). Our studies revealed that there was a 15-fold stimulation in the rate of cytochrome b_5 reduction when it was reduced by NADPH-cytochrome P450 reductase in the presence of cytochrome P450 at low ionic strength (lo I), as compared to the reduction in the absence of cytochrome P450. Along with the stimulation in cytochrome b_5 reduction rate the turnover of benzphetamine was enhanced almost six-fold at low ionic strength. At high ionic strength (hi I), when cytochrome b_5 affinity for cytochrome P450 is low[16] there was no effect on cytochrome b_5 reduction by the presence of cytochrome P450 (Table 2), but the reduction of cytochrome P450 was slowed, due to competition with cytochrome b_5 for reductase.[35] Since cytochrome b_5, when complexed to cytochrome P450, was unable to accept electrons from other redox partners,[34] the enhanced rate of electrons reaching the cytochrome b_5 had to be coming from the NADPH-cytochrome P450 reductase via cytochrome P450. Our conclusions were that in the monooxygenase reaction the cytochrome b_5 functions as an electron buffer, stimulating the monooxygenase reaction by accepting an electron from ferrous cytochrome P450 and returning it to oxyferrous cytochrome P450.[35] This would entail only one interaction with the reductase, during which two electrons would be transferred. In the absence of cytochrome b_5 the cytochrome P450 would require two interactions with the reductase, since the heme is a one electron acceptor. It was suggested that this is the basis for the stimulation afforded by cytochrome

Table 3. Reduction of CYP3A4 with cytochrome b5 and Fp_T

Substrate	Form of Cyt b5	K (min)
Testosterone	Holo b5	730
Testosterone	Apo b5	960
Testosterone	None	<1
None	Holo b5	7
Ethylmorphine	None	660

Data from [Yamazaki, 1996 #14182]

b_5, since the rate of formation of a functional complex between NADPH-cytochrome P450 reductase and cytochrome P450 has been shown to be a limiting factor.[36]

In contrast to this suggestion, a recent study using CYP3A4, a human form of cytochrome P450, indicated that while cytochrome b_5 afforded a 3-fold stimulation of testosterone 6β-hydroxylation,[37] the same stimulation was obtained when the apoprotein of cytochrome b_5 was added to the medium. The investigators concluded that the stimulation by cytochrome b_5 of this monooxygenase reaction did not involve electron transport properties of cytochrome b_5. At first glance these results are in direct opposition to those of Morgan and Coon,[28] who observed that addition of apo-cytochrome b_5 to the assay medium containing rabbit CYP2B4 was, if anything, slightly inhibitory of the metabolism of benzphetamine and 7-ethoxycoumarin; apo-cytochrome b_5 was also inhibitory of the metabolism of acetanilide when added to the assay containing rabbit CYP1A2. Metabolism studies with rat cytochrome P450B1, for which cytochrome b_5 is necessary, also showed an inability of apo-cytochrome b_5 to replace cytochrome b_5 in the turnover of p-nitroanisole.[38] The difference may lie in the apparently obligatory role for cytochrome b_5 in the monooxygenations by CYP3 family enzymes. Rat CYP3A1, for example has been shown to have a requirement for phospholipid, detergent and cytochrome b_5 in the hydroxylation of testosterone.[32] This is similar to the requirement of rat cytochrome P450B1 for the metabolism of p-nitroanisole.[30] The human ortholog, CYP3A4 has been shown to have the same cytochrome b_5 and detergent requirements for testosterone hydroxylation.[33] Interestingly the data by Yamazaki and coworkers[37] showed that both apo-cytochrome b_5 and holo-cytochrome b_5 were stimulatory of the rate of reduction of the CYP3A4 by NADPH and NADPH-cytochrome P450 reductase (Table 3) when testosterone was the substrate. So it would appear that with this substrate cytochrome b_5, which is generally concluded to provide a rate-limiting second electron (but not the first electron[11]) for monooxygenations, is having some other effect on CYP3A4, since it stimulated input of the first electron with this protein (Table 3). It is possible that the interaction of family 3A proteins with cytochrome b_5 results in some structural modification, like those shown earlier to increase the affinity for substrates, and to result in shifts in the spin equilibrium of the cytochrome P450 toward the high spin configuration.[14,15] As seen in Table 3, when another compound, ethylmorphine, was added as the substrate it alone enabled the reduction of the CYP3A4 without the addition of either cytochrome b_5 or its apo-protein. The nature of cytochrome b_5 interaction with the 3A family of cytochrome P450 and its response to steroids will need considerably more study.

3. TOPOLOGY OF THE CYTOCHROME b_5-BINDING DOMAIN OF CYTOCHROME P450

The mass of the heme prosthetic group is relatively small (616 Da) compared with the mass of the cytochrome b_5 (17 kDa), and of the cytochrome P450 (50 kDa) hemoproteins. Consequently, one might expect a precise alignment of redox centers between the proteins is needed in order for optimal efficiency of electron transfer. A number of studies have indicated that conserved arrangements of charged amino acids are found on different protein redox partners that might aid in the orientation of their redox centers. These could function to align the proteins by a process of complementary charge pairing (salt bridges). The nonheme iron protein, adrenodoxin, for example, interacts with adrenal mitochondrial cytochrome P450$_{scc}$ by complementary charge pairing, an interaction involving three carboxyl residues on the adrenodoxin. Loss of those anionic residues inhibited redox interac-

Table 4. Alignment of putative cytochrome b_5 docking domain on P450

101	E	Q	R	Q	F	R^{112}	A	L	A
2B4	R	W	R	A	L	R^{125}	R	F	S
2C11	Q	W	K	E	I	R	R	F	S
2A5	R	A	K	Q	L	R	R	F	S
2A4	R	A	K	Q	L	R	S	F	S
2E1	T	W	K	D	T	R	R	F	S
3A1	E	W	K	R	Y	R	A	L	L
3A4	E	W	K	R	L	R	S	L	L
3A6	D	W	K	R	V	R	T	L	L
102	N	W	K	K	A	H	N	I	L

tions between the two proteins and precluded formation of EDC-catalyzed complexes between them.[39] The carboxyl residues on adrenodoxin are analogous to corresponding carboxyl residues on bacterial putidaredoxin, the nonheme iron electron transfer protein involved in electron transfer from putidaredoxin reductase to CYP101. When those carboxyl residues were neutralized, it resulted in loss of ability to be reduced by putidaredoxin reductase.[40] As indicated above, cytochrome b_5, too, has very highly conserved carboxyl residues, which aid in its orientation with redox partners.

A very large number of cytochrome P450 genes exist (see: drnelson.utmem.edu/genesperspecies.html for list of over 450 forms). At least 60 different forms have been reported for rat and about 40 genes are identified for humans. In alignments of the primary sequences of the proteins, a high degree of phylogenetically conserved residues have been observed for members of the cytochrome P450 superfamily,[41,42] allowing an analogy between their structures and that of the first crystallized form, bacterial cytochrome CYP101. Other bacterial forms of cytochrome P450 have since been crystallized, including CYP108 (P450$_{terp}$) and CYP102 (P450$_{BM-3}$), and although they possess very low sequence identity between them, their tertiary structures have been shown to be very similar to that of CYP101.[43] Much better sequence identity occurs between the approximately 200 forms of mammalian cytochrome P450 (http://drnelson.utmem.edu/p450bpub237.html). In key regions of the tertiary structure of these forms, regions that are suggested from x-ray crystallography to be involved in heme anchoring, oxygen activation and substrate binding, high levels of sequence identity are seen.

In 1983 it was noted that some forms of cytochrome P450 have an amino acid sequence corresponding to the substrate recognition sequence, RRXS, of the cyclic AMP-dependent protein kinase, where X represents a hydrophobic residue such as phenylalanine. Incubation of one of these forms, CYP2B4, with cAMP and the protein kinase resulted in phosphorylation of the hemoprotein.[44] Proteolytic digests of the phosphorylated cytochrome P450 showed that the substrate recognition sequence, RRFS[128], contained a phosphorylated serine.[45] Sixty-two forms of cytochrome P450 have the substrate recognition sequence of the cAMP-dependent protein kinase, and a number of these were shown to undergo phosphorylation by the cAMP-dependent protein kinase.[46,47] All of these forms are in family 2 and are in the 2A, 2B, 2C, 2D, 2E, and 2G sub-families (Table 4). While studying the phosphorylation it was observed that after phosphorylation of CYP2B4 the hemoprotein would no longer bind to cytochrome b_5.[29] A competitive inhibition by cytochrome b_5 of cytochrome P450 phosphorylation by the cAMP-dependent protein kinase could be demonstrated.[47] Based upon these observations, it was suggested that the cytochrome b_5-binding domain on CYP2B4 overlapped a region that contained a phosphorylatable serine, RRFS[128].[47] In a subsequent study it was found that mutation of the second arginine in the RRFS sequence of mouse cyp2a-5 to serine (this is the sequence in mouse cyp2a-4, see Table 4) resulted in loss of ability of cytochrome b_5 to bind to the cytochrome P450.[48] Mouse cyp2a-4 does not bind cytochrome b_5. In the table are also shown the alignments for three forms of cytochrome P450 from the 3A subfamily, one from rat, one from rabbit and one from human. These forms were mentioned above as requiring cytochrome b_5 for expression of substrate metabolism. None of these have the cAMP-dependent protein kinase substrate recognition motif, possessing only one of the R groups , but have an additional arginyl residue two positions away.

Cytochrome b_5 was found to competitively inhibit the binding of the acidic protein putidaredoxin and its ability to transfer electrons to CYP101, its physiological electron acceptor.[49] This was taken to indicate that the two acidic proteins bind to a similar site on CYP101. Since the crystal structures of cytochrome b_5 and CYP101 had both been determined (the crystal structure of putidaredoxin has not been determined), this permitted computer modeling of the interaction and a postulation of the putidaredoxin docking site on CYP101. A good model was obtained with electrostatic contact involving cytochrome b_5 anionic residues E^{44}, a heme propionate side chain, E^{48}, and D^{60}, with a roughly circular patch of cationic residues on the proximal surface of CYP101 (see Figure 7). Residues K^{344}, R^{72}, R^{112} and R^{364}, respectively, on CYP101 could charge-pair with cytochrome b_5 in that energy minimization model.[49] In that model, the heme of cytochrome b_5 was oriented perpendicular to that of CYP101, as compared with the parallel orientation of the hemes indicated for the interaction between cytochrome b_5 and cytochrome c.[3] The surface of the CYP 101 in space-filling model showing the relationship of the four proposed charge-pairing residues is seen in Figure 8. Support for this model was provided by site-directed mutagenesis of acidic residues on cytochrome b_5 [50] and of two of the above cationic residues of cytochrome CYP101.[51] Mutation of residues $R^{72} \rightarrow Q$ and $K^{344} \rightarrow Q$ on CYP101 resulted in increased K_m values for putidaredoxin binding. Residue R^{112} on CYP101 aligns with residue R^{125} of CYP2B4 of rabbit liver microsomes when a large number of different forms of cytochrome P450 are aligned [52] (See Table 5, below, and http://drnelson.utmem.edu/genesperspecies.html). Residue R^{112} of CYP101 also forms a hydrogen bond to a heme propionate and may be involved in electron transfer.[53] Mutation of $R^{112} \rightarrow C$ caused a large increase in the K_m for putidaredoxin and greatly decreased the fast phase reduction rate constant, possibly by increasing the efficiency of binding. Other mutations to this residue caused an increase in the midpoint potential of the hemoprotein and increased the dissociation constant for the interaction with putidaredoxin, as well as inhibited the electron transfer between the two proteins, suggesting R^{112} normally hydrogen bonds to the heme propionate.[54] In the Nelson alignment residue K^{344} of CYP101 aligns with L^{414} of CYP2B4 and leucine of a large number of mammalian forms (see Table 5). However, in the alignment of Chang and coworkers,[55] an alignment based on 3D structural conservation, residue K^{421} of CYP2B4 (and many other mammalian forms) is aligned with K^{344} and this may be a more accurate alignment. CYP3A4 is not one of those forms that show a cationic residues by either alignment, suggesting that its different response to cytochrome b_5 may be due to binding at some other site on CYP3A4. A number of attempts have been made to discern the site of interaction on mammalian cytochrome P450s of NADPH-cytochrome P450 reductase, using chemical modification of lysyl and arginyl residues[56,57] and by using site directed mutagenesis.[58] With these techniques a large number of cationic residues have been identified. Some of these residues overlap the region we have designated as the cytochrome b_5-binding domain. Since the binding of cytochrome b_5 to CYP2B4 blocked phosphorylation of the cytochrome P450, but did not influence the K_m

Table 5. Alignment of mammalian cytochrome P450s with CYP101 putative and potential docking residues for cytochrome b$_5$

CYP101	R72	R79	R109	R112A	K344	H352	H361	R364
h2C9	K	G	K	RR	L	A	A	R
R2C11	K	G	K	RR	L	A	A	R
h2E1	K	Y	K	RR	L	T	G	R
R2B4	R	Q	R	RR	L	L	G	R
m2a-5	K	A	K	RR	L	I	G	R
m2a-4	K	A	K	RS	L	I	G	R
h2A6	K	A	K	**LR**	L	I	G	R
h2A7	K	A	K	RS	L	I	G	R
r2C13	K	G	K	**RH**	L	A	S	R
r2C12	K	G	K	RR	L	A	G	**S**
h3A4	K	C	**K**	RS	S	S	R	L

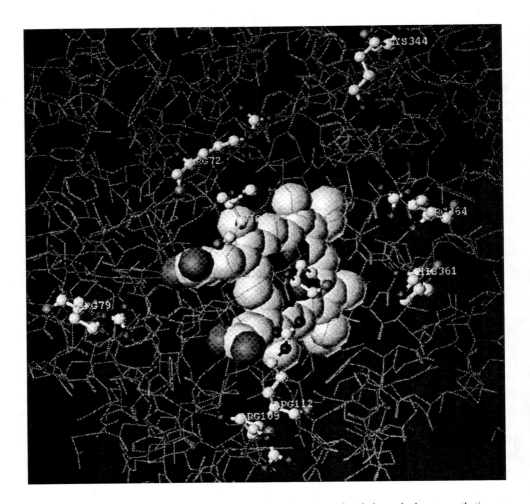

Figure 7. Proximal surface of CYP101 showing cationic residues around and above the heme prosthetic group. The heme is shown in space filling model and the cationic residues above it in ball and stick model using a Silicon Graphics computer and SYBYL 6.3.

Figure 8. The proposed putidaredoxin binding site of CYP101. The space-filling model was drawn using a Silicon Graphics computer with SYBYL 6.3. Cationic residues R72, R112, K344 and R364 are labeled.

for NADPH-cytochrome P450 reductase, it is clear that this region does not contain the docking site for NADPH-cytochrome P450 reductase.

Recently, a form of bacterial cytochrome P450 has been isolated and crystallized from Bacillus megaterium (CYP102) and resembles the mammalian endoplasmic reticulum cytochrome P450s in utilizing NADPH-cytochrome P450 reductase.[59] Because of its use of NADPH-cytochrome P450 reductase rather than a nonheme iron protein CYP102 has been suggested to be a prototype of the microsomal cytochrome P450s [59] and it has been suggested that it should be a better model of the microsomal forms. However, most alignments for the purpose of modeling tertiary structure of mammalian forms still make use of CYP101. We note that the newly crystallized enzyme does not align as well with mammalian forms of cytochrome P450 as CYP101 with respect to the region of interaction with cytochrome b_5 (see Table 4). Perhaps this is because CYP101 makes use of the acidic nonheme iron protein, putidaredoxin, for electron transport, and, as indicated above, a number of the

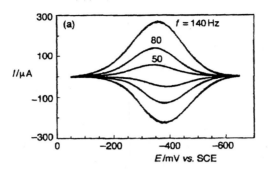

Figure 9. Square Wave Voltammetry of CYP101. From[73] with permission.

mammalian forms make use of the acidic cytochrome b_5. Also, as indicated above, cytochrome b_5 has been shown to be competitive with putidaredoxin for binding to CYP101, suggesting that the topology for its interaction on mammalian forms may be similar. This cytochrome b_5 binding domain is possibly evolutionarily related to the putidaredoxin binding site on CYP101 and serves as an additional electron input domain, at least for some forms of cytochrome P450. As seen in the table, CYP102 lacks the cationic R^{364} and has an acidic residue in place of the K^{344}. It does, however, have a cationic residue aligned with R^{72} of CYP101, and there is a histidyl residue aligned with the R^{112}.

4. ROUTE OF ELECTRON FLOW FROM THE SURFACE TO THE HEME OF P450

One of the fundamental processes in biological systems is electron transfer, and theories explaining the mechanism of the process are still being developed.[60] The most difficult factor to delineate is the role of specific amino acids which lie between and influence the redox centers. The concepts of how such electron transfer occurs include direct electron transfer from prosthetic group to prosthetic group,[61] to tunneling via a mixture of covalent and hydrogen bond orbitals[62] to suggestions that electrons are carried through proteins via aromatic residues, as via W^{106} of putidaredoxin to CYP101[63,64] by a process of covalent switching.[65] With cytochrome b_5, at least, it would appear that the exposed heme edge is capable of electron transfer.[66,67] At the interface between the donor and acceptor sites of cytochrome b_5 and cytochrome c is a phylogenetically conserved aromatic residue on cytochrome c, F^{82}, which may possibly be involved in heme contact and has an influence on the rate constant for electron transfer.[68] Site-directed mutation of this residue to variants with an aliphatic residue at position 82 generally resulted in rates of electron transfer 10,000 times slower.[69] A similar observation involving the role of W^{106} of putidaredoxin in redox coupling with CYP101 has been made.[64] However, subsequent studies indicated a non-electron conductance role for this aromatic residue, but rather a role in enhanced affinity of the reduced putidaredoxin for the oxidized hemoprotein.[70] The heme moiety of CYP101 is buried in the molecule, as, presumably, is that of mammalian forms of cytochrome P450. Figure 7 shows the proximal surface of the molecule with the heme prosthetic group in space filling model and the cysteinyl residue in ball and stick model connecting to the heme iron. The heme is in a shallow depression of the proximal surface with C^{357} above it. For oxygen activation to occur electrons must reach the heme, and

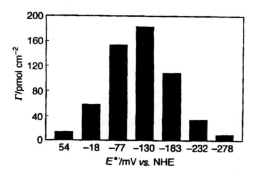

Figure 10. Frequency distribution of formal potentials of CYP101. From[73] with permission.

there is a need for conductance of electrons from the surface of the protein to the heme iron. The proximal surface of the hemoprotein is the shortest distance to the heme, hence our interest in the topology of this region.

There are data in the literature suggesting in some redox proteins the presence of multiple redox sites. For example, in the interaction between two redox proteins multiple manners of interaction may occur. Different pathways of electron flow may result, depending on the orientation to each other of the redox centers of two proteins or on the redox surfaces between the two redox centers.[67,71] This would imply a process of conformational gating,[72] in which different interacting conformers of the proteins can influence rates obtained of electron flux. Some of our own data might suggest that cytochrome P450 forms, too, might have more than one route of electron input to the heme. For example, in studying the electrochemical reduction of CYP101 and CYP2B4 in lipid films on electrodes, we noted that the half peak width was more than double the theoretical value of 90 mV for ideal reversible electron transfer, although the hemoprotein could be completely reduced and oxidized multiple times without harm to it.[73] With CYP101 it was not possible to make use of the ideal Lavirons theory[74] to describe the voltammetry peak shapes nor to consistently estimate the rate constants for the electron transfer from the cyclic voltammetry. For this reason we turned to square wave voltammetry (Figure 9). But here too, the half peak widths were significantly broader than the ideal case.[73] We hypothesized that the broadening of the peak widths might be the result of a dispersion of midpoint potentials ($E^{0'}$) for the cytochrome P450 on the electrode thin film. When the dispersion model was combined with the square wave theory for confined species, it was possible to fit the voltammograms, using non-linear regression, by considering a population of redox centers, *i.e.,* by calculating distribution concentrations of populations of the P450 with different midpoint potentials. The number of cytochrome P450 redox proteins that would allow a good fit of the square wave voltammogram was between 5 and 7 (Figure 10). We can rationalize this as being indicative of different populations of cytochrome P450 confined on the electrode surface with different orientations relative to the plane of the electrode normal. If this is correct, then it might indicate that measurement of a redox potential provides an average value. The values determined by the regression analyses of square wave voltammograms onto the dispersion model may be indicative of different routes electrons take from the electrode through the hemoprotein surface to reach the buried heme of cytochrome P450. This would suggest it is possible for electrons to funnel into the cytochrome P450 heme via different routes, and might explain the different rates of reduction of different forms of cytochrome P450. These have different amino acid sequences, yet must interact with NADPH-cytochrome P450 reductase. To date, despite much speculation the binding site of NADPH-cytochrome P450 reductase on cytochrome P450 forms has yet to be elucidated. It may be that docking between these two proteins, in contrast to interactions between cytochrome b_5 and cytochrome P450, does not involve a well-defined site with a highly conserved sequence.

REFERENCES

1. Mathews F, Levine M, Argos P (1972) Three-dimensional Fourier synthesis of calf liver cytochrome b5 at 2.8 A resolution. *J. Mol. Biol.* **64**:449–464.
2. Mathews FS, Argos P, Levine M (1971) The structure of cytochrome b5 at 2.0 Å resolution. *Cold Springs Harbor Symposia on Quantitative Biology* **36**:387–395.
3. Salemme F (1976) An hypothetical structure for an intermolecular electron transfer complex of cytochromes c and b5. *J. Mol. Biol.* **102**:563–568.
4. Poulos T, Mauk A (1983) Models for the complexes formed between cytochrome b5 and the subunits of methemoglobin. *J. Biol. Chem.* **258**:7369–7373.
5. Passon P, Hultquist D (1972) Soluble cytochrome b5 reductase from human erythrocytes. *Biochim. Biophys. Acta* **275**:62–73.
6. Loverde A, Strittmatter P (1968) The role of lysyl residues in the sturcture and reactivity of cytochrome b5 reductase. *J. Biol. Chem.* **243**:5779–5787.
7. Dailey H, Strittmatter P (1979) Modification and identification of cytochrome b5 carboxyl groups involved in protein-protein interaction with cytochrome b5 reductase. *J. Biol. Chem.* **254**:5388–5396.
8. Dailey H, Strittmatter P (1980) Characterization of the interaction of amphipathic cytochrome b5 with stearyl coenzyme A desaturase and NADPH: cytochrome P-450 reductase. *J. Biol. Chem.* **255**:5184–5189.
9. Conney A, Brown R, Miller J, Miller E (1957) The metabolism of methylated aminoazo dyes. VI. Intracellular distribution and properties of the demethylase system. *Cancer Res.* **17**:628–633.
10. Cohen BS, Estabrook RW (1971) Microsomal electron transport reactions. III. Cooperative interactions between reduced diphosphopyridine nucleotide and reduced triphosphopyridine nucleotide linked reactions. *Arch. Biochem. Biophys.* **143**:54–65.
11. Hildebrandt A, Estabrook R (1971) Evidence for the participation of cytochrome b5 in hepatic microsomal mixed-function oxidation reactions. *Arch. Biochem. Biophys.* **143**:66–79.
12. Mannering GJ, Kuwahara S, Omura T (1974) Immunochemical evidence for the participation of cytochrome b5 in the NADH synergism of the NADPH-dependent mono-oxidase system of hepatic microsomes. *Biochem. Biophys. Res. Commun.* **57**:476–481.
13. Jansson I, Schenkman JB (1977) Studies on three microsomal electron transfer systems. Specificity of electron flow pathways. *Arch. Biochem. Biophys.* **178**:89–107.
14. Chiang J (1981) Interaction of purified microsomal cytochrome P-450 with cytochrome b5. *Arch. Biochem. Biophys.* **211**:662–673.
15. Tamburini P, White R, Schenkman J (1985) Chemical characterization of protein-protein interactions between cytochrome P-450 and cytochrome b5. *J. Biol. Chem.* **260**:4007–4015.
16. Tamburini P, Schenkman J (1986) Mechanism of interaction between cytochromes P-450 RLM5 and b5: evidence for an electrostatic mechanism involving cytochrome b5 heme propionate groups. *Arch. Biochem. Biophys.* **245**:512–522.
17. Voznesensky A, Schenkman J (1994) Quantitative analysis of electrostatic interactions between NADPH-cytochrome P450 reductase and cytochrome P450 enzymes. *J. Biol. Chem.* **269**:15724–15731.
18. Rodgers KK, Sligar SG (1991) Mapping electrostatic interactions in macromolecular associations. *J. Mol. Biol.* **221**:1453–1460.
19. Mueller GC, Miller JA (1949) The reductive cleavage of 4-dimethylaminoazobenzene by rat liver: the intracellular distribution of the enzyme system and its requirement for triphosphopyridine nucleotide. *J. Biol. Chem.* **180**:1125–1136.
20. LaDu B, Gaudette L, Trousof N, Brodie B (1955) Enzymatic dealkylation of aminopyrine (pyramidon) and other alkylamines. *J. Biol. Chem.* **214**:741–752.
21. Axelrod J (1955) The enzymatic demethylation of ephedrine. *J. Pharmacol. Exp. Ther.* **114**:430–438.
22. Nilsson A, Johnson B (1963) Cofactor requirements of the o-demethylating liver microsomal enzyme system. *Arch. Biochem. Biophys.* **101**:494–498.

23. Correia M, Mannering G (1973) Reduced diphosphopyridine nucleotide synergism of the reduced triphosphopyridine nucleotide-dependent mixed-function oxidase system of hepatic microsomes. II. Role of the type I drug-binding site of cytochrome P-450. *Mol. Pharmacol.* **9**:470–485.

24. Lu A, West S, Vore M, Ryan D, Levin W (1974) Role of cytochrome b5 in hydroxylation by a reconstituted cytochrome P-450-containing system. *J. Biol. Chem.* **249**:6701–6709.

25. Hrycay E, Estabrook R (1974) The effect of extra bound cytochrome b5 on cytochrome P-450- dependent enzyme activities in liver microsomes. *Biochem. Biophys. Res. Commun.* **60**:771–778.

26. Jansson I, Schenkman J (1973) Evidence against participation of cytochrome b5 in the hepatic microsomal mixed function oxidase reaction. *Mol. Pharmacol.* **9**:840–845.

27. Lu AYH, Levin W, Selander H, Jerina DM (1974) Liver microsomal electron transport systems. III. The involvement of cytochrome b5 in the NADPH-supported cytochrome P-450-dependent hydroxylation of chlorobenzene. *Biochem. Biophys. Res. Commun.* **61**:1348–1355.

28. Morgan E, Coon M (1984) Effects of cytochrome b5 on cytochrome P-450-catalyzed reactions. *Drug Metab. Dispos.* **12**:358–364.

29. Jansson I, Schenkman JB (1987) Influence of cytochrome b5 on the stoichiometry of the different oxidative reactions catalyzed by liver microsomal cytochrome P450. *Drug Metab. Dispos.* **15**:344–348.

30. Sugiyama T, Miki N, Yamano T (1979) The obligatory requirement of cytochrome b5 in the p-nitroanisole o-demethylation reaction catalyzed by cytochrome P-450 with a high affinity for cytochrome b5. *Biochem. Biophys. Res. Commun.* **90**:715–720.

31. Miki N, Sugiyama T, Yamano T, Miyake Y (1981) Characterization of a highly purified form of cytochrome P-450B1. *Biochem. Int.***3**:217–223.

32. Eberhart D, Parkinson A (1991) Cytochrome-P450 IIIA1 (P450p) requires cytochrome-b5 and phospholipid with unsaturated fatty acids. *Arch. Biochem. Biophys.* **291**:231–240.

33. Shet MS, Faulkner KM, Holmans PL, Fisher CW, Estabrook RW (1995) The effects of cytochrome b5, NADPH-P450 reductase, and lipid on the rate of 6β-hydroxylation of testosterone as catalyzed by a human P450 3A4 fusion protein. *Arch. Biochem. Biophys.* **318**:314–321.

34. Tamburini P, Schenkman J (1987) Purification to homogeneity and enzymological characterization of a functional covalent complex composed of cytochromes P-450 isozyme 2 and b5 from rabbit liver. *Proc. Natl. Acad. Sci. USA* **84**:11–15.

35. Schenkman JB, Voznesensky AI, Jansson I (1994) Influence of ionic strength on the P450 monooxygenase reaction and role of cytochrome b5 in the process. *Arch. Biochem. Biophys.* **314**:234–241.

36. Backes W, Eyer C (1989) Cytochrome P-450 LM2 reduction. Substrate effects on the rate of reductase-LM2 association. *J. Biol. Chem.* **264**:6252–6259.

37. Yamazaki H, Johnson WW, Ueng Y-F, Shimada T, Guengerich FP (1996) Lack of electron transfer from cytochrome b5 in stimulation of catalytic activities of cytochrome P450 3A4. Characterization of a reconstituted cytochrome P450 3A4/NADPH-cytochrome P450 reductase system and studies with apo-cytochrome b5. *J. Biol. Chem.* **271**:27438–27444.

38. Sugiyama T, Miki Y, Miyake Y, Yamano T (1982) Interaction and electron transfer between cytochrome b5 and cytochrome P-450 in the reconstituted p-nitroanisole O-demethylase system. *J. Biochem.* **92**:1793–1803.

39. Lambeth J, Geren L, Millett F (1984) Adrenodoxin interaction with adrenodoxin reductase and cytochrome P-450scc. Cross-linking of protein complexes and effects of adrenodoxin modification by 1-ethyl-3-(3-dimethylaminopropyl)carbodiimide. *J. Biol. Chem.* **259**:10025–10029.

40. Geren L, Tuls J, O'Brien P, Millett F, Peterson J (1986) The involvement of carboxylate groups of putidaredoxin in the reaction with putidaredoxin reductase. *J. Biol. Chem.* **261**:15491–15495.

41. Nelson D, Strobel H (1989) Secondary structure prediction of 52 membrane-bound cytochromes P450 shows a strong structural similarity to P450cam. *Biochemistry* **28**:656–660.

42. Gotoh O, Fujii-Kuriyama Y (1989) Evolution, structure, and gene regulation of cytochrome P-450. in *Frontiers in Biotransformation*Vol. 1 (Ruckpaul K, Rein H, eds.), pp. 195–243, Akademie-Verlag, Berlin

43. Hasemann CA, Kurumbail RG, Boddupalli SS, Peterson JA, Deisenhofer J (1995) Structure and function of cytochromes P450: a comparative analysis of three crystal structures. *Structure* **2**:41–62.

44. Pyerin W, Wolf C, Kinzel V, Kubler D, Oesch F (1983) Phosphorylation of cytochrome P-450-dependent monooxygenase components. *Carcinogenesis* **4**:573–576.

45. Muller R, Schmidt W, Stier A (1985) The site of cyclic AMP-dependent protein kinase catalyzed phosphorylation of cytochrome P-450 LM2. *FEBS Letters* **187**:21–24.

46. Jansson I, Epstein P, Bains S, Schenkman J (1987) Inverse relationship between cytochrome P-450 phosphorylation and complexation with cytochrome b5. *Arch. Biochem. Biophys.* **259**:441–448.

47. Epstein P, Curti M, Jansson I, Huang C-K, Schenkman J (1989) Phosphorylation of cytochrome P-450: regulation by cytochrome b5. *Arch. Biochem. Biophys.* **271**:424–432.

48. Juvonen R, Iwasaki M, Negishi M (1992) Roles of residue-129 and residue-209 in the alteration by cyto-chrome-b5 of hydroxylase activities in mouse-2A-P450s. *Biochemistry* **31**:11519–11523.

49. Stayton P, Poulos T, Sligar S (1989) Putidaredoxin competitively inhibits cytochrome b5-cytochrome P-450cam association: A proposed molecular model for a cytochrome P-450cam electron-transfer complex. *Biochemistry* **28**:8201–8205.

50. von Bodman S, Schuler M, Jollie D, Sligar S (1986) Synthesis, bacterial expression, and mutagenesis of the gene coding for mammalian cytochrome b5. *Proc. Natl. Acad. Sci. USA* **83**:9443–9447.

51. Stayton P, Sligar S (1990) The cytochrome P-450cam binding surface as defined by site-directed mu-tagenesis and electrostatic modeling. *Biochemistry* **29**:7381–7386.

52. Nelson D, Strobel H (1988) On the membrane topology of vertebrate cytochrome P-450 proteins. *J. Biol. Chem.* **263**:6038–6050.

53. Koga H, Sagara Y, Yaoi T, Tsujimura M, Nakamura K, Sekimizu K, Makino R, Shimada H, Ishimura Y, Yura K and others (1993) Essential role of the Arg112 residue of cytochrome- P450cam for electron trans-fer from reduced putidaredoxin. *FEBS Letters* **331**:109–113.

54. Nakamura K, Horiuchi T, Yasukochi T, Sekimizu K, Hara T, Sagara Y (1994) Significant contribution of arginine-112 and its positive charge of Pseudomonas putida cytochrome P-450(cam) in the electron trans-port from putidaredoxin. *Biochim. Biophys. Acta* **1207**:40–48.

55. Chang YT, Stiffelman OB, Loew GH (1996) Computer modeling of 3D structures of cytochrome P450s. *Biochimie* **78**:771–779.

56. Shen S, Strobel H (1992) The role of cytochrome-P450 lysine residues in the interaction between cyto-chrome-P450IA1 and NADPH- cytochrome-P450 reductase. *Arch. Biochem. Biophys.* **294**:83–90.

57. Shen S, Strobel H (1993) Role of lysine and arginine residues of cytochrome- P450 in the interaction be-tween cytochrome-P4502B1 and NADPH-cytochrome P450 reductase. *Arch. Biochem. Biophys.* **304**:257–265.

58. Shimizu T, Tateishi T, Hatano M, Fujii-Kuriyama Y (1991) Probing the role of lysines and arginines in the catalytic function of cytochrome P450d by site-directed mutagenesis. *J. Biol. Chem.* **266**:3372–3375.

59. Ravichandran K, Boddupalli S, Hasemann C, Peterson J, Deisenhofer J (1993) Crystal structure of hemo-protein domain of P450BM-3, a prototype for microsomal P450's. *Science* **261**:731–736.

60. Chuev GN, Lakhno VD (1993) A polaron model for electron transfer in globular proteins. *J. Theor. Biol.* **163**:51–60.

61. Poulos T, Kraut J (1980) A hypothetical model of the cytochrome c peroxidase-cytochrome c electron transfer complex. *J. Biol. Chem.* **255**:10322–10330.

62. Beratan D, Onuchic J, Winkler J, Gray H (1992) Electron-tunneling pathways in protein. *Science* **258**:1740–1741.

63. Davies M, Qin L, Beck J, Suslick K, Koga H, Horiuchi T, Sligar S (1990) Putidaredoxin reduction of cyto-chrome P-450cam-dependence of electron transfer on the identity of putidaredoxin C-terminal amino acid. *J. Amer. Chem. Soc.* **112**:7396–7398.

64. Stayton PS, Sligar SG (1991) Structural microheterogeneity of a tryptophan residue required for efficient biological electron transfer between putidaredoxin and cytochrome P-450cam. *Biochemistry* **30**:1845–1851.

65. Baldwin J, Morris G, Richards W (1991) Electron transport in cytochromes-P-450 by covalent switching. *Proc. Royal Soc. B* **245**:43–51.

66. Davies D, Lawther J (1989) Kinetics and mechanism of electron transfer from dithionite to microsomal cy-tochrome b5 and to forms of the protein associated with charged and neutral vesicles. *Biochem. J* **258**:375–380.

67. Lederer F (1994) The cytochrome b5-fold: an adaptable module. *Biochimie* **76**:674–692.

68. Guillemette JG, Barker PD, Eltis LD, Lo TP, Smith M, Brayer GD, Mauk AG (1994) Analysis of the bi-molecular reduction of ferricytochrome c by ferrocytochrome b5 through mutagenesis and molecular mod-elling. *Biochimie* **76**:592–604.

69. Liang N, Mauk A, Pielak G, Johnson J, Smith M, Hoffman B (1988) Regulation of interprotein electron transfer by residue 82 of yeast cytochrome c. *Science* **240**:311–313.

70. Davies M, Sligar S (1992) Genetic variants in the putidaredoxin cytochrome-P- 450(cam) electron-transfer complex - identification of the residue responsible for redox-state-dependent conformers. *Biochemistry* **31**:11383–11389.

71. Hazzard J, McLendon G, Cusanovich M, Tollin G (1988) Formation of electrostatically-stabilized complex at low ionic strength inhibits interprotein electron transfer between yeast cytochrome c and cytochrome c peroxidase. *Biochem. Biophys. Res. Commun.* **151**:429–434.

72. Hoffman BM, Ratner MA (1987) Gated electron transfer: when are observed rates controlled by conforma-tional interconversion? *J. Am. Chem. Soc.* **109**:6237–6243.

73. Zhang Z, Nassar A, Lu Z, Schenkman JB, Rusling JF (1997) Direct electron injection from electrodes to cytochrome P450cam in biomembrane-like films. *J. Chem. Soc., Faraday Trans.* **93**:1769–1774.
74. Laviron E (1979) General expression of the linear potential sweep voltammogram in the case of diffusionless electrochemical systems. *J. Electrochem. Chem.* **101**:19–28.

3

STRUCTURES OF MITOCHONDRIAL P450 SYSTEM PROTEINS

Israel Hanukoglu*

E. Katzir Biotechnology Program
The Research Institute
The College of Judea and Samaria
Ariel, 44837 Israel

ABSTRACT

Mitochondrial P450 type enzymes are generally involved in the metabolism of cholesterol derived steroidal compounds. The reactions catalyzed by these enzymes include cholesterol conversion to pregnenolone, 11-beta and 18 hydroxylation reactions in adrenal steroid biosynthesis, C-27 hydroxylation of cholic acid in bile acid metabolism, and 1alpha and 24 hydroxylations of vitamin D. These P450 mediated reactions require molecular oxygen and two electrons donated by NADPH. The electrons of NADPH are transferred to P450 by an electron transfer system that includes a specific flavoprotein, adrenodoxin reductase, and an iron-sulfur protein, adrenodoxin. These proteins are not specific for individual P450s and serve as electron donors for different P450 in different tissues. This review presents an overview of the major sequence and structural characteristics of the mitochondrial P450 system proteins.

1. INTRODUCTION

Mitochondrial P450 type enzymes are generally involved in the metabolism of cholesterol derived steroidal compounds. The reactions catalyzed by these enzymes include cholesterol conversion to pregnenolone, 11β and 18 hydroxylation reactions in steroid biosynthesis, C-27 hydroxylation of cholic acid in bile acid metabolism, and 1α and 24'hydroxylations of 25-OH-vitamin D (Table 1).[1-6] These reactions are catalyzed by specific P450s and follow the usual monooxygenation stoichiometry.[7,8]

---

* E-mail: israel@mail.com

Molecular and Applied Aspects of Oxidative Drug Metabolizing Enzymes,
edited by Arınç *et al*. Kluwer Academic / Plenum Publishers, New York, 1999.

Table 1. Reactions catalyzed by the mitochondrial P450 systems

P450	Gene	Major reaction	Highest levels in
P450scc	CYP11A1	Cholesterol side chain cleavage	Steroidogenic cells in adrenal cortex and gonads
P450c11	CYP11B1	Steroid 11β-hydroxylation	Zona fasciculata of adrenal cortex
P450c18	CYP11B2	Steroid C-18 hydroxylation	Zona glomerulosa of adrenal cortex
P450cc24	CYP24	25-OH-vitamin D3-24 hydroxylation	Kidney tubules
P450c27	CYP27A	Sterol C-27 hydroxylation Vitamin D3-25 hydroxylation	Liver
P4501?	CYP27B	25-OH-vitamin D3-1α hydroxylation	Kidney

$$R-H + NADPH + H^+ + O_2 \longrightarrow R-OH + NADP^+ + H_2O$$

In these reactions NADPH donates two electrons which are transferred to P450 via two electron transfer proteins, adrenodoxin reductase, which is an FAD containing flavoenzyme, and adrenodoxin, which is a [2Fe-2S] ferredoxin type iron-sulfur protein.[1,2,9,10] FAD of adrenodoxin reductase accepts two electrons from NADPH, and these are transferred one at a time to adrenodoxin which is a one electron carrier. There is evidence that the availability of NADPH to the mitochondrial P450 systems is regulated in coordination with steroid biosynthesis.[11] The mitochondrial P450 electron transfer system is similar to that of some bacterial P450's such as P450cam from Pseudomonas putida that includes a ferredoxin reductase and a ferredoxin (named putidaredoxin) as the electron transfer proteins.[9,12]

All three proteins of mitochondrial P450 systems are located on the matrix side of the inner mitochondrial membrane.[1,13] Whereas the mitochondrial P450s are tightly bound to the membrane, the electron transfer proteins are soluble in the matrix. All three proteins are encoded as larger precursors, and their signal peptides are cleaved during transfer into mitochondria.[14,15]

In contrast to the multiplicity of P450 forms,[12,16] there is generally only one form of adrenodoxin reductase, and adrenodoxin, encoded by one or two similar nuclear genes in all animal species.1 Thus, the electron transfer proteins are not specific to individual P450s and serve as electron donors for different cytochromes P450 in different tissues. Adrenodoxin reductase and adrenodoxin are expressed in all human tissues examined.[1,16] Their highest levels of expression are observed in steroidogenic cells especially in adrenal cortex and ovarian corpus luteum.[1] The levels of these proteins show no significant sex, or interindividual variation in bovine adrenal cortex.[17]

Previous studies established several general principles of function for P450 system electron transport chains:[7–10] 1) The reductases are generally expressed at much lower levels than P450s, there being only one molecule of reductase per about 10 or more molecules of P450. 2) The protein components are independently mobile and do not form static multicomponent complexes. 3) Proteins that are redox partners form transient high affinity 1:1 complexes during their random diffusions, in accordance with the principles of mass action. Dissociation constants of these protein-protein complexes are strongly influenced by the redox states of the proteins and other molecules in the environment, such as P450 substrate, ions, and phospholipids. 4) The transfer of an electron between two redox partners depends on the formation of a specific high affinity 1:1 complex between the two proteins. In P450 systems electron transfer is not always coupled to substrate monooxygenation. P450s and their electron transfer proteins may transfer electrons to other acceptors, such as O_2. This type of "uncoupling" or "leaky electron transport" is observed in both mitochondrial and microsomal systems.[7,9,11] The regulation of protein-protein complex formation generally enhances productive associations for monooxygenase activities and helps to minimize uncoupled reactions that produce harmful free radicals.

The sections below present a general overview of the major sequence and structural characteristics of adrenodoxin reductase, adrenodoxin and mitochondrial P450s. As the extensive number of references are already listed in the sequence databases, the codes of the proteins are provided in the tables below for retrieval of individual references.

2. STRUCTURE OF ADRENODOXIN REDUCTASE

In both human and bovine genomes there is only one gene that encodes for adrenodoxin reductase.[1] The sequences of human, bovine, rat, and mouse adrenodoxin reductase have been deduced from cloned DNAs (Table 1). Whereas the signal peptides of these proteins share 55–82% identity, the mature peptide sequences show 88–95% identity among the enzymes from four different species (Figure 1). In addition to these, a yeast gene homologous to the mammalian adrenodoxin reductase gene has been identified, yet its function remains uncharacterized.[18] Despite functional similarities, adrenodoxin reductase shows no sequence homology with the bacterial P450cam system putidaredoxin reductase or other types of oxidoreductases.[19]

In immunoelectron microscopy of adrenal cells, adrenodoxin reductase appears as membrane associated.[20] However, its sequence does not have a hydrophobic membrane spanning segment (Fig. 1) and, it probably functions as a peripheral membrane protein.

The FAD and NAD(P) binding sites of adrenodoxin reductase were identified using an ADP dinucleotide binding site consensus motif.[19] FAD and NAD(P) both have ADP as a common part of their structures. In most FAD or NAD(P) binding enzymes, the sites that bind this ADP portion also share a similar conformation of a βαβ-fold. The most highly conserved sequence in this fold is Gly-X-Gly-X-X-Gly/Ala which forms a tight turn between the first β-strand and the α-helix.[19,21,22] Analysis of adrenodoxin reductase sequence led to the discovery that in NADP binding sites of this type, there is an Ala instead of the third Gly residue, and it was proposed that this is a major determinant of NADP vs. NAD specificity of enzymes.[19] This hypothesis was verified for glutathione reductase.[23,24]

The full-length sequence of adrenodoxin reductase shows no similarity to that of the bacterial putidaredoxin reductase, yet, the FAD and NAD(P) motifs appear in both enzymes at nearly identical positions:[9] the FAD site at the amino terminus, and the NAD(P) site at 146–151 residues from the amino terminus (Fig. 1). Similar spacing of the FAD and NADP sites is also observed in many other flavoenzymes.[19] The sequence of the FAD binding amino terminus is highly conserved across species.[25] A bovine adrenodoxin reductase cDNA expressed in yeast with an extra segment encoding four additional residues at the N-terminus did not yield an active enzyme, suggesting that the four residues disrupted the incorporation of FAD into the apoprotein.[26] One alternative splicing product of adrenodoxin reductase gene encodes six extra residues in both bovine and human genes.[27,28] However, this form represents ~1% of the total reductase mRNA population,[28] and expression of its cDNA in E. coli did not yield active enzyme, suggesting that the structure of the enzyme.[29]

Whereas the bovine adrenodoxin reductase was reported to be glycosylated,[30,31] a recent study could not confirm previous evidence for functional glycosylation of the bovine

```
                 |         20        |         40        |         60
Human  : MASRCWRWWGWSAWPRTRLPPAGSTPS--FCHHFSTQEKTPQICVVGSGPAGFYTAQHLL  58
Bovine : ..P......P..S.T......SR.IQN--.GQ..G....Q..........G..G.....G...  58
Rat    : ..P.....S....GV.PL.SR...TPG..KK.....T.................G...  60
Mouse  : ..P...H..R....SGL.PS.SR...TPG..QK....................G...  60
                                                          FAD binding

                 |         80        |        100        |        120
Human  : K-HPQAHVDIYEKQPVPFGLVRFGVAPDHPEVKNVINTFTQTAHSGRCAFWGNVEVGRDV 117
Bovine : .H.SR........L.....................R.D....Y......... 118
Rat    : .H.TR........L.....................R.D....R...V..... 120
Mouse  : .H.TH........L.....................R.D....Q...V..... 120

                 |        140        |        160        |        180
Human  : TVPELQEAYHAVVLSYGAEDHRALEIPGEELPGVCSARAFVGWYNGLPENQELEPDLSCD 177
Bovine : ..Q...D.............Q..D.........F..............R..A...... 178
Rat    : S....R.............QP.........V.............K.A...... 180
Mouse  : S....R.............QP.G.......V..............A...... 180

                 |        200        |        220        |        240
Human  : TAVILCQCNVALDVARILLTPPEHLERTDITKAALGVLRQSRVKTVWLVGRRGPLQVAFT 237
Bovine : .....................D...K...E....A.........I........... 238
Rat    : .....................K...EV.............I........... 240
Mouse  : .....................K...E....A.........I........... 240
          NADP binding

                 |        260        |        280        |        300
Human  : IKELREMIQLPGARPILDPVDFLGLQDKIKEVPRPRKRLTELLLRTATEKPGPAEAARQA 297
Bovine : ..........T..M...A......R...AA.....M...........VE....R. 298
Rat    : ..........TQ.....S.......R..D...............VE....R. 300
Mouse  : ..........T.....S.......R..D.....R..........VE..... 300

                 |        320        |        340        |        360
Human  : SASRAWGLRFFRSPQQVLPSPDGRRAAGVRLAVTRLEGVDEATRAVPTGDMEDLPCGLVL 357
Bovine : ........................I.........IG.........V........ 358
Rat    : L.............T.....V..I.....G.S......V......L. 360
Mouse  : L.............T...Q.V..I.....S....G.S......V......L. 360

                 |        380        |        400        |        420
Human  : SSIGYKSRPVDPSVPFDSKLGVIPNVEGRVMDVPGLYCSGWVKRGPTGVIATTMTDSFLT 417
Bovine : ........I......P....V..M....V...............T........ 418
Rat    : ..V......I......P...I...T....VNA...............T........ 420
Mouse  : ..V......I......P........T....VN...............T........ 420

                 |        440        |        460        |        480
Human  : GQMLLQDLKAGLLPSGPRPGYAAIQALLSSRGVRPVSFSDWEKLDAEEVARGQGTGKPRE 477
Bovine : ..I........H.....S.F.K...D....W...........S...AS..... 478
Rat    : S.V..K.............T.....D................S.......... 480
Mouse  : S.A..E.............V......N................S.......... 480

                 |
Human  : KLVDPQEMLRLLGH 491
Bovine : ..L.......... 492
Rat    : ....RR...Q.... 494
Mouse  : ....RR........ 494
```

Figure 1. Alignment of adrenodoxin reductase sequences from vertebrate species. The amino terminal Serine of the mature peptides, and the conserved glycines and alanines in the FAD and NADP binding sites are marked by background shading. Dashes in sequence represent gaps inserted for alignment.

Table 2. Characteristics of adrenodoxin reductase and adrenodoxin amino acid sequences determined from cloned cDNA or gene sequences

Species	SWISSPROT code	Prepeptide	Signal peptide	Mature peptide Length	Mature peptide MW
Adrenodoxin reductase					
Human	adro_human	491	32	459	49967
Bovine	adro_bovin	492	32	460	50296
Rat	adro_rat	494	34	460	50316
Mouse	adro_mouse	494	34	460	50128
Yeast	adro_yeast	493			
Adrenodoxin					
Human	adx_human	184	60	124	13561
Bovine	adx1_bovin	186	58	128	14048
Sheep	adx_sheep	ND			
Pig	adx_pig	186	58	128	14012
Rat	adx_rat	188	64	124	13588
Mouse	adx_mouse	188	64	124	13617
Chick	adx_chick	ND		124	13558

ND: Not determined.

enzyme.[32] The porcine enzyme is free of carbohydrate.[33] Adrenodoxin reductase expressed in E. coli functioned as well as the native enzyme in a reconstituted mitochondrial P450scc system. Thus, the apoprotein may be assembled to active holoenzyme without eukaryotic posttranslational modifications.[34]

Adrenodoxin reductase has been crystallized in several laboratories.[35,36,37] The elucidation of its crystal structure is necessary to increase our understanding beyond the sequence analyses, to identify in detail the cofactor pockets and the adrenodoxin binding sites, and to elucidate the routes of electron transfer.

3. STRUCTURE OF ADRENODOXIN

The sequence of adrenodoxin has been deduced from cloned DNAs in six mammalian species and in chicken (Table 2, Figure 2). Similar to adrenodoxin reductase, the mature peptide sequence of adrenodoxin is much more conserved than its signal peptide: Whereas the homology of the mature peptide ranges between 81–97% among these species, the homology of the signal peptide ranges between 31–67% (except for rat and mouse 82%). Adrenodoxin sequence shows significant homology with bacterial putidaredoxin and their sequences can be aligned over their entire lengths with only a few gaps.[38] A [2Fe-2S] ferredoxin from E. coli also shares 36% identity with adrenodoxin and putidaredoxin.[39] Waki et al.[40] reported the purification of a [2Fe-2S] protein from bovine liver mitochondria that can support C-25 and C-27 hydroxylations of steroids, but with an amino terminus sequence completely different from adrenodoxin.

Bovine adrenodoxin is translated from multiple species of mRNA encoded by a single gene.[41] The protein sequences encoded by these mRNAs differ only in the C-terminus of the signal peptide and the first two residues of the mature sequence.[42] In the human genome there are two genes, but both encode the same protein product.[43] In polyacrylamide gel electrophoresis, adrenodoxin purified from different tissues two major bands or a broad band may be observed around ~12–14 kDa.[44,45] This heterogeneity was considered

```
              |       20      |       40      |       60
Human  : MAAAGGARLLRAASAVLG--GPAGRWLHHAGSRAGSSGLLRNRGP--GGSAEASRSLSVS  56
Bovine: ..-----.....V...A..--DT....RLL.RP...AG..RGS...GL..G.V.T.T....  54
Sheep  : ------------------------------------------------------------   -
Pig    : ..-----V....V...A..--DT.V..QPLV.P...NR.PGGSIWLGL..R.A.A.T..L.  54
Rat    : ....P........C.SVAFR.LDC.R.LVC.T...PAVPQWTPS.HTLAE.GPG.P....  60
Mouse  : ....P........C.SVPFR.LDRCR.LVC.TG..TAISPWTPS.RLHAE.GPG.P....  60
Chick  : ----------------------------------------------CS--AVAVRTL.P..L.  15


              |       80      |      100      |      120
Human  : ARARSSSEDKITVHFINRDGETLTTKGKVGDSLLDVVVENNLDIDGFGACEGTLACSTCH 116
Bovine: G..Q.................................I.........Q...........  114
Sheep  : ----------V..N............................................  50
Pig    : ...W........................K....Q..............I..........  114
Rat    : ............V....K..............................I..........  120
Mouse  : ................K..............I.........I..........  120
Chick  : ...AC...............DK..A...P............................  75


              |      140      |      160      |      180
Human  : LIFEDHIYEKLDAITDEENDMLDLAYGLTDRSRLGCQICLTKSMDNMTVRVPETVADARQ 176
Bovine: ....Q..F...E.................|......A.........DA.S...E 174
Sheep  : ....Q....E..................|......A.........DA.S...E 110
Pig    : .......F...E................|......A..........A.....E 174
Rat    : .....................F...N....|.V....A..........A...V.. 180
Mouse  : .....................F.......|.V....A..........A...V.. 180
Chick  : .......F...........M........ET...|....K............A...... 135


              |
Human  : SIDVGKTS      184
Bovine: ...M.MN.SKIE  186
Sheep  : ...M.MN.SKIE  122
Pig    : ...L..N.SKLE  186
Rat    : .V.MS.N.      188
Mouse  : .V.MS.N.      188
Chick  : .V.LS.N.      143
```

Figure 2. Alignment of the adrenodoxin sequences from six vertebrate species. The amino terminal Serine of the mature peptides, and the cysteines that are involved in the formation of [2Fe-2S] complex are marked by background shading.

to represent a multiplicity of tissue specific forms of adrenodoxin. However, it apparently results from proteolytic cleavage of up to 14 residues from the carboxy terminus of the mature adrenodoxin during the purification process.[46,47] Trypsin treatment of purified adrenodoxin produces a truncated form of adrenodoxin (des 116–128) which shows a lower Km (higher affinity) in supporting P450 activity.[48]

In the [2Fe-2S] center two iron atoms are coordinated to four cysteines and two labile sulfur atoms. Mature bovine adrenodoxin sequence includes five cysteines. Chemical modification and site-directed mutagenesis studies indicated that Cys-46, 52, 55, and 92 are involved in iron-sulfur coordination, whereas Cys-95 is free.[38,49,50]

A cluster of negatively charged residues of bovine adrenodoxin have been implicated in complex formation with both adrenodoxin reductase and mitochondrial P450s in studies employing different approaches.[51–54] Site directed mutation of Asp-76 and Asp-79 showed that these residues are essential for adrenodoxin binding to both adrenodoxin reductase and P450scc.[54] In contrast, modification of all lysine or arginine residues did not affect adrenodoxin interactions with either of its redox partners, suggesting that these are not located at

Table 3. Characteristics of mitochondrial P450 amino acid sequences determined from cloned cDNA or gene sequences

Species	SWISSPROT Code	Prepeptide	Signal peptide	Mature peptide	
				Length	MW
P450scc					
Human	cpm1_human	521	39	482	56117
Bovine	cpm1_bovin	520	39	481	56398
Sheep	cpm1_sheep	520	39	481	56399
Goat	cpm1_caphi	520	39	481	56304
Pig	cpm1_pig	520	39	481	56240
Rabbit	q28827				
Rat	cpm1_rat	526	36	490	56946
Trout	cpm1_oncmy	514	39	475	55324
P450c11					
Human	cpn1_human	503	24	479	54889
Baboon	cpn1_papha	503	24	479	55004
Bovine	cpn1_bovin	503	24	479	55087
Sheep	cpn1_sheep	503	24	479	55028
Pig	cpn1_pig	503	24	479	54647
Cavia	cpn1_cavpo	500	24	476	55146
Rat	cpn1_rat	499	24	475	54612
Mouse	cpn1_mouse	500	24	476	54468
Hamster	cpn1_mesau	499	24	475	54015
Frog	cpn1_ranca	517	45	472	54492
P450c18					
Human	cpn2_human	503	24	479	54934
Rat	cpn2_rat	500	24	476	54274
Mouse	q64661	500	24	476	54568
Hamster	cpn2_mesau	500	24	476	54526
P450c24					
Human	cp24_human	513	35	478	54935
Rat	cp24_rat	514	35	479	55537
Mouse	cp24_mouse	514	35	479	55427
P450c27A					
Human	cp27_human	531	33	498	56908
Rabbit	cp27_rabit	535	32	503	57033
Rat	cp27_rat	533	32	501	57185
P450c27B					
Human		508			
Rat		501			

the binding site for either protein.[52] These findings support the conclusions based on kinetic studies[9,10,55,56] that the binding sites of these two enzymes on adrenodoxin overlap. Yet, the differential effects of carboxy-terminal truncation at Arg-115 on interactions with adrenodoxin reductase and P-450 suggest that the sites are not identical.[48] Mutation of Tyr-82 did not affect reductase binding but changed Km values with P450scc and P450c11, suggesting that Tyr-82 may affect the binding of P450.[57] A mutated form of adrenodoxin missing six amino terminal residues supported only 60% of the activity of P450scc suggesting that the amino terminal residues may also play a role in P450scc binding or electron transfer.[58] Similarly deletion studies of C-terminus of bovine adrenodoxin indicate that this region, and especially Pro-108 are essential for the structural integrity of the protein.[59] Site-directed mutagenesis was also used in assigning the NMR signals from His residues.[60,61]

Table 4. Percent identity of P450c11 amino acid sequences from ten species. The bold numbers represent the number of amino acids that are identical

	Human	Baboon	Bovine	Sheep	Pig	Cavia	Rat	Mouse	Hamster	Frog
Human	**503**	96%	72%	74%	73%	63%	63%	67%	61%	44%
Baboon	485	**503**	72%	74%	73%	62%	63%	67%	61%	44%
Bovine	367	366	**503**	95%	80%	61%	59%	64%	58%	47%
Sheep	374	374	482	**503**	82%	62%	60%	65%	58%	47%
Pig	368	371	407	415	**503**	61%	61%	65%	58%	45%
Cavia	320	316	310	313	310	**500**	58%	65%	57%	42%
Rat	318	318	299	304	308	294	**499**	80%	74%	41%
Mouse	339	339	324	329	329	330	402	**500**	74%	45%
Hamster	310	307	296	294	293	289	370	372	**499**	38%
Frog	232	231	246	246	236	222	218	236	202	**517**

Bovine and chick adrenodoxins can be phosphorylated. This may affect their interaction with P450 and hence the activity of P450.[62,63] Adrenodoxin cDNAs expressed in E. coli encode proteins functionally as active as the native protein, indicating that the [2Fe-2S] centers of these proteins can be properly assembled in bacteria, and that eukaryote specific posttranslational modifications are not necessary for activity.[49,50,54,64]

The structure of adrenodoxin crystals currently available cannot be solved because of their complexity.[65] The elucidation of the complete structure of adrenodoxin awaits formation of different crystals suited for crystallographic analysis. Proton NMR studies indicate that the structure of adrenodoxin is similar to that of Spirulina platensis ferredoxin.[66]

4. STRUCTURES OF MITOCHONDRIAL P450s

The hydrophobic character of the mitochondrial P450s have hindered the crystallization of these proteins for crystallographic structure analysis. The structures of several soluble bacterial P450s have been determined.[12,67,68] Yet, the very low sequence similarity (<20%) between the mitochondrial and bacterial P450 sequences does not permit an exact alignment of these sequences along their entire lengths. A model for the 3-D structure of P450scc has been suggested based on the crystal structure of P450cam and alignment of the two sequences.[69] This model remains speculative because of the low similarity between these sequences. Thus, our present understanding of mitochondrial P450 structure is based on sequence comparisons and biochemical analyses of these proteins.

To date the sequences of six different types mitochondrial P450s have been determined from cloned DNAs from vertebrate species (Tables 1, and 3 and Figure 3). These P450s contain 24–39 residue long signal peptides. The sequences of the P450c11 and P450c18 specific for 11?- and 18-hydroxylation of steroids are most similar showing 91% amino acid sequence identity in humans. The sequences of other mitochondrial P450s in humans share between 24–37% sequence identity. The intron-exon organization of P450scc, P450c11 and P450c18 are highly similar, clearly indicating a common evolutionary origin at least for these three mitochondrial P450s.[70] Recently an insect P450 sequence has been isolated that shows strong homology to the vertebrate mitochondrial P450s, yet the function of this P450 remains to be determined.[71] There is immunological evidence that insect ecdysone 20-monooxygenase may be related to the vertebrate mitochondrial P450s.[72]

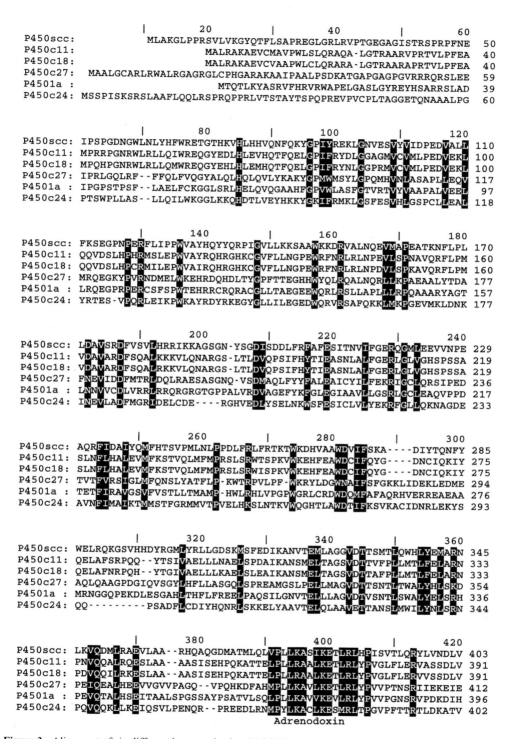

Figure 3. Alignment of six different human mitochondrial P450 sequences. Shaded regions mark identical residues. The positions of putative adrenodoxin binding and heme binding regions are marked below the sequences.

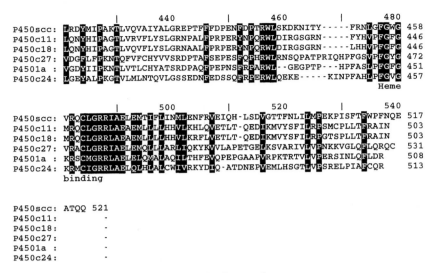

Figure 3. (*Continued*)

For each form of P450, interspecies homology varies in general according to the evolutionary relatedness of the species. The complete sequence of P450c11 has been determined in ten species (Table 3). Human and baboon P450c11 sequences share 96% identity. Whereas, frog P450c11 shares 38–47% sequence identity with other species (Table 4). If conservative substitutions are considered, (such as Thr for Ser, or Glu for Asp) then the percent similarity increases by up to 20% over the figures noted in Table 4.

Mitochondrial P450s behave as integral membrane proteins, as they are strongly hydrophobic and their isolation requires treatment with detergents. In contrast to the microsomal P450s, mitochondrial P450s do not have a hydrophobic membrane spanning segment. Thus, their mode of association with the inner mitochondrial membrane is still not understood. A mitochondrial P450c27 that was engineered to contain the microsomal targeting signal of bovine P450c17 front the mature form of rat P450c27 was localized in the microsomes and could utilize the microsomal NADPH-P450 reductase as an electron donor.[73] Yet, purified human P450c27 does not have the capability to utilize microsomal P450 reductase as an electron donor.[74] Thus, mitochondrial P450s may bind to the microsomal membranes and that there is no intrinsic requirement for the mitochondrial membrane environment for the function of the mitochondrial P450s. Mitochondrial P450s can also function in purified form in the absence of a membrane environment as well as in phospholipid vesicles.[75] The signal sequences of P450s are apparently not completely species specific as, in vitro synthesized bovine P450scc precursor can be imported into isolated soybean cotyledon mitochondria and processed therein to the mature size product.[76]

Since all the mitochondrial P450s bind heme and interact with adrenodoxin, the binding sites for these molecules should be conserved. All forms of P450, including bacterial and microsomal P450s show strong homology in the heme binding segment before the L-helix close to the carboxy termini of the enzymes.[68,77] All mitochondrial P450 sequences noted in Table 3 carry the consensus signature sequence FGxGxRxCxG in this region (Fig. 3).

The adrenodoxin binding site of P450scc was identified by chemical modification studies.[78] The sequence of this site is also conserved in P450scc from all species[79] as well as other mitochondrial P450s (Fig. 3). The labeling of Cys264 of P450scc affected the interaction of P450scc with adrenodoxin and significantly inhibited its enzymic activity.[80] Labeled and unlabelled enzymes were cleaved by trypsin and split into two fragments. It has been suggested that the hinge connecting the two domains in the region Arg250-Asn257 is exposed to the surface of the membrane and involved in the interaction of P450scc with adrenodoxin.[80]

In contrast to the conservation of binding sites for heme and adrenodoxin, the substrate binding sites of P450s would be expected to show divergence in accordance with the diversity of the substrates. Indeed among bacterial and microsomal P450s the sequences of structurally known or predicted sites of substrate binding show great diversity.[67,77] A study utilizing a suicide substrate for P450scc suggested that the substrate binding site may be located at its amino terminus.[81] Among the mitochondrial P450s the amino terminus is the least conserved segment (Fig. 3). However, a recent study showed that P450c11 could be converted into an enzyme with the activity of P450c18 by two residue substitutions of Ser288Gly and Val320Ala.[82] Thus, even if the amino terminus is involved in the formation of a substrate channel or binding site, critical residues are located in the middle of the polypeptide chain.

As noted for the electron transfer proteins, the elucidation of the structure of the mitochondrial P450s awaits isolation of enzyme crystals suitable for X-ray diffraction analysis.

REFERENCES

1. I. Hanukoglu, 1992, Steroidogenic enzymes: structure, function, and regulation of expression, J. Steroid Biochem. Mol. Biol. 43:779–804.
2. J.D. Lambeth, 1991, Enzymology of mitochondrial side-chain cleavage by cytochrome P-450scc, Frontiers Biotransformation 3:58–100.
3. N.R. Orme-Johnson, 1990, Distinctive properties of adrenal cortex mitochondria. Biochim. Biophys. Acta 1020:213–231.
4. P.C. White, K.M. Curnow, and L. Pascoe, 1994, Disorders of steroid 11 beta-hydroxylase isozymes. Endocr. Rev. 15:421–438.
5. K.I. Okuda, 1994, Liver mitochondrial P450 involved in cholesterol catabolism and vitamin D activation. J. Lipid Res. 35:361–372.
6. K. Okuda, E. Usui, Y. Ohyama, 1995, Recent progress in enzymology and molecular biology of enzymes involved in vitamin D metabolism. J. Lipid Res. 36:1641–1652.
7. A.I. Archakov, and G.I. Bachmanova, 1990, Cytochrome P-450 and active oxygen. Taylor and Francis, Hants, U.K.
8. J.B. Schenkman, and H. Greim (eds.), 1993, Cytochrome P450. (Handbook Expl. Pharmacol. vol. 105), Springer-Verlag, Berlin.
9. I. Hanukoglu, 1996, Electron transfer proteins of cytochrome P450 systems. Adv. Mol. Cell Biol. 14:29–55.
10. J.D. Lambeth, D.W. Seybert, J.R. Lancaster, J.C. Salerno, and H. Kamin, 1982, Steroidogenic electron transport in adrenal cortex mitochondria. Mol. Cell. Biochem. 45:13–31.
11. I. Hanukoglu, and R. Rapoport, 1995, Routes and regulation of NADPH production in steroidogenic mitochondria. Endocrine Res. 21:231–241.
12. K.N. Degtyarenko, 1995, Structural domains of P450-containing monooxygenase systems. Protein Engineering 8:737–747.
13. K. Ishimura, and H. Fujita, 1997, Light and electron microscopic immunohistochemistry of the localization of adrenal steroidogenic enzymes. Microsc. Res. Tech. 36:445–453.
14. T. Omura, 1993, Localization of cytochrome P450 in membranes: Mitochondria. Handb. Expl. Pharmacol. 105:61–69.

15. N. Hachiya, K. Mihara, K. Suda, M. Horst, G. Schatz, and T. Lithgow, 1995, Reconstitution of the initial steps of mitochondrial protein import. Nature 376:705–709.

16. D.R. Nelson, L. Koymans, T. Kamataki, J.J Stegeman, R. Feyereisen, D.J. Waxman, M.R. Waterman, O. Gotoh, M.J. Coon, R.W. Estabrook, I.C. Gunsalus, and D.W. Nebert, 1996, P450 superfamily: update on new sequences, gene mapping, accession numbers, and nomenclature. Pharmacogenetics 6:1–42.

17. T. Lacour, and B. Dumas, 1996, A gene encoding a yeast equivalent of mammalian NADPH-adrenodoxin oxidoreductases. Gene 174:289–292.

18. I. Hanukoglu, and T. Gutfinger, 1989, cDNA sequence of adrenodoxin reductase: Identification of NADP binding sites in oxidoreductases. Eur. J. Biochem. 180:479–484.

19. F. Mitani, 1979, Cytochrome P450 in adrenocortical mitochondria. Mol. Cell. Biochem. 24:21–43.

20. M.G. Rossman, A. Liljas, C.-I. Branden, and L.J. Banaszak, 1975, Evolutionary and structural relationships among dehydrogenases. Enzymes 9:61–102.

21. R.K. Wierenga, M.C.H. DeMaeyer, and W.G.J Hol, 1985, Interaction of pyrophosphate moieties with ?-helices in dinucleotide binding proteins. Biochemistry 24:1346–1357.

22. P.R.E. Mittl, A. Berry, N.S. Scrutton, R.N. Perham, and G.E. Schulz, 1993, Structural differences between wild-type NADP-dependent glutathione reductase from Escherichia coli and a redesigned NAD-dependent mutant. J. Mol. Biol. 231:191–195.

23. N.S. Scrutton, A. Berry, and R.N. Perham, 1990, Redesign of the coenzyme specificity of a dehydrogenase by protein engineering. Nature 343:38–43.

24. M. Yamazaki, and Y. Ichikawa, 1990, Crystallization and comparative characterization of reduced nicotiṇamide adenine dinucleotide phosphate-ferredoxin reductase from sheep adrenocortical mitochondria. Comp. Biochem. Physiol. 96B:93–100.

25. M. Akiyoshi-Shibata, T. Sakaki, Y. Yabusaki, H. Murakami, and H. Ohkawa, 1991, Expression of bovine adrenodoxin and NADPH-adrenodoxin reductase cDNAs in Saccharomyces cerevisiae. DNA Cell Biol. 10:613–621.

26. Y. Sagara, Y. Takata, T. Miyata, T. Hara, and T. Horiuchi, 1987, Cloning and sequence analysis of adrenodoxin reductase cDNA from bovine adrenal cortex. J. Biochem. 102:1333–1336.

27. S.B. Solish, J. Picado-Leonard, Y. Morel, R.W. Kuhn, T.K. Mohandas, I. Hanukoglu, and W.L. Miller, 1988, Human adrenodoxin reductase - Two mRNAs encoded by a single gene on chromosome 17cen-q25 are expressed in steroidogenic tissues. Proc. Natl. Acad. Sci. U.S.A. 85:7104–7108.

28. S.T. Brentano, S.M. Black, D. Lin, and W.L. Miller, 1992, cAMP post-transcriptionally diminishes the abundance of adrenodoxin reductase mRNA. Proc. Natl. Acad. Sci. U.S.A. 89:4099–4103.

29. M.E. Brandt, and L.E. Vickery, 1992, Expression and characterization of human mitochondrial ferredoxin reductase in Escherichia coli. Arch. Biochem. Biophys. 294:735–740.

30. A. Hiwatashi, Y. Ichikawa, N. Maruya, T. Yamano, and K. Aki, 1976, Properties of crystalline reduced nicotinamide adenine dinucleotide phosphate-adrenodoxin reductase from bovine adrenocortical mitochondria. I. Physicochemical properties of holo- and apo- NADPH-adrenodoxin reductase and interaction between non-heme iron proteins and the reductase. Biochemistry 15:3082–3090.

31. K. Suhara, K. Nakayama, O. Takikawa, and M. Katagiri, 1982, Two forms of adrenodoxin reductase from mitochondria of bovine adrenal cortex. Eur. J. Biochem. 125:659–664.

32. R.J. Warburton, and D.W. Seybert, 1995, Structural and functional characterization of bovine adrenodoxin reductase by limited proteolysis. Biochim. Biophys. Acta 1246:39–46.

33. A. Hiwatashi, and Y. Ichikawa, 1978, Crystalline reduced nicotinamide adenine dinucleotide phosphate-adrenodoxin reductase from pig adrenocortical mitochondria. Essential histidyl and cysteinyl residues of the NADPH binding site and environment of the adrenodoxin-binding site. J. Biochem. 84:1071–1086.

34. Y. Sagara, A. Wada, Y. Takata, M.R. Waterman, K. Sekimizu, and T. Horiuchi, 1993, Direct expression of adrenodoxin reductase in Escherichia coli and the functional characterization. Biol. Pharm. Bull. 16:627–630.

35. Y. Nonaka, S. Aibara, T. Sugiyama, T. Yamano, and Y. Morita, 1985, A crystallographic investigation on NADPH-adrenodoxin oxidoreductase. J. Biochem. 98:257–260.

36. R.-J. Kuban, A. Marg, M. Resch, and K. Ruckpaul, 1993, Crystallization of bovine adrenodoxin-reductase in a new unit cell and its crystallographic characterization. J. Mol. Biol. 234:245–248.

37. I. Hanukoglu, R. Rapoport, S. Schweiger, D. Sklan, L. Weiner, and G. Schulz, 1992, Structure and function of the mitochondrial P450 system electron transfer proteins, adrenodoxin reductase and adrenodoxin. J. Basic Clin. Physiol. Pharmacol. 3(Suppl.):36–37.

38. J.R. Cupp, and L.E. Vickery, 1988, Identification of free and [Fe2S2]-bound cysteine residues of adrenodoxin, J. Biol. Chem. 263:17418–17421. (erratum in J. Biol. Chem. 264:7760, 1989.)

39. D.T. Ta, and L.E. Vickery, 1992, Cloning, sequencing, and overexpression of a [2Fe-2S] ferredoxin gene from Escherichia coli. J. Biol. Chem. 267:11120–11125.

40. N. Waki, A. Hiwatashi, and Y. Ichikawa, 1986, Purification and biochemical characterization of hepatic ferredoxin (hepatoredoxin) from bovine liver mitochondria. FEBS Lett. 195:87–91.

41. Y. Sagara, H. Sawae, A. Kimura, Y. Sagara-Nakano, K. Morohashi, K. Miyoshi, and T. Horiuchi, 1990, Structural organization of the bovine adrenodoxin gene. J. Biochem. 107:77–83.

42. T. Okamura, M. Kagimoto, E.R. Simpson, and M.R. Waterman, 1987, Multiple species of bovine adrenodoxin mRNA. Occurrence of two different mitochondrial precursor sequences associated with the same mature sequence. J. Biol. Chem. 262:10335–10338.

43. C.Y. Chang, D.A. Wu, T.K. Mohandas, and B.C. Chung, 1990, Structure, sequence, chromosomal location, and evolution of the human ferredoxin gene family. DNA Cell Biol. 9:205–212.

44. W.J. Driscoll, and J.L. Omdahl, 1986, Kidney and adrenal mitochondria contain two forms of NADPH-adrenodoxin reductase dependent iron-sulfur proteins. J. Biol. Chem. 261:4122–4125.

45. C.R. Bhasker, T. Okamura, E.R. Simpson, and M.R. Waterman, 1987, Mature bovine adrenodoxin contains a 14-amino-acid COOH-terminal extension originally detected by cDNA sequencing. Eur. J. Biochem. 164:21–25.

46. A. Hiwatashi, N. Sakihama, M. Shin, and Y. Ichikawa, 1986, Heterogeneity of adrenocortical ferredoxin. FEBS Lett. 209:311–315.

47. N. Sakihama, A. Hiwatashi, A. Miyatake, M. Shin, and Y. Ichikawa, 1988, Isolation and purification of mature bovine adrenocortical ferredoxin with an elongated carboxyl end. Arch. Biochem. Biophys. 264:23–29.

48. J.R. Cupp, and L.E. Vickery, 1989, Adrenodoxin with a COOH-terminal deletion (des 116–128) exhibits enhanced activity. J. Biol. Chem. 264:1602–1607.

49. H. Uhlmann, V. Beckert, D. Schwarz, and R. Bernhardt, 1992, Expression of bovine adrenodoxin in E. Coli and site-directed mutagenesis of /2Fe-2S/ cluster ligands. Biochem. Biophys. Res. Commun. 188:1131–1138.

50. B. Xia, H Cheng, V. Bandarian, G.H. Reed, and J.L. Markley, 1996, Human ferredoxin: overproduction in Escherichia coli, reconstitution in vitro, and spectroscopic studies of iron-sulfur cluster ligand cysteine-to-serine mutants. Biochemistry 35:9488–9495.

51. J.D. Lambeth, L.M. Geren, and F. Millett, 1984, Adrenodoxin interaction with adrenodoxin reductase and cytochrome P-450scc. Cross-linking of protein complexes and effects of adrenodoxin modification by 1-ethyl-3-(3-dimethylaminopropyl)carbodiimide. J. Biol. Chem. 259:10025–10029.

52. J. Tuls, L. Geren, J.D. Lambeth, and F. Millett, 1987, The use of a specific fluorescence probe to study the interaction of adrenodoxin with adrenodoxin reductase and cytochrome P-450scc. J. Biol. Chem. 262:10020–10025.

53. T. Hara, and T. Miyata, 1991, Identification of a cross-linked peptide of a covalent complex between adrenodoxin reductase and adrenodoxin. J. Biochem. 110:261–266.

54. V.M. Coghlan, and L.E. Vickery, 1992, Electrostatic interactions stabilizing ferredoxin electron transfer complexes. Disruption by "conservative" mutations. J. Biol. Chem. 267:8932–8935.

55. I. Hanukoglu, and C.R. Jefcoate, 1980, Mitochondrial cytochrome P-450scc: Mechanism of electron transport by adrenodoxin. J. Biol. Chem. 255:3057–3061.

56. I. Hanukoglu, V. Spitsberg, J.A. Bumpus, K.M. Dus, and C.R. Jefcoate, 1981, Adrenal mitochondrial cytochrome P-450scc: Cholesterol and adrenodoxin interactions at equilibrium and during turnover. J. Biol. Chem. 256:4321–4328.

57. V. Beckert, R. Dettmer, and R. Bernhardt, 1994, Mutations of tyrosine 82 in bovine adrenodoxin that affect binding to cytochromes P45011A1 and P45011B1 but not electron transfer. J. Biol. Chem. 269:2568–2573.

58. Y. Sagara, T. Hara, Y. Ariyasu, F. Ando, N. Tokunaga, and T. Horiuchi, 1992, Direct expression in Escherichia coli and characterization of bovine adrenodoxins with modified amino-terminal regions. FEBS Lett. 300:208–212.

59. H. Uhlmann, S. Iametti, G. Vecchio, F. Bonomi, and R. Bernhardt, 1997, Pro108 is important for folding and stabilization of adrenal ferredoxin, but does not influence the functional properties of the protein. Eur. J. Biochem. 248:897–902.

60. V. Beckert, H. Schrauber, R. Bernhardt, A.A. Van Dijk, C. Kakoschke, and V. Wray, 1995, Mutational effects on the spectroscopic properties and biological activities of oxidized bovine adrenodoxin, and their structural implications. Eur. J. Biochem. 231:226–235.

61. B. Xia, H. Cheng, L. Skjeldal, V.M. Coghlan, L.E. Vickery, and J.L. Markley, 1995, Multinuclear magnetic resonance and mutagenesis studies of the histidine residues of human mitochondrial ferredoxin. Biochemistry 34:180–187.

62. N. Monnier, G. Defaye, and E.M. Chambaz, 1987, Phosphorylation of bovine adrenodoxin. Structural study and enzymatic activity. Eur. J. Biochem. 169:147–153.

63. M.L. Mandel, B. Moorthy, and J.G. Ghazarian, 1990, Reciprocal post-translational regulation of renal 1α- and 24-hydroxylases of 25-hydroxyvitamin D3 by phosphorylation of ferredoxin. Biochem. J. 266:385–392.

64. C. Tang, and H.L. Henry, 1993, Overexpression in Escherichia coli and affinity purification of chick kidney ferredoxin. J. Biol. Chem. 268:5069–5076.

65. A. Marg, R-J. Kuban, J. Behlke, R. Dettmer, and K. Ruckpaul, 1992, Crystallization and X-ray examination of bovine adrenodoxin. J. Mol. Biol. 227:945–947.

66. S. Miura, and Y. Ichikawa, 1991, Proton nuclear magnetic resonance investigation of adrenodoxin. Assignment of aromatic resonances and evidence for a conformational similarity with ferredoxin from Spirulina platensis. Eur. J. Biochem. 197:747–757.

67. K.G. Ravichandran, S.S. Boddupalli, C.A. Hasemann, J.A. Peterson, and J. Deisenhofer, 1993, Crystal structure of hemoprotein domain of P450BM-3, a prototype for microsomal P450's. Science 261:731–736.

68. C.A. Hasemann, K.G. Ravichandran, S.S. Boddupalli, J.A. Peterson, and J. Deisenhofer, 1995, Structure and function of cytochromes P450: A comparative analysis of three crystal structures, Structure 3:41–62.

69. S. Vijayakumar, and J.C. Salerno, 1992, Molecular modeling of the 3-D structure of cytochrome P-450scc. Biochim. Biophys. Acta 1160:281–286.

70. T. Omura, and K. Morohashi, 1995, Gene regulation of steroidogenesis, J. Steroid Biochem. Molec. Biol. 53:19–25.

71. P.B. Danielson, and J.C. Fogleman, 1997, Isolation and sequence analysis of cytochrome P45012B1: the first mitochondrial insect P450 with homology to 1α,25 dihydroxy-D3 24-hydroxylase. Insect Biochem. Mol. Biol. 27:595–604.

72. J.H. Chen, T. Hara, M.J. Fisher, and H.H. Rees, 1994, Immunological analysis of changes in ecdysone 20-mono-oxygenase expression in the cotton leaf, Spodoptera littoralis. Biochem. J. 299:711–717.

73. T. Sakaki, S. Kominami, K. Hayashi, M. Akiyoshi-Shibata, and Y. Yabusaki, 1996, Molecular engineering study on electron transfer from NADPH-P450 reductase to rat mitochondrial P450c27 in yeast microsomes. J. Biol. Chem. 271:26209–26213.

74. I.A. Pikuleva, I. Bjorkhem, and M.R. Waterman, 1997, Expression, purification, and enzymatic properties of recombinant human cytochrome P450c27 (CYP27), Arch. Biochem. Biophys. 343:123–130.

75. I. Hanukoglu, V. Spitsberg, J.A. Bumpus, K.M. Dus, and C.R. Jefcoate, 1981, Adrenal mitochondrial cytochrome P-450scc: Cholesterol and adrenodoxin interactions at equilibrium and during turnover. J. Biol. Chem. 256:4321–4328.

76. V.N. Luzikov, L.A. Novikova, J. Whelan, M. Hugosson, and E. Glaser, 1994, Import of the mammalian cytochrome P450(scc) precursor into plant mitochondria. Biochem. Biophys. Res. Commun. 199:33–36.

77. S.E. Graham-Lorence, and J.A. Peterson, 1996, P450s: Structural Similarities and Functional Differences. FASEB J. 10:206–214.

78. M. Tsubaki, Y. Iwamoto, A. Hiwatashi, and Y. Ichikawa, 1989, Inhibition of electron transfer from adrenodoxin to cytochrome P-450scc by chemical modification with pyridoxal 5'-phosphate: identification of adrenodoxin-binding site of cytochrome P-450scc. Biochemistry 28: 6899–6907.

79. E. Okuyama, T. Okazaki, A. Furukawa, R.-F. Wu, and Y. Ichikawa, 1996, Molecular cloning and nucleotide sequences of cDNA clones of sheep and goat adrenocortical cytochromes P450scc (CYP11A1). J. Steroid Biochem. Molec. Biol. 57:179–185.

80. A. Chernogolov, S. Usanov, R. Kraft, and D. Schwartz, 1994, Selective chemical modification of Cys264 with diiodofluorescein iodacetamide as a tool to study the membrane topology of cytochrome P450scc (CYP11A1). FEBS Lett. 340:83–88.

81. M. Tsujita, and Y. Ichikawa, 1993, Substrate-binding region of cytochrome P-450(scc) (P-450XIA1). Identification and primary structure of the cholesterol binding region in cytochrome P-450(scc). Biochim. Biophys. Acta 1161:124–130.

82. K.M. Curnow, P. Mulatero, N. Emeric-Blanchouin, B. Aupetit-Faisant, P. Corvol, and L. Pascoe, 1997, The amino acid substitutions Ser288Gly and Val320Ala convert the cortisol producing enzyme, CYP11B1, into an aldosterone producing enzyme. Nat. Struct. Biol. 4:32–35.

83. A. Bairoch, R. Apweiler, 1998, The SWISS-PROT protein sequence data bank and its supplement TrEMBL in 1998. Nucl. Acids Res. 26:38–42.

4

BIOCHEMICAL ASPECTS OF FLAVIN-CONTAINING MONOOXYGENASES (FMOs)

Ernest Hodgson,[1] Nathan J. Cherrington,[1] Richard M. Philpot,[2] and Randy L. Rose[1]

[1]Department of Toxicology
North Carolina State University
Raleigh, North Carolina 27695-7633
[2]National Institute for Environmental Health Sciences
Research Triangle Park, North Carolina 27709

1. INTRODUCTION

1.1. Historical

Xenobiotics are metabolized by many enzymes, including a number of isoforms of cytochrome P450 (P450), the flavin-containing monooxygenases (FMO), prostaglandin synthetase, alcohol and aldehyde dehydrogenases, molybdenum hydroxylases, esterases, as well as a number of transferases, particularly the glutathione S-transferases, the glucuronyl transferases and the sulfotransferases.[1,2] Both activation and detoxication reactions can be catalyzed by any of these enzymes but P450 is the most important with regard to activation of toxicants. While many of the reactions carried out by the FMO and P450 are similar there are differences between them. The most significant of these differences are that FMO does not, so far as is known, carry out oxidations at carbon atoms,[3] prefering soft nucleophiles as substrates and carrying out oxidations at heteroatoms, such as nitrogen, sulfur, selenium and phosphorus, in organic molecules.

Oxygen and NADPH-dependent microsomal oxidations were attributed, in a landmark 1963 paper by Estabrook et al.[4] to P450. This attribution has proven to be correct as P450 isoforms are now known to oxidize a huge array of both xenobiotics and endogenous metabolites through a large number of different chemical reactions. At the same time, workers studying the oxidation of N,N-dimethylaniline and other secondary and tertiary amines determined that another enzyme, differing from P450 in thermal stability and sensitivity to carbon monoxide, was present in hog liver microsomes.[5–7] This enzyme, termed

Molecular and Applied Aspects of Oxidative Drug Metabolizing Enzymes,
edited by Arinç *et al.* Kluwer Academic / Plenum Publishers, New York, 1999.

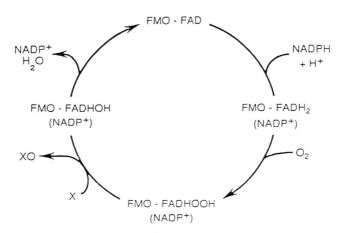

Figure 1. Catalytic cycle of FMO.

amine oxidize and other names, including "Ziegler's enzyme", was purified from hog liver and shown to contain flavin-adenine dinucleotide (FAD).[8]

It was subsequently determined that this "amine oxidase" had a wider range of substrates, and a more appropriate name, the flavin-containing monooxygenase (usually abbreviated to FMO)(EC 1.14.13.8), was adopted. Among organic compounds metabolized by FMO are those containing nitrogen, sulfur or phosphorus heteroatoms.[9–13] More recently, boron and selenium containing compounds have been identified as FMO substrates.[14,15]

1.2. Cytochrome P450 and FMO—Mechanisms Compared

In mammals, both FMO and P450, particularly the P450 isoforms involved in the oxidation of xenobiotics, have similar distribution in the endoplasmic reticulum as well as similar organ distribution, being at high levels in the liver but broadly distributed in such tissues as the nervous system, lungs, kidney and skin. However, for particular substrates the relative importance may vary due to differential distribution of isoforms among organs. Both FMO and P450 require NADPH and O_2 but only in the case of P450 is a reductase required to transfer electrons from NADPH to the monooxygenase. The biggest difference between FMO and P450 lies in the mechanism of oxidation. The FAD prosthetic group of FMO first reacts with NADPH and then molecular oxygen to give rise to the enzyme bound hydroperoxyflavin responsible for the oxidation of suitable substrates. These initial reactions occur in the absence of substrate and, in the resting state, the enzyme exists primarily in the hydroperoxyflavin form. The consequence, in terms of substrate level kinetics, of this series of events (Figure 1) is that all substrates, with very few exceptions, have the same Vmax, although Km may vary.[16]

The mechanism by which P450 oxidizes xenobiotic substrates is quite different. The initial step is the binding of substrate to the oxidized enzyme followed by a one electron reduction catalyzed by the NADPH-P450 reductase to form a reduced cytochrome-substrate complex. The next several steps include reaction with molecular oxygen to form a ternary oxygenated complex, followed by a second electron transfer, the electron usually derived from the P450 reductase but, in some cases, from cytochrome b_5 to form, eventually, a complex that breaks down to yield the oxygenated product, water and the oxidized cytochrome (Figure 2). The consequence of this sequence, relative to substrate level kinetics, is that both Vmax and Km may vary.[16]

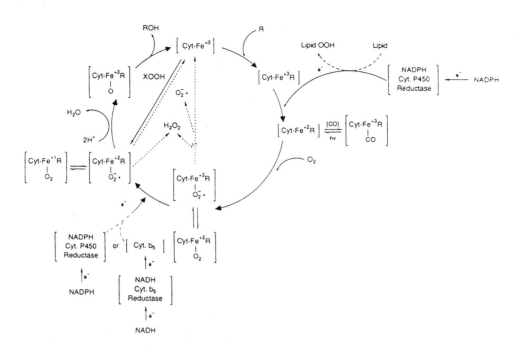

Figure 2. Catalytic cycle of P450.

Similarities and differences between FMOs and P450s, relative to the oxidation of xenobiotics are summarized in Table 1.

1.3. FMO Nomenclature

A uniform FMO nomenclature based on inferred amino sequences was adopted in 1994[17] in order to alleviate the confusion caused by laboratory-specific trivial names. The system of nomenclature is based on evidence of five isoforms that fit into a single gene family. Isoforms are 50–58% identical compared to each other and all five orthologs are over 80% identical across species. The system designates the mammalian flavin-containing monooxygenase gene family as FMO and individual genes as FMO followed by an arabic numeral. These names, as well as earlier names used in the original publications, are shown in Table 2. No sequences are known for FMOs from non-mammalian vertebrates or from invertebrates, thus the true extent of the FMO gene family is not readily apparent.

2. SUBSTRATE SPECIFICITY

2.1. Examples of N, S, P, and Se FMO Monooxidations

Examples of FMO substrates are shown in Table 3 and Figure 3. Several SAR studies have been carried out with FMO. Since our laboratory has had a long term interest in the metabolism of pesticides, particularly the organophosphorus insecticides (OPs), they

Table 1. Similarities and differences between the flavin-containing monooxygenase (FMO) and cytochrome P450 (P450)

	FMO	P450
Co-substrates	NADPH, O_2	NADPH, O_2
Coupled enzymes	None	NADPH-P450 reductase Cytochrome b_5*
Cellular location	Microsomes	Microsomes, mitochondria
Inducers	None known	Many, e.g. phenobarbital, TCDD.
Inhibitors	Competitive substrates only	Competitive substrates and mechanism based inhibitors, e.g. SKF 525A, piperonyl butoxide
Isoforms	Few, only 5 known currently	Many, over 500 known currently
Substrates	Some inorganics, organic compounds with N, S, Se, P heteroatoms.	Organic compounds with and without N, S, Se, P heteroatoms.

*Cytochrome b_5 is required only for some P450 isoforms and for some substrates

will be used as an example of SAR studies carried out either on mammalian microsomes or on purified liver FMO. Two types of reaction take place when OPs are metabolized by FMO. Those containing thioether groups such as phorate, disulfoton, sulprofos are metabolized to sulfoxides but not to their oxons, while phosphonates, such as fonofos, undergo an oxidative attack on the phoshorus atom to yield the oxon.[34] The SAR for metabolism at the thioether sulfur are as follows:

1. Thiono and thiolo sulfur atoms are not attacked at appreciable rates;
2. The sulfoxides formed are not further metabolized to sulfones by FMO, although they are by P450;

Table 2. Reported sequences of mammalian FMOs

Isoform*	Name used**	Species	Source	Reference
FMO1	Liver	Pig	cDNA	Gasser et al., 1990[18]
	1A1	Rabbit	cDNA	Lawton et al., 1990[19]
	Form 1	Rabbit	Protein	Ozols, 1990[20]
	FMO1	Human	cDNA	Dolphin et al., 1991[21]
	RFMO1	Rat	cDNA	Itoh et al., 1993[22]
	MFMO1	Mouse	cDNA	GenBank
	FMO1	Mouse	cDNA	Cherrington et al., 1998[23]
FMO2	1B1	Rabbit	cDNA	Lawton et al., 1990[19]
	Lung	Rabbit	Protein	Guan et al., 1991[24]
	1B1	Guinea pig	cDNA	Nikbakht et al., 1992[25]
FMO3	Form 2	Rabbit	Protein	Ozols, 1991[26]
	FMO II	Human	cDNA	Lomri et al., 1992[27]
	1D1	Rabbit	cDNA	Burnett et al., 1994[28]
		Human	cDNA	GenBank, 1993
	FMO3	Mouse	cDNA	Falls et al., 1997[29]
FMO4	FMO2	Human	cDNA	Dolphin et al., 1992[30]
	1E1	Rabbit	cDNA	Burnett et al., 1994[28]
FMO5	1C1	Rabbit	cDNA	Atto-Asafo-Adjei et al., 1993[31]
	Form 3	Rabbit	Protein	Ozols. 1994[32]
	FMO5	Human	cDNA	Overby et al., 1995[33]
	FMO5	Guinea pig	cDNA	Overby et al., 1995[33]
	FMO5	Mouse	cDNA	Cherrington et al., 1998[23]

*Current agreed nomenclature. **Name used at the time of publication.

Table 3. FMO substrates*

Inorganic
HS⁻, S_8, I⁻, IO⁻, I_2, CNS⁻
Organic nitrogen compounds
Secondary and tertiary acyclic and cyclic amines
N-alkyl- and N,N-dialkylarylamines
Hydrazines
Primary amines
Organic sulfur compounds
Thiols and disulfides
Cyclic and acyclic sulfides
Mercapto-purines, -pyrimidines, and -imidazoles
Dithio acids and dithiocarbamides
Thiocarbamides and thioamides
Organic phosphorus compounds
Phosphines
Phosphonates
Other substrates
Boronic acids
Selenides
Selenocarbamides

*Adapted from Ziegler.[11]

3. Steric hindrance at the thioether sulfur atom is important. For example $RSCH_2SCH_3$ is a much better FMO substrate than $RSCH_2SC_6H_5$;

4. Variations around the phosphorus atom affect FMO oxidation at remote thioether sulfur atoms. For example, substitution of thiono and thiolo sulfur atoms by oxygen reduces activity.

SAR for oxon formation are as follows:

1. At least one bond to the phosphorus atom must be a phosphonate (C-P) bond, thus phosphorodithioates, such as parathion, are not metabolized to their oxons by FMO while fonofos, a phosphonate insecticide, is metabolized to fonofos oxon;

2. The presence of the sulfur atom is not essential since trivalent phosphorus compounds, phosphines, are metabolized to their pentavalent phosphine oxides. Thus diethylphenylphosphine is an excellent FMO substrate with a K_m lower than 2.5 μM.

2.2. Fate of Substrates Common to P450 and FMO

Although FMO and P450 have many substrates in common, as illustrated above, the products may be different and have different toxic potencies. Thus the relative contributions of the two pathways to the metabolism of a particular substrate is important. In contrast to FMO, several P450 isoforms are easily induced, thus changing the relative contributions with the conditions of exposure. Although FMOs prefer soft nucleophiles as substrates and P450s hard nucleophiles, with the exception of compounds attacked at carbon atoms this applies only to their relative ability to serve as substrates for one enzyme or the other since many organic compounds with N, S and/or P heteroatoms are substrates for both enzymes.

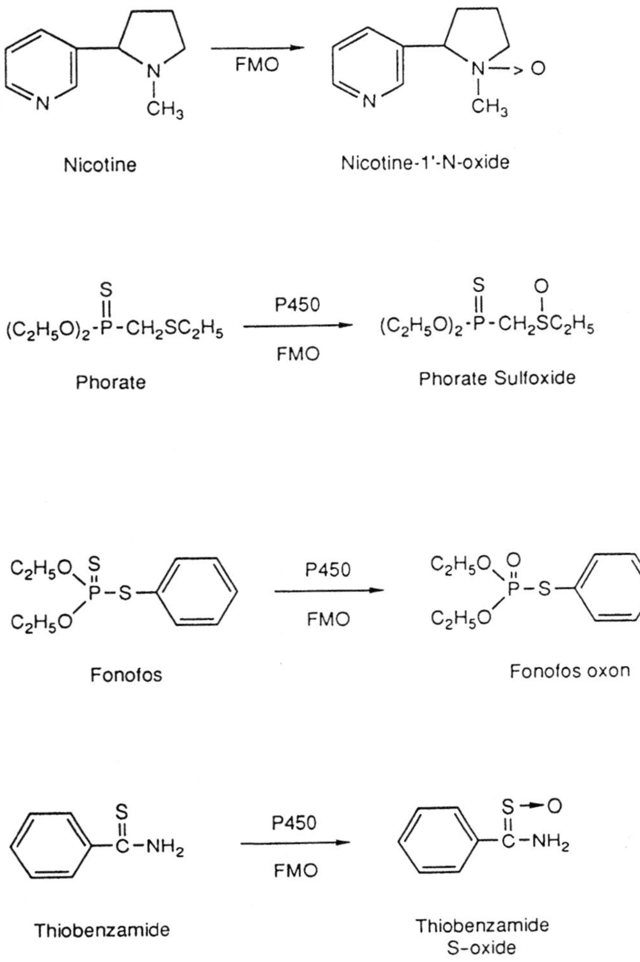

Figure 3. Examples of FMO-catalyzed oxidations.

Substrates may have complex oxidation patterns involving both FMO and P450 isoforms and may show regioselectivity in the site on the molecule attacked. As a result substrates may yield different products (e.g., the drugs thioridazine and tamoxifen) or different isomers of the same product (e.g., the insecticide phorate).

Thioridazine, an antipsychotic drug (Figure 4), is subject to oxidative metabolism at the 2- and 5- sulfur atoms or at a nitrogen atom.[35] In mouse liver, FMO is solely responsible for the formation of the N-oxide while P450 is responsible for oxidations at the sulfur atoms.[35] Experiments with recombinant human P450 isoforms suggest that CYP 2D6 forms the 2- and 5-sulfoxides but that CYP 3A4 does not, although it is probable that other CYP isoforms are involved in the overall metabolism of thioridazine. In rat liver the anti-mammary cancer, antiestrogenic, drug tamoxifen is also metabolized to its N-oxide by FMO and, at the same time metabolized to its other major metabolites, the desmethyl and 4-hydroxy derivatives, by P450.[36,37] Formation of tamoxifen N-oxide by FMO in human liver microsomes was also indicated. The insecticide phorate, on the other hand, is oxi-

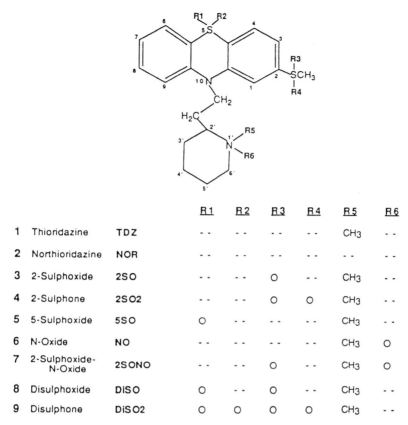

			R1	R2	R3	R4	R5	R6
1	Thioridazine	TDZ	- -	- -	- -	- -	CH₃	- -
2	Northioridazine	NOR	- -	- -	- -	- -	- -	- -
3	2-Sulphoxide	2SO	- -	- -	O	- -	CH₃	- -
4	2-Sulphone	2SO2	- -	- -	O	O	CH₃	- -
5	5-Sulphoxide	5SO	O	- -	- -	- -	CH₃	- -
6	N-Oxide	NO	- -	- -	- -	- -	CH₃	O
7	2-Sulphoxide-N-Oxide	2SONO	- -	- -	O	- -	CH₃	O
8	Disulphoxide	DISO	O	- -	O	- -	CH₃	- -
9	Disulphone	DiSO2	O	O	O	O	CH₃	- -

Figure 4. Thioridazine and oxidative metabolites of thioridazine.

dized to its sulfoxide by both P450 and FMO. In this case, however, the reactions are stereospecific (see below—Section 2.3).

2.3. Relative Contributions of FMO and Cytochrome P450

A number of methods are available for determining the relative contributions of FMO and P450 to the overall oxidation of a common substrate, including the manipulation of microsomes in which both enzymes are found. The technique in which mild heat treatment (50C for 90 sec.) of the microsomes is used to inactivate FMO or an antibody to the NADPH-P450 reductase is used to inactivate P450 has proven useful in our hands,[38–40] particularly in the case of hepatic enzymes.

The initial studies[38] on the relative contributions of FMO and P450 to the total oxidation of a substrate common to both enzymes involved the oxidation of thiobenzamide by microsomes from the liver and lung of mouse and rat. The ratios of the contribution of P450 to FMO toward the overall oxidation varied from 1:1 in the case of mouse liver to 1:4 in the case of mouse lung, with the ratios for the rat liver (1:2) and lung (2:3) falling bewteen these two values.

The insecticide phorate has been used extensively in our laboratory as a model for the study of FMO and P450. It undergoes a complex series of oxidations (Figure 5). The

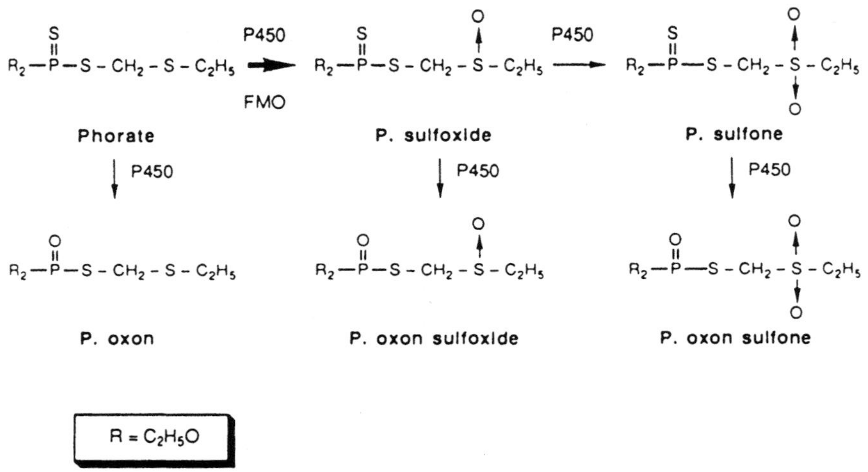

Tissue	Sex	Control[b]	+AR[b,c]	% FMO	% P450
Liver	M	12.7	2.8	21.7	78.3
Liver	F	14.4	3.7	24.0	76.1
Lung	M	3.3	1.9	59.1	41.3
Lung	F	5.7	3.1	54.0	46.0
Kidney	M	1.6	1.2	72.0	28.1
Kidney	F	1.9	1.7	89.5	10.5
Liver-Pb[d]	M	69.7	10.1	14.3	85.5

Relative Contributions of FMO and P450 to Microsomal Oxidation of Phorate in Mouse[a]

[a]From Kinsler *et al.* (1988, 1990).
[b]Phorate sulfoxide formed, nmols/min per mg protein.
[c]Antibody to P450 reductase.
[d]Phenobarbital-treated mice.

Figure 5. Oxidative metabolism of phorate and relative contributions of FMO and P450 to the formation of phorate sulfoxide.

products are generally more toxic than phorate and thus the reaction sequence is an activation process. For example the I_{50}s for the inhibition of acetylcholinesterase are some three orders of magnitude lower in the case of the oxons of the sulfoxide and the sulfone than the I_{50} for the oxon of the parent compound. FMO forms only one product, the sulfoxide, while P450 forms the sulfoxide and additional products. The sulfoxidation reaction is stereospecific with FMO producing the (-)-sulfoxide and several P450 isoforms producing the (+)-sulfoxide. While both sulfoxide isomers are substrates for further metabolism by P450 isoforms, the (+)-sulfoxide is preferred to to the (-)-sulfoxide.[34] The relative contribution of FMO to sulfoxide formation is higher in female than in male mice, in agreement with the gender studies discussed below (Section 3.1). Although overall sulfoxide formation is higher in the liver than in any extra-hepatic tissue, the contribution of FMO relative to P450 is higher in the lung, kidney and skin, being as high as 90% of the total metabolism in renal microsomes from female mice. In the liver, on the other hand, P450 is responsible for 50% of phorate sulfoxidation in female mice and 70% in male mice (Fig. 5). In consideration of potential interactions between xenobiotics it is of interest that the contribution of P450 to phorate sulfoxidation relative to that of FMO is increased following treatment with P450 inducers such as phenobarbital and reduced by P450 inhibitors such as piperonyl butoxide.[39,40]

2.4. Endogenous Substrates

Relatively few endogenous substrates have been identified as substrates for FMOs. Cysteamine undergoes oxidation to the disulfide, cystamine,[41] and cysteine-S-conjugates may also be metabolized by FMO.[42] However, the Kms for these substrates are some three orders of magnitude higher than many xenobiotic substrates, such as methimazole, suggesting that these reactions may be of little physiological significance.

2.5. Specific FMO Isoforms

2.5.1. Characterization of Specific FMO Isoforms. Differences in the rate and optimum conditions required for the oxidation of amine substrates by microsomes from different tissues[41] as well as differences in the sensitivity to mercury[43,44] and in substrate specificity[45–47] all suggested that there might be different forms of FMO with different tissue distributions. The demonstration that the lungs and liver of the rabbit contained immunologically distinct forms of FMO provided more direct evidence of multiple forms[48,49] a finding that was soon extended to other species.[50]

The first purified FMO was obtained from pig liver microsomes;[51,52] subsequently procedures were developed for isolation of FMOs from the liver of rat,[53] mouse,[54] and rabbit.[49] Rabbit lung FMO was first purified by Tynes et al.[49] and Williams et al.[48] It was the earlier indirect studies of substrate and inhibitor specificity (described above) and the purification of "liver" and "lung" forms that led to the conclusion that different isoforms of FMO must be present in the same species.

The techniques of molecular biology have been extensively used for the characterization of individual FMO isoforms. In 1990, Gasser et al.[18] screened a pig liver cDNA library with antibodies to the pig liver FMO and identified a clone with a coding region for 532 amino acids and a calculated molecular weight of 58.8 kDa. Subsequent studies by Lawton et al., 1990[19] revealed an orthologous form in rabbit liver. A high degree of similarity was noted with peptide sequences.[20,26,32] In studies involving rabbit, rat, mouse, guinea pig and human carried out during the next several years,[21,22,25,28,31] it became clear that there were five mammalian FMOs, now named FMO1 through FMO5.

Because of the importance of the mouse in toxicology studies and the almost complete lack of information on mouse FMOs our studies have concentrated on this species. The methods used in our laboratory for cloning, sequencing and expression of FMO3 are summarized below as an example of a typical protocol. These methods are not original to our laboratory and the original references are cited, in full, in Falls et al.[29] The methods used for other FMO isoforms[23] differ in detail but not in principle.

cDNA was reverse transcribed from hepatic RNA isolated from adult female CD-1 mice, size fractionated from 1.5–6.5 KB, and libraries constructed using the λZAP Express Vector Kit (Stratagene Cloning Systems). The cDNA library was screened with a 3′*Hind*III-*Eco*R1 fragment (720 bases of rabbit FMO3 cDNA radiolabed with ^{32}P using a random primer kit (Boehringer Manheim) under low stringency conditions. Clones of interest were isolated and screened to purity. Full length clones were selected by PCR using

T7 and BK reverse primers. Putative full length clones were excised using exAssist helper phage as the BK phagemid, transfected into XLOLR *E. coli* and sequenced using the dideoxy mediated chain termination method.

Full-length clones were expressed in E. coli using the vector pJL-2, a derivative of pKK233–2(Pharmacia, LKB Biotechnology, Inc.) The cloning sites in this vector are *Xba*I, *Nco*I, *Eco*RV, and *Hind*III. For mouse FMO3, high fidelity PCR was used to add *Xba*I and *Hind*III restriction sites at the 5' and 3'-ends of the cDNA. Because of the presence of an internal *Xba*I site, the PCR product was cut and the two fragments ligated sequentially into pJL-2 and the orientation of the inserts confirmed by restriction analysis.

For expression in *E. coli*, XL1-Blue cells were transformed with recombinant pJLl-FMO3, and recombinants selected on LB plates containing ampicillin. Single colonies were grown to log phase overnight in LB in the presence of ampicillin, induced with IPTG, incubated overnight at 30°C and subcellular fractions prepared as described by Lawton and Philpot.[55] Membrane fractions of recombinant *E. coli* were subjected to SDS-PAGE electrophoresis, and Western blot analysis performed using antibody to rabbit FMO3.

We have also used these methods for the cloning, sequencing and expression of FMO1 and FMO5 from the mouse and for the expression, from clones obtained from other laboratories, of human FMO3 and FMO5 and rabbit FMO1, FMO3, and FMO5.

For the analysis of mouse FMO3 mRNA and proteins, RNA and liver microsomes were prepared from tissues of adult male and female CD-1 mice. RNAs for FMO1, FMO3 and FMO5 were determined by Northern blot analysis using the appropriate cDNA probes. Microsomes were prepared and subjected to Western blot analysis using specific antibodies for FMO1, FMO3, and FM05. FMO activity of recombinant E. coli membranes and mouse liver microsomes were determined using methimazole and other substrates.

Using the above techniques the sequence of mouse FMO3 was obtained from several clones isolated from the mouse liver cDNA library.[29] The nucleotide sequence of FMO3 was 2020 bases in length containing 37 bases in the 5'-flanking region, 1602 in the coding region and 381 in the 3'-flanking region. The derived protein sequence consisted of 534 amino acids including the FAD and NADP$^+$ pyrophosphate binding sites starting at positions 9 and 191, respectively. Mouse protein sequence homologies were 79 and 82% identical to the human and rabbit forms, respectively. Mouse FMO3 was expressed in *E. coli* and compared to *E. coli* expressed human FMO3. The FMO3 proteins were identical in molecular weight (~ 58 kDa) as identified by SDS-PAGE and immunoblotting. The expressed FMO3 enzymes reacted in a similar manner toward heat, metal ions and detergent. Catalytic activities of mouse and human FMO3 were high toward the substrate methimazole. In addition, trimethylamine and thioacetamide both inhibited, by more than 85%, the FMO-dependent oxidation of methimazole by either enzyme, indicating that they are also FMO3 substrates. Thiourea and thiobenzamide, and to a lesser degree, *N,N*-dimethylaniline also inhibited methimazole oxidation. FMO3 transcripts were detected in hepatic mRNA samples from females, but not in samples from males. In contrast, FMO3 was detected in mRNA samples from male and female mouse lung; FMO3 message was not detected in the kidney of either gender.

Full-length cDNA clones encoding FMO1 and FMO5 were also isolated from the same cDNA library. The sequence for the FMO1 clone contained 2310 bases, 1596 in the coding region, encoding a protein of 532 amino acids with a predicted molecular weight of of 59.9 kDa, an 83.3% identity to human FMO1 and 83–94% homology to other FMO homologs. The FMO5 clone consisted of 3168 bases with 1599 in the coding region, encoding a protein of 533 amino acids, a molecular weight of 60.0 kDa, 84.1% homology and 83–84% homology to other FMO5 homologs.

2.5.2. Substrate Specificity of Specific Isoforms. Few of the substrate specificity studies carried out on microsomes or purified 'native' proteins have been repeated on specific recombinant FMO isoforms and, as a result, their substrate specificities are not yet well understood.

Purified FMO1 from pig liver oxidized all of a series of phenothiazines with side chain lengths varying from 2 to 7 methylene carbons whereas purified FMO2 from rabbit lung oxidized only the C6 and C7 derivatives.[56] Prochlorperazine and trifluperazine are metabolized by FMO1 and FMO2; however, the structurally similar compounds chlorpromazine and imipramine have shorter side chains and are substrates of FMO1 but not substrates of FMO2.[48,57] As a result of these studies Nagata et al.[56] proposed that the substrate binding site of FMO2 is more restrictive than that of FMO1.

The studies of Lomri et al.[27,58] using cDNA expressed FMO1, FMO2 and FMO3 supported the above conclusion[56] and also led to the conclusion that the substrate binding site of FMO3 more closely resembled that of FMO2 than that of FMO1. It has also been demonstrated that FMO2 and FMO3 convert (S)-nicotine exclusively to trans-(S)-nicotine N-1'-oxide, whereas FMO1 produces a 1:1 mixture of cis- and trans N-oxides of (S)-nicotine.[59]

Rettie et al.[60] synthesized a series of n-alkyl-substituted p-tolyl sulfides with alkyl side chains ranging from 1 to 4 carbon atoms long for use as substrates for four expressed rabbit FMO isoforms. Each isoform displayed a different profile of oxidation with these compounds. FMO1 catalyzed stereoselective oxidation to R sulfoxides, FMO2 oxidized only the methyl derivative stereoselectively. Based on these results and others a speculative model for the FMO binding site was proposed.[60–62] The proposed site consists of a substrate channel, and a compartment containing the hydroperoxyflavin with binding pockets for the substrate and a lipophilic amine activator on either side. This speculative model is of interest but the data is still lacking to provide a definitive model of the active site of any of the FMO isoforms.

The characteristics of the expressed enzymes are often difficult to compare to the "native" enzymes purified from the tissues of the same species. For, example, the "native" enzyme from mouse liver may include the cumulative activities from FMO1, FMO3 and FMO5. FMO1 and FMO3 may vary with gender but when expressed are often at a higher level than FMO5.[29] Methimazole, the substrate commonly used to measure FMO activity is an aproximately 3-fold better substrate for FMO1 than for FMO3, thus many previous results on microsomal FMO activity determined with methimazole as substrate are primarily the activity of FMO1, with FMO3 making a smaller contribution. This is shown in the Km for FMO activity toward methimazole in mouse liver microsomes, 17.0 µM, which is much closer to that of FMO1, 6.4 µM, than of FMO3, 56. 2 µM, even though more FMO3 is present in the microsomes.[29,49,63]

Preliminary studies[63] on the substrate specificity of FMO1, FMO3, and FMO5 from the mouse show that, with pesticide substrates, many similarities exist between the mouse isoforms and the corresponding forms from humans and other species. As has been observed with other substrates, each ortholog shares a very similar list of substrates across species while different isoforms show the same differences in substrate specificity regardless of species. Phorate appears to be a preferred substrate for FMO1, while fonofos and imidacloprid may be substrates for FMO5. This is in contrast to methimazole, a substrate for all isoforms from all three species (mouse, rabbit and human) examined. Methimazole is, however, a much better substrate for FMO1 than for FMO3 with even lower activity as a substrate for FMO5.

3. PHYSIOLOGICAL EFFECTS ON FMO EXPRESSION

3.1. Gender

Recently[29,64,65] we have examined the role of gender in the expression of FMO isoforms in mouse liver. While it has long been known that the FMO activity toward several substrates was higher in the liver of female than of male mice, these studies were carried out before it was known that several isoforms of FMO were present in mouse liver, only substrate oxidation was measured and microsomes were used as the enzyme source. In initial studies we demonstrated that the hepatic FMO activity in microsomes from all strains of mice examined were higher in females than males. Based on protein (western blots) and mRNA (northern blots) levels it was shown that this difference was due to differences in FMO1 and FMO3 levels, FMO5 being the same in the liver of either gender. FMO1 was 2–3 times higher in the females than in that of males while the level of FMO3, aproximately equivalent to that of FMO1 in the female liver was absent from the liver of males. Thus in the liver of adult mice there is a gender-independent isoform, FMO5, a gender-dependent isoform, FMO1, and a gender-specific isoform, FMO3. This effect, while dramatic, is tissue dependent. FMO1, FMO3, and FMO5 are all expressed at similar levels in the lung and kidney of both male and female CD-1 mice.

The roles of testosterone, 17 β-estradiol and progesterone in the regulation of hepatic FMOs have been examined in gonadectomized CD-1 mice, normal mice receiving hormonal implants and gonadectomized mice receiving various hormonal treatments. Following castration of males, overall hepatic FMO activities were significantly increased and serum testosterone levels significantly decreased; however, administration of physiological levels of testosterone to castrated animals returned FMO and testosterone to control levels. When sexually intact or ovariectomized female mice were treated with testosterone, concomitant with high serum testosterone levels their overall hepatic FMO activities were reduced to those of their male counterparts. In males, based on both protein and mRNA measurements, castration was shown to dramatically increase FMO1 and FMO3 expression and testosterone replacement to castrated males to bring about a reduction in FMO1 and ablation of FMO3 expression. None of the above treatments affected the level of FMO5 expression, indicating a sex hormone-independent mechanism for regulation of this isoform.

Although the expression of FMO isoforms in rat liver is also gender dependent, the situation is dramatically different. FMO1 is expressed in the liver of both males and females but, in contrast to the mouse, the level is higher in the male. The expression of FMO3 in rat liver shows no difference between genders, again in sharp contrast to mouse liver. FMO5 in rat liver, as in the mouse, appears unaffected by gender.

3.2. Development

We[66] have also investigated changes in expression of FMO1, FMO3, and FMO5 in mouse liver during development. In the case of FMO1 the isoform is already expressed by gestation day 13, expression is higher after 2 days of postnatal development and is still equivalent in males and females at 2 weeks of age. At four weeks of age the gender difference is apparent with higher expression in the females; at 6 weeks the gender difference is fully apparent. FMO3, on the other hand, is not expressed at any time in the fetus and at 2 weeks post-partum is expressed only at low levels in both males and females. At 4 weeks of age FMO3 is expressed in both males and females at levels equivalent to those seen in the adult female. At six weeks of age FMO3 is no longer expressed in the male due to the repressive effects of testosterone. The developmental pattern of expression of FMO3 in the female mouse is thus very similar to that in humans in which FMO1 is expressed in the fetal liver and FMO3 in the adult liver.[21,30] The gender effect is not, however, seen in adult humans and FMO3 is the principal isoform in both males and females.[66]

4. SUMMARY AND CONCLUSIONS

During the last three decades, the FMOs have been firmly established as important enzymes, in mammals, for the oxidation of xenobiotics of many chemical and use classes. In recent years, following the demonstration that multiple forms of FMO are expressed in mammals, progress has been made on their molecular biology, on the factors that control their expression, and on the substrate specificity of the individual isoforms. Further progress may be expected in all of these areas. The extent of the FMO gene family remains to be determined; although FMO-like activity has been detected in vertebrates other than mammals and, with less probability, in invertebrates, no significant molecular characterization has been carried out in any group of animals other than the Mammalia. Whether or not any or all of the FMOs have an endogenous function also remains an open question. Finally, it is clear that with recent and probable future developments, knowledge of FMOs will play a role in the development of drugs, agricultural chemicals and other commercial xenobiotics.

ACKNOWLEDGMENTS

The studies carried out at NCSU were all supported, in part, by NIEHS grants ES-00044 and 2T32ESO7046 and the North Carolina Agricultural Research Service.

REFERENCES

1. E. Hodgson, R. L. Rose, D-Y. Ryu, J. G. Falls, and P. E. Levi, 1995, Pesticide-metabolizing enzymes, *Toxicol. Lett.* **82/83:** 73–81.
2. E. Hodgson, and P. E. Levi, 1996, Pesticides: an important, but underused, model for the environmental health sciences, *Environ. Hlth. Perspect.* **104 (Suppl. 1):** 97–106.
3. D. M. Ziegler, 1991, Unique properties of the enzymes of detoxication, *Drug Metabol. Disp.,* **19:** 847–852.
4. R. W. Estabrook, D. Y. Cooper, and D. Rosenthal, 1963, The light reversible carbon monoxide inhibition of the C-21 hydroxylase system of adrenal cortex, *Biochem.,* **338:** 741–755.
5. J. M. Machinist, W. H.Orme-Johnson, and D. M. Ziegler, 1966, Microsomal oxidases II. Properties of a pork liver microsomal N-oxide dealkylase, *Biochem.* **5:** 2939–2943.
6. J. M. Machinist, E. W. Dehner, and D. M. Ziegler, 1968, Microsomal oxidases III. Comparison of species and organ distribution of dialkylarylamine-N-oxide dealkylase and dialkylamine-N-oxidase, *Arch. Biochem. Biophys.* **125 ;** 858–864.
7. D. M. Ziegler, and F. H. Pettit, 1966, Microsomal oxidases I. The isolation and dialkylarylamine oxygenase activity of pork liver microsomes, *Biochem.,* **5:** 2932–2938.
8. D. M. Ziegler, C. J. Mitchell, and D. Jollow, 1969, Properties of a purified hepatic microsomal mixed-function amine oxidase, in:*Microsomes and Drug Oxidations,* (J. R. Gillette, A. H. Conney, R. W. Estabrook, J. R. Fouts, and G. J. Mannering, eds.) pp. 173–188, Academic Press, NY.
9. D. M. Ziegler, 1980, Microsomal flavin-containing monooxygenase: Oxygenation of nucleophilic nitrogen and sulfur compounds, in: *Enzymatic Basis of Detoxication, Vol. I,* (W. B. Jakoby, ed.) pp. 201–227, Academic Press, NY.

10. D. M. Ziegler, 1988, Flavin-containing monooxygenases: Catalytic mechanism and substrate specificity. *Drug Metabol. Rev.,* **6**: 1–32.

11. D. M. Ziegler, 1990, Flavin-containing monooxygenases: Enzymes adapted for multisubstrate specificity, *Trends Pharmacol. Sci.,* **11**: 321–324.

12. E. Hodgson, and P. E. Levi, 1988, Species, organ and cellular variation in the flavin-containing monooxygenases. *Drug Metabol. Drug Interact.,* **6**: 219–233.

13. P. E. Levi, and E. Hodgson, (1989) Monooxygenations: Interactions and expression of toxicity, in: *Insecticide Action, From Molecule to Organism,* (T. Narahashi and J. E. Chambers, eds.), Plenum Press, NY.

14. K. C. Jones and D. P. Ballou, 1986, Reactions of the 4a-hydroperoxide of liver microsomal flavin-containing monooxygenase with nucleophilic and electrophilic substrates, *J. Biol. Chem.,* **261**: 2552–2559.

15. D. M. Ziegler, P. Graf, L. Poulsen, H. Sies, and W. Stal, 1992, NADPH-dependent oxidation of reduced ebselen, 2-selenylbenzanilide, and of 2-(methylseleno)benzanilide catalyzed by pig liver flavin-containing monooxygenase. *Chem. Res. Toxicol.,* **5**: 163–166.

16. E. Hodgson, and P. E. Levi, 1994, Metabolism of toxicants: Phase I reactions, in: *Introduction to Biochemical Toxicology, 2nd Edition,* (E. Hodgson and P. E. Levi, eds.) Appleton and Lange, Norwalk, CT.

17. M. R. Lawton, J. R. Cashman, T. Cresteil, C. T. Dolphin, A. A. Elfarra, R. N. Hines, E. Hodgson, T. Kimura, J. Ozols, I. R. Phillips, R. M. Philpot, L. L. Poulsen, A. E. Rettie, E. A. Shepard, D. E. Williams, and D. M. Ziegler, 1994, A nomenclature for the mammalian flavin-containing monooxygenase gene family based on amino acid sequence identities, *Arch. Biochem. Biophys.,* **308**: 254–257.

18. R. Gasser, R. E. Tynes, M. P. Lawton, K. K. Korsmeyer, D. M. Ziegler, and R. M. Philpot, 1990, The flavin-containing monooxygenase expressed in pig liver: Primary sequence, distribution, and evidence for a single gene. *Biochem.* **29**: 119–124.

19. M. P. Lawton, R. Gasser, R. E. Tynes, E. Hodgson, and R. M. Philpot, 1990, the flavin-containing monooxygenase enzymes expressed in rabbit and lung are products of related, but distinctly different genes, *J. Biol. Chem.,* **265**:5855–5861.

20. J. Ozols, 1990, Covalent structure of liver microsomal flavin-containing monooxygenase form 1. *J. Biol. Chem.,* **265**:10289–10299.

21. C. T. Dolphin, E. A. Shepard, S. Povey, C. N. A. Palmer, D. M. Ziegler, R. Ayesh, R. L. Smith and I. R. Phillips, 1991, Cloning, primary sequence and chromosomal mapping of a human flavin-containing monooxygenase (FMO1), *J. Biol. Chem.* **266**:12379–12385.

22. K. Itoh, T. Kimura, T. Yokoi, S. Itoh, and T. Kamataki, 1993, Rat liver flavin-containing monooxygenase (FMO): cDNA cloning and expression in yeast, *Biochim. Biophys. Acta,* **1173**:165–171.

23. N. J. Cherrington, J. G. Falls, R. L. Rose, K. M. Clements, R. M. Philpot, P. E. Levi, and E. Hodgson, 1998, Molecular cloning, sequence, and expression of mouse flavin-containing monooxygenases 1 and 5 (FMO1 and FMO5), *J. Biochem. Molec. Toxicol. in press.*

24. S. Guan, A. M. Falick, D. E. Williams, and J. R. Cashman, 1991, Evidence for complex formation between rabbit lung flavin-containing monooxygenase and calreticulin. *Biochem.* **30**:9892–9900.

25. K. N. Nikbakht, M. P. Lawton, and R. M. Philpot, 1992, Guinea pig or rabbit lung flavin-containing monooxygenases with distinct mobilities in SDS-PAGE are allelic variants that differ in primary structure at only two positions, *Pharmacogenetics,* **2**:207–216.

26. J. Ozols, 1991, Multiple forms of liver microsomal flavin-containing monooxygenases: Complete covalent structure of form 2, *Arch. Biochem. Biophys.,* **290**:103–115.

27. N. Lomri, J. Thomas, and J. R. Cashman, 1993, Expression in Escherichia coli of the cloned flavin-containing monooxygenase from pig liver, *J. Biol. Chem.,***268**:5048–5059.

28. V. L. Burnett, M. P. Lawton, and R. M. Philpot, 1994, Cloning and sequencing of flavin-containing monooxygenases FMO3 and FMO4 from rabbit and characterization of FMO3. *J. Biol. Chem.,* **269**:14314–14322.

29. J. G. Falls, N. J. Cherrington, K. M. Clements, R. M. Philpot, P. E. Levi, R. L. Rose, and E. Hodgson, 1997, Molecular cloning, sequencing, and expression in Escherichia coli of mouse flavin-containing monooxygenase 3 (FMO3): Comparison with the human isoform, *Arch. Biochem. Biophys.,* **347**:9–18.

30. C. T. Dolphin, E. A. Shepard, S. Povey, R. L. Smith and I. R. Phillips, 1992, Cloning, primary sequence and chromosomal localization of human FMO2, a new member of the flavin-containing monooxygenase family, *Biochem. J.,* **287**:261–267.

31. E. Atta-Asafo-Adjei, M. P. Lawton, and R. M. Philpot, 1993, Cloning, sequencing, distribution, and expression in Escherichia coli of flavin-containing monooxygenase 1C1. *J. Biol. Chem.,* **268**:9681–9689.

32. J. Ozols, 1994, Isolation and structure of a third form of liver microsomal monooxygenase. *Biochem.,* **33**:3751–3757.

33. L. H. Overby, A. R. Buckpitt, M. P. Lawton, E. Atta-Asafo-Adjei, J. Schulze, and R. M. Philpot, 1995, Characterization of flavin-containing monooxygenase 5 (FMO5) cloned from human and guinea pig: Evi-

dence that the unique catalytic properties of FMO5 are not confined to the rabbit ortholog, *Arch. Biochem. Biophys.,* **317**:275–284.

34. P. E. Levi, 1992, Metabolism of organophosphorus compounds by the flavin-containing monooxygenase, in: *Organophosphates: Chemistry, Fate, and Effects,*(J. E. Chambers and P. E. Levi, eds.) pp 141–154.

35. B. L. Blake, R. L. Rose, R. B. Mailman, P. E. Levi and E. Hodgson, 1995, Metabolism of thioridazine by microsomal monooxygenases: Relative roles of P450 and flavin-containing monooxygenase, *Xenobiotica,* **25**:377–393.

36. C. Mani, H. V. Gelboin, S. S. Park, R. Pearce, A. Parkinson and D. Kupfer, 1993, Metabolism of the anti-mammary cancer antiestrogenic agent tamoxifen I. Cytochrome P450-catalyzed N-demethylation and 4-hydroxylation, *Drug Metabol. Disp.,* **21**:645–656.

37. C. Mani, E. Hodgson, and D. Kupfer, 1993, Metabolism of the antimammary cancer antiestrogenic agent tamoxifen II Flavin-containing monooxygenase-mediated N-oxidation, *Drug Metabol. Disp.,* **21**:657–661.

38. R. E. Tynes and E. Hodgson, 1983, Oxidation of thiobenzamide by the FAD-containing monooxygenase and cytochrome P-450-dependent monooxygenases of liver and lung microsomes, *Biochem. Pharmacol.* **32**:3419–3428.

39. S. P. Kinsler, P. E. Levi and E. Hodgson, 1988, Hepatic and extrahepatic microsomal oxidation of phorate by cytochrome P-450 and FAD-containing monooxygenase systems in the mouse. *Pestic. Biochem. Physiol.,* **31**: 54–60.

40. S. P. Kinsler, P. E. Levi and E. Hodgson, 1990, Relative contributions of the cytochrome P450 and flavin-containing monooxygenases to the microsomal oxidation of phorate following treatment of mice with phenobarbital, hydrocortisone, acetone and piperonyl butoxide. *Pestic. Biochem. Physiol.,* **37**: 174–181.

41. L. L. Poulsen, R. Hyslop and D. M. Ziegler, 1974, S-oxidation of thioureylenes catalyzed by a microsomal flavoprotein mixed-function oxidase, *Biochem. Pharmacol.* **23**: 3431–3440.

42. A. A. Elfarra, 1995, Potential role of the flavin-containing monooxygenases in the metabolism of endogenous compounds, *Chem. Biol. Interact.,* **96**:47–56.

43. T. R. Devereux and J. R. Fouts, 1974, N-oxidation and demethylation of N,N-dimethylaniline by rabbit liver and lung microsomes: Effect of age and metals, *Chem. Biol. Interact.* **8**: 91–105.

44. T. R. Devereux, R. M. Philpot and J. R. Fouts, 1977, The effect of Hg^{2+} on rabbit hepatic and pulmonary solubilized, partially purified N,N-dimethylaniline N-oxidases, *Chem. Biol. Interact.,* **18**: 277–287.

45. Y. Ohmiya and H. M. Mehendale, 1980, Uptake and metabolism of chlorpromazine by rat and rabbit lungs, *Drug Metabol. Disp.,* **8**:313–318.

46. Y. Ohmiya and H. M. Mehendale, 1981, Pulmonary metabolism of imipramine in the rat and rabbit, *Pharmacology,* **22**:172–182.

47. Y. Ohmiya and H. M. Mehendale, 1983, N-Oxidation of N,N-dimethylaniline in the rabbit and rat lung. *Biochem. Pharmacol.,* **32**:1281–1285.

48. D. E. Williams, S. E. Hale, A. S. Muerhoff and B. S. S. Masters, 1985, Rabbit lung flavin-containing monooxygenase: Purification, characterization and induction during pregnancy, *Mol. Pharmacol.,* **28**:381–390.

49. R. E. Tynes and E. Hodgson, 1985, Catalytic activity and substrate specificity of the flavin-containing monooxygenase in microsomal systems: Characterization of the hepatic, pulmonary and renal enzymes of the mouse, rabbit and rat. *Arch. Biochem. Biophys.,* **240**:77–93.

50. R. E. Tynes and R. M. Philpot, 1987, Tissue- and species-dependent expression of multiple forms of mammalian microsomal flavin-containing monooxygenases, *Mol. Pharmacol.,* **31**:569–574.

51. D. M. Ziegler, and C. H. Mitchell, 1972, Microsomal oxidase IV: Properties of a mixed-function amine oxidase isolated from pig liver microsomes. *Arch. Biochem. Biophys.,* **150**:116–125.

52. M. S. Gold and D. M. Ziegler, 1973, Dimethylaniline N-oxidase and aminopyrine N-demethylase activities of human liver tissue, *Xenobiotica,* **3**:179–189.

53. T. Kimura, M. Kodama and C. Nagata, 1983, Purification of a mixed function amine oxidase from rat liver microsomes, *Biochem. Biophys. Res. Commun.,* **110**:640–645.

54. P. J. Sabourin, B. P. Smyser and E. Hodgson, 1984, Purification of the flavin-containing monooxygenase from mouse and pig liver microsomes. *Int. J. Biochem.,* **16**:713–720.

55. M. R. Lawton and R. M. Philpot, 1995, Emergence of the flavin-containing monooxygenase gene family. *Rev. Biochem. Toxicol.,* **11**:1–27.

56. T. Nagata, D. E. Williams and D. M. Ziegler, 1990, Substrate specificities of rabbit lung and porcine liver flavin-containing monooxygenases: Differences due to substrate size, *Chem, Res. Toxicol.,* **3**:372–376.

57. L. L. Poulsen, K. Taylor, D. E. Williams, B. S. S. Masters and D. M. Ziegler, 1986, Substrate specificity of the rabbit lung flavin-containing monooxygenase for amines: Oxidation products of primary alkylamines, *Mol. Pharmacol.,* **30**:680–685.

58. N. Lomri, Z.-C. Yang, and J. R. Cashman, 1993, Expression in Escherichia coli of the cloned flavin-containing monooxygenase D (form II) from adult human liver: Determination of a distinct tertiary amine substrate specificity, *Chem. Res. Toxicol.,* **6:**425–429.

59. R. N. Hines, J. R. Cashman, R. M. Philpot, D. E. Williams and D. M. Ziegler, 1994, The mammalian flavin-containing monooxygenases: Molecular characterization and regulation of expression. *Toxicol. Appl. Pharmacol.,* **125:**1–6.

60. A. E. Rettie, M. P. Lawton, A. J. M. Sadeque, G. P. Meier, and R. M. Philpot, 1994, Prochiral sulfoxidation as a probe for multiple forms of the microsomal flavin-containing monooxygenase: Studies with FMO1, FMO2, FMO3 and FMO5 expressed in Escherichia coli, *Arch. Biochem. Biophys.,* **311:**269–377.

61. J. R. Cashman, L. D. Olsen, C. E. Lambright and M. J. Presas, 1990, Enantioselective S-oxygenation of p-methoxy phenyl-1,3-dithiolane by various tissue preparations: Effects of estradiol. *Mol. Pharmacol.* **37:**319–327.

62. J. R. Cashman, 1995, Structural and catalytic properties of the mammalian flavin-containing monooxygenase. *Chem. Res. Toxicol.,* **8:**165–181.

63. N. J. Cherrington, R. L. Rose, J. G. Falls, R. M. Philpot, P. E. Levi and E. Hodgson, 1997, Functional characterization of cDNA expressed flavin-containing monooxygenases FMO1 and FMO5 from mouse and comparative pesticide metabolism by FMOs 1, 3 and 5 from mouse, human and rabbit. *in preparation.*

64. J. G. Falls, B. L. Blake, Y. Cao, P. E. Levi and E. Hodgson, 1995, Gender differences in hepatic expression of flavin-containing monooxygenase isoforms (FMO1, FMO3 and FMO5) in mice, 1995, *J. Biochem. Toxicol.,* **10:**171–177.

65. J. G. Falls, D-Y. Young, Y. Cao, P. E. Levi and E. Hodgson, 1997, Regulation of mouse liver flavin-containing monooxygenases 1 and 3 by sex steroids, *Arch. Biochem. Biophys.* **342:**212–223.

66. N. J. Cherrington, Y. Cao, J. A. Cherrington, R. L. Rose and E. Hodgson, 1997, Physiological factors affecting protein expression of flavin-containing monooxygenases 1, 3 and 5. *in preparation.*

5

EXPRESSION AND REGULATION OF FLAVIN-CONTAINING MONOOXYGENASES

Richard M. Philpot, Christine P. Biagini, Geraldine T. Carver, Lila H. Overby, M. Keith Wyatt, and Kiyoshi Itagaki

Molecular Pharmacology Section
Laboratory of Signal Transduction
National Institute of Environmental Health Science
Research Triangle Park, North Carolina 27709

1. INTRODUCTION

Humans are exposed on a more or less continual basis to a myriad of synthetic and natural chemicals with the potential to cause deleterious effects. Most exogenous chemicals undergo oxidative metabolism as a prelude to excretion. This "biotransformation" creates polar products that can be eliminated directly or following conjugation. However, the same metabolic machinery can form reactive metabolites that may bind covalently to macromolecules in the cell. This "activation" process is a required step in the expression of many carcinogenic, mutagenic, and other toxic responses associated with many environmental contaminants.

Oxidative metabolism of xenobiotics is catalyzed primarily by members of two gene families—cytochromes P450 (P450) and the flavin-containing monooxygenases (FMO). Members of the P450 and FMO gene families exhibit exceptionally broad substrate specificities; thousands of chemicals are known to be metabolized by these enzymes. Tissue- and species-specific regulation of expression of these "phase I" enzymes may be related to tissue- and species-specific toxic responses. Likely, many site-selective toxic responses are initiated by localized metabolism, and characteristics of metabolism may actually play a role in determining specificity. Of the two phase I gene families, the FMO is the lesser studied.

Although FMO-catalyzed activity was first described in 1964 and an FMO enzyme was purified from pig liver in 1972,[1] the FMO was not recognized as a gene family for nearly twenty years. Some indications of a distinct pulmonary form of the enzyme were provided as early as 1974[2,3] but conclusive evidence for multiple genes was not reported until 1990.[4] Now it is clear that there are five isoforms of the FMO (FMO1- 5), each with a different substrate specificity.[5] Three areas of FMO research currently under investigation in our laboratory are; FMO and human drug-metabolism, expression and characterization of FMO4, and modulation of FMO activity by tricyclic antidepressants.

Molecular and Applied Aspects of Oxidative Drug Metabolizing Enzymes,
edited by Arınç *et al.* Kluwer Academic / Plenum Publishers, New York, 1999.

2. FMO AND HUMAN DRUG-METABOLISM

2.1. Introduction

The most thoroughly studied of the FMO isoforms are FMO1, the major form expressed in liver of adult pigs,[6] and FMO2, the major form in lung of adult rabbit.[7] However, neither of these isoforms appears to be expressed to any significant extent in liver of adult human.[8] On the basis of catalytic, immunochemical and expression data, it has been suggested that FMO3 makes the greatest contribution to FMO-mediated drug metabolism in adult human liver.[8-11] However, FMO3 has not been quantitated in human samples nor has its specific activity (catalytic constant) been determined.

In addition to FMO3, FMO5 has also been detected in human liver.[12] While the activity of FMO5 appears to be limited by low substrate affinity, its concentration in human liver and its ability to metabolize known FMO drug substrates have not been investigated. We have cloned human FMO3 and human FMO5, and compared the properties of the expressed enzymes with those present in hepatic microsomal samples from adult humans.

2.2. Characterization of Recombinant Human FMO3 and FMO5

Both FMO3 and FMO5 catalyzed the S-oxidation of methimazole. However, FMO3 (Km = 35 mM, Vmax = 50 nmol product $\times$ min^{-1} $\times$ nmol FMO3^{-1}) was about 5×10^3 times more efficient than FMO5 (Km = 6 mM, Vmax = 1.4 nmol $\times$ min^{-1} $\times$ nmol FMO5^{-1}) in this reaction. Km values determined from rates of FMO3-catalyzed oxidation of NADPH in the presence of ranitidine or cimetidine were ~ 2mM and ~ 4 mM, respectively (Figure 4). The Km values for FMO5-mediated NADPH oxidation with ranitidine and cimetidine could not be determined accurately because of substrate solubility problems but were clearly in excess of 10 mM. The metabolism of ranitidine (1 mM) catalyzed by recombinant FMO3 and FMO5 was assessed by HPLC. The N-oxide and S-oxide metabolites were both formed but no N-demethylated product was detected. Rates of formation of the N-oxide and S-oxide with FMO3 were 5.10 ± 0.07 and 0.98 ± 0.02 nmoles $\times$ min^{-1} $\times$ nmole enzyme^{-1}. Rates of N-oxidation with FMO5 were less than 0.05 nmole $\times$ min^{-1} $\times$ nmole enzyme^{-1} (2–3 times background rates in incubations containing non-recombinant *E. coli*).

2.3. Characterization of Human Hepatic Microsomal FMO3 and FMO5

2.3.1. Quantitation of FMO3 and FMO5 in Human Liver Samples. Known amounts of recombinant FMO3 and FMO5, determined on the basis of flavin concentrations, were detected by immunoblotting. Standard curves were constructed and used to quantitate FMO3 and FMO5 in human hepatic microsomal preparations. Positive identification of FMO3 or FMO5 was provided by monospecific antibodies that detected single protein bands with mobilities matching those of expressed FMO3 or FMO5. Calculations were based on slopes obtained from results of immunodetection with three concentrations of microsomal protein. The concentrations of FMO3 and FMO5 in five human hepatic microsomal preparations were highly variable—12.5 to 117 pmol/mg protein for FMO3 and 3.5 to 34 for FMO5 (Table 1), and only a weak relationship between FMO3 and FMO5 contents was apparent (r = 0.75). Relative amounts of mRNA for FMO3 and FMO5 in samples of RNA from the five human livers were also highly variable. However, the mRNA levels were, if anything, somewhat inversely related to the protein concentrations

Table 1. FMO protein concentrations, mRNA contents and activities in human hepatic microsomal samples

Sample	FMO content (pmol/mg prot.)		mRNA content (relative)		Metabolism (nmol product/mg prot./min.)	
	FMO3	FMO5	FMO3	FMO5	Methimazole	Ranitidine
437	72.0	7.5	3411	1059	3.3	0.35
405	22.0	11.5	4426	2777	1.6	0.29
300	12.5	3.5	1778	2193	0.9	0.13
8	78.0	34.0	1074	945	4.1	0.42
58	117.0	30.0	534	1450	5.7	0.76

(r = -0.58 for FMO3 and -0.52 for FMO5). Little or no agreement between the relative amounts of mRNA for FMO3 and FMO5 was observed (r = 0.52).

2.3.2. FMO-Catalyzed Metabolism in Human Hepatic Microsomal Samples. FMO-catalyzed metabolism of methimazole (S-oxidation) and ranitidine (N-oxidation) varied significantly among the five human hepatic microsomal samples; 0.9 to 5.7 and 0.13 to 0.76 nmol $\times$ mg protein^{-1} $\times$ min^{-1}, respectively (Table 1). However, agreement between the rates of metabolism in each sample was excellent (r = 0.95), and each rate was directly proportional to the FMO3 concentration (r = 0.99 with methimazole and 0.93 with ranitidine). The kinetic parameters (K_m = 25 mM; V_{max} = 51 nmol $\times$ nmol FMO3^{-1} $\times$ min^{-1}) were nearly identical to those calculated with recombinant FMO3. No evidence for the participation of a second enzyme in the reaction was observed even though sample 8, which was used for the kinetic determinations, contained the highest relative concentration of FMO5 (34 pmol/mg *vs* 78 for FMO3).

2.4. Role of FMO3 and FMO5 in Human Hepatic Drug Metabolism

Livers of adult, male humans appear to contain highly variable amounts of FMO3 and FMO5. FMO3 content is greater than FMO5 content but the ratio of the two varies from 2:1 to 10:1. Variability in FMO-dependent methimazole S-oxidation is also observed. This variability appears to be entirely a function of FMO3 content (r= .99). Regardless of the FMO3/FMO5 ratio, no evidence for the participation of FMO5 in the metabolism of methimazole is observed, which is consistent with the poor catalytic efficiency of FMO5 determined with the recombinant enzyme. Like S-oxidation of methimazole, N-oxidation of ranitidine also correlates very well with FMO3 content. The conclusion that FMO3 is likely the major contributor to FMO-mediated activity in adult human liver has been reached in a number of studies.[9–11,13,14] Our results offer quantitative proof that this is indeed the case for the S-oxidation of methimazole and the N-oxidation of ranitidine. Given the general lack of activity of FMO5[12,15] this will likely hold true for the majority of substrates.

3. EXPRESSION AND CHARACTERIZATION OF FMO4

3.1. Introduction

Our understanding of the human FMO isoforms, in particular FMO3 and FMO5, is based primarily on studies of enzymes expressed in *E. coli* or insect cells (see above). In contrast to these successes, attempts to express rabbit FMO4 in *E. coli*, yeast or COS-1 cells, or human FMO4 in *E. coli* or insect cells, have proven futile. Thus, the catalytic properties and distribution of FMO4, for which no purified enzyme or antibodies are available, have not been determined, although distribution of mRNA suggests that FMO4 may play a role in the brain.[16]

A comparison of the human and rabbit FMO4 transcripts with those encoding all other FMO isoforms described reveals a difference that might be associated with the inability to express FMO4 in heterologous systems. The coding regions of the FMO4 transcripts are 60 to 75 nucleotides longer than the coding regions of the other isoforms owing to a shift in the stop codon to the 3' of the consensus position. Elongation of FMO4 is due to alteration of the FMO consensus stop codon by a single base change.

3.2. Expression of FMO4

A stop codon was introduced at the consensus position in the cDNA for human FMO4 and a truncated enzyme (FMO4*), which contained 531, rather than 558 amino acids, was expressed in *E. coli*. SDS-PAGE and Western Blot analysis with a cross-reacting antibody to FMO3 showed that the expressed enzyme is detected as a protein of approximately 60kDa.[17] The effect on expression of the 3' region of the cDNAs encoding other FMO isoforms was also examined with FMO3. The coding region of the cDNA for FMO3 was extended to the same length as that of FMO4. This change decreased somewhat but the level was still far above that minimum required for detection. Comparisons of *E. coli* transcript levels with normal and modified cDNAs showed no differences. These results indicate that the lack of expression of FMO4 is related to a problem associated with translation rather than transcription. Recent results demonstrate that translation may be blocked by an RNA/RNA hybridization that involves a sequence of only seven bases. Three silent mutations are sufficient to allow for translation and expression of full-length, unaltered, FMO4.

3.3. Characterization of Recombinant FMO4

The 100,000g particulate fractions were used to examine the activities of FMO4 with methimazole as the substrate. The activity of recombinant FMO4 at a methimazole concentration of 1 mM was slightly greater than 1 nmol product $\times$ min^{-1} $\times$ mg protein^{-1} (n=4) with a specific activity of about 2.5 nmol product $\times$ min^{-1} $\times$ nmol FMO4^{-1}. The apparent Km for the reaction was 3.3 mM and the Vmax was about 4 nmol product $\times$ min^{-1} $\times$ nmol FMO4^{-1}. The effects of a number of factors on the metabolism of methimazole catalyzed by FMO4 were examined. The pH optimum for the reaction was found to be near 10.2. The responses of FMO3 and FMO3* to pH were nearly identical with pH optima near 9.5. The activity was found to be moderately temperature sensitive (more stable than has been noted for FMO1, but less stable than FMO2). The effects of n-octylamine (2mM), sodium cholate (1%), magnesium chloride (10mM) or imipramine (0.3mM) on the activity of FMO4* were all inhibitory (25%, 54%, 91% and 14%).

3.4. Transcription of FMO3 and FMO4 in *E. coli*

Lack of expression of FMO4 in *E. coli*, and reduced expression of FMO3 following elongation to form FMO3*, was examined to determine whether the fault could be localized to problems associated with transcription or translation. RNA was isolated from *E.*

coli transformed with FMO3, FMO3*, FMO4, FMO4*, or pJL vector alone and examined with ^{32}P-labeled cDNA probes for FMO3 and FMO4. Patterns and amounts of RNA detected were very similar when results obtained for FMO3 and FMO3* or FMO4 and FMO4* were compared. No evidence of abnormal or absent transcript was seen in the case of FMO4 and no decrease in transcript amount was apparent with FMO3*.

4. MODULATION OF FMO ACTIVITY BY TRICYCLIC ANTIDEPRESSANTS

4.1. Introduction

Isoform-dependent activation of catalytic activity by certain amines is one of the more interesting characteristics of the FMOs. Both modulator-induced and substrate-induced (self-activation) activation have been described. Ziegler et al.[18] showed that some short-chain amines, in particular n-octylamine, can increase the rate of oxidation of dimethylaniline catalyzed by pig FMO1. It is of interest that n-octylamine is not a substrate for pig FMO1. Self-activation in the metabolism of some substrates (N,N-dimethyl octylamine and other N,N-dimethyl alkylamines) but not others (dimethylaniline) was also observed.[10] With FMO2, however, markedly different results are obtained. First, n-octylamine is not an activator for FMO2,[7] but rather is a substrate.[19,20] Second, the activity of FMO2, but not FMO1, is enhanced by tricyclic antidepressants, compounds that are substrates for FMO1 but not for FMO2.[21] These results and others[22] support Ziegler's conclusion[18] that the FMO contains distinct catalytic and regulatory sites.

4.2. Modulation of FMO Activity by Imipramine

4.2.1. The Effect of Imipramine on FMO1 and FMO2. Although modulators of FMO activity are classified as either inhibitors or activators with a given isoform of the enzyme, that is only partially correct. Imipramine, for example, can be an inhibitor or an activator of the same FMO isoform depending upon the substrate concentration and the FMO isoform (Figure 1). With both inhibition and activation the extent of the effect is related directly to the concentration of imipramine. With rabbit FMO1 and FMO2, higher concentrations of methimazole result in activation and lower concentrations in inhibition. At a methimazole concentrations of ~100 uM or 50 uM no effect (activation = inhibition) is observed with FMO1 or FMO2, respectively. On the other hand, the activity of pig FMO1 is inhibited by imipramine at all concentrations of methimazole. The inhibition appears competitive with an I_{50} that is less than the Km for methimazole; reactions containing equimolar concentrations of imipramine and methimazole are inhibited by greater than 50%. The same Km (~3 uM) for methimazole is obtained with both pig FMO1 and rabbit FMO1, whereas the Km with FMO2 is about 300 μM.[7]

4.2.2. Characterization of Chimeric Proteins Containing Pig and Rabbit FMO1. Random chimerigensis[23] was used to produce chimeric transcripts that contained varying amounts of rabbit FMO1 at the 5' and pig FMO1 at the 3'. Active enzymes from 32 chimeras were expressed in *E. coli.* and eight were characterized in detail. The chimeras formed by random chimeragenesis and characterized were: R157P532 (rabbit protein from residue 1 to residue 157, pig protein from residue 157 to 532), R205P532, R235P532, R298P532, R380P535, R432P535, R465P535, R507P535. The metabolism of methima-

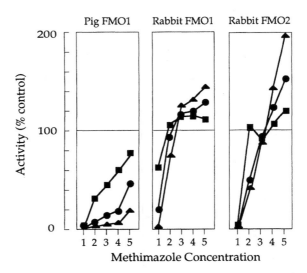

Figure 1. The effect of imipramine on the metabolism of methimazole catalyzed by pig FMO1, rabbit FMO1 and rabbit FMO2. Metabolism was determined at five concentrations of methimazole (1 = 1 µM; 2 = 50 µM; 3 = 100 µM; 4 = 300 µM; and 5 = 1 mM) and three concentrations of imipramine (square = 50 µM, circle = 300 µM, and triangle = 750 µM). Data are reported as percent of activity obtained in the absence of imipramine.

zole catalyzed by pig FMO1, rabbit FMO1 and rabbit/pig chimeras showed the same Km for methimazole (~3 µM). In contrast to the results with methimazole, metabolism of imipramine by different forms of the enzyme was highly variable. Data with pig FMO1 were linear with a Km of ~0.5 mM whereas the data with rabbit FMO1 were markedly curvilinear with disproportional increases in activity with increasing substrate concentration. The

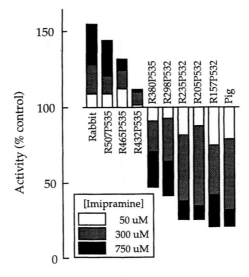

Figure 2. The effect of imipramine on the metabolism of methimazole catalyzed by pig FMO1, rabbit FMO1 and chimeric enzymes containing rabbit FMO1 and pig FMO1. The data are reported as percent of activity obtained in the absence of imipramine.

extent of the curvilinearity was inversely related to the amount of pig FMO1 included at the carboxy end of the chimeric proteins. Linear results, similar to those with pig FMO1, were obtained with R157P532. The effects of imipramine (50, 300 and 750 μM) on metabolism of methimazole (1 mM) catalyzed by various forms of FMO1 are shown in Figure 2. Under these conditions rabbit FMO1 is activated and pig FMO1 is inhibited, both in a concentration dependent manner. The responses of the chimeras relative to the two wild-type enzymes can be described in one of four ways: first, little alteration of the rabbit FMO 1 response—R507P535; two, marked reduction in activation, but no inhibition—R465P535 and R432P535; third, elimination of activation and introduction of significant inhibition—R380P535 and R298P532; fourth, inhibition characteristic of pig FMO1—R235P532, R205P532, and R157P532.

Similar results were obtained for metabolism of 100 μM methimazole except for the change with R432P535 which was now clearly greater than that with R465P535. The results at 100 uM confirmed that no differences were apparent between R380P535 and R298P532 or among R235P532, R205P532, R157P532 and pig FMO1. Amino acids 381–432 and 433–465 were identified from the properties of the simple chimeras as regions of FMO1 with major influence on the imipramine interaction. Minor effects could be attributed to regions 236–298 and 466–507, whereas regions 1–235, 299–380, and 508–535 appeared to have no influence on the imipramine effect. Of the seventy amino acid differences between pig and rabbit FMO1, twelve are in the 381–432 region, seven in the 433–465 region and a total of twelve in the 236–298 and 466–507 regions.

4.2.3. Specific Amino Acids Involved in the Imipramine Interaction. Residues in rabbit FMO1 were changed to the corresponding residues found in pig FMO1 by site-directed mutagenesis. A total of 20 mutants containing from 1 to 9 amino acid changes were characterized. The following substitutions, made singly or in conjunction with other changes, had no effect on the activation of methimazole (100 μM) metabolism by imipramine (750 μM) observed with rabbit FMO1: T236M, E386D, Y391W, Q394R, F396L, K410Q, E414T, S445Y, N447D, and L452M, L461H, M466I, and S472T. However, several single changes (I400N, H420P, and M381L) did result in mutants whose activities were only 72%, 63%, and 59%, respectively, of the activity with rabbit FMO1.

Results with mutants containing multiple changes indicated the importance of two additional positions—186 and 433. The activity of mutant #12 containing R186S and L452M (negative) was only 46% of rabbit FMO1 (Table 2), and the addition of A433S and F396L (negative) to a mutant containing M381L, and I400N decreased activity from 54% to 29% of the control (Table 2; mutants 9 and 15). Combination mutants showed that the effects of M381L, I400N, and A433S were at least partially independent (Table 2, mutants 9 and 15). This was also the case for R186S when added to M381L and I400N (decrease from 54% to 36% of control activity; mutants 9 and 14). However, addition of R186S to M381L, I400N, and A433S had little or no effect (mutants 15 and 16). In contrast, addition of H420P to R186S, M381L, I400N, and A433S (mutants 16 and 17) decreased activity from 25% to 15% of control. The effect of imipramine on metabolism of methimazole with combination mutants (#17–20) containing R186S, M381L, I400N, H420P, and A433S (~15% of control activity) approached that observed with chimera R157P532 or pig FMO1 (~7.5% of control activity). In addition, the kinetics of methimazole metabolism catalyzed by these mutants were the same as for the wild type rabbit and pig FMO1 orthologs, whereas the kinetics for imipramine metabolism most closely resembled those of pig FMO.

Table 2. The effects changing residues present in rabbit FMO1 to those present in pig FMO1 on the modulation of methimazole (100 μM) metabolism by imipramine (750 μM)

Mutant	Mutations (Rabbit aa - Residue # - Pig aa)	Activity
Rabbit	complete rabbit FMO1 sequence	100
1	Y391W	100
2	K410Q	100
3	N447D	100
4	F396L, L452M	100
5	K410Q, E414T	100
6	I400N	75
7	H420P	64
8	M381L	60
9	M381L, I400N	54
10	M381L, E386D, Q394R, F396L, I400N, K410Q, E414T	52
11	M381L, Q394R, F396L, I400N, K410Q, E414T, L452M	51
12	R186S, L452M	47
13	R186S, M381L, I400N, S445Y	36
14	R186S, M381L, I400N	35
15	M381L, F396L, I400N, A433S	29
16	R186S, M381L, F396L, I400N, A433S	25
17	R186S, M381L, F396L, I400N, A433S, H420P	14
18	R186S, M381L, F396L, I400N, A433S, H420P, L461H, S472T	13
19	R186S, T236N, M381L, F396L, I400N, A433S, H420P, L461H,	13
20	R186S, M381L, F396L, I400N, A433S, H420P, L461H, M466I	13
Pig	complete pig FMO1 sequence	8

4.3. Discussion

Elimination of the activation of FMO1 activity by imipramine as the structure of the enzyme is changed from that of the rabbit ortholog to that of the pig ortholog appears to involve two simultaneous changes, decreases in activation and increases in inhibition. These changes are consistent with a decrease in the ability of imipramine to activate its own metabolism and a simultaneous decrease in the Km for metabolism of imipramine. We speculate that activation and inhibition by imipramine involves an equilibrium between two orientations in a single "binding site". Inhibition of methimazole may be due to competition for the reactive oxygen rather than for a binding site. Likely, activation by imipramione involves an increae in the ratelimiting step which is thought to be release of oxidized cofactor or water form the enzyme.

REFERENCES

1. D.M. Ziegler and C.H. Mitchell, 1972, Microsomal Oxidase IV: Properties of a mixed-function amine oxidase isolated from pig liver microsomes, *Arch. Biochem. Biophys.* **150**: 116–125.
2. T.R. Devereux, R.M. Philpot and J.R. Fouts, 1977, The effect of Hg^{+2} on rabbit hepatic and pulmonary solubilized, partially purified N,N-dimethylaniline N-oxidases, *Chem.-Biol. Interactions* **18**: 277–287.
3. T.R. Devereux and J.R. Fouts, 1974, N-oxidation and demethylation of N,N-dimethyaniline by rabbit liver and lung microsomes, *Chem.-Biol. Interactions* **8**: 91–105.

4. M.P. Lawton, R. Gasser, R.E. Tynes, E. Hodgson, and R.M. Philpot, 1990, The flavin-containing monooxygenase enzymes expressed in rabbit liver and lung are products of related but distinctly different genes, *J. Biol. Chem.* **265:** 5855–5861.

5. M. Lawton and R.M. Philpot, 1995, Emergence of the Flavin-containing Monooxygenase Gene Family, in: *Rev. Biochem. Toxicol.,* Volume 11 (E. Hodgson, J.R. Bend and R.M. Philpot, eds.) pp. 1–27, Toxicology Communications, Inc, Raleigh, NC.

6. D.M. Ziegler, 1988, Flavin-containing monooxygenases: Catalytic mechanism and substrate specificity, *Drug Metab. Rev,* **19:** 1–32.

7. M.P. Lawton, T. Kronbach, E.F. Johnson, and R.M. Philpot, 1991, Properties of expressed and native flavin-containing monooxygenases: Evidence of multiple forms in rabbit liver and lung, *Mol. Pharmacol.* **40:** 692–698.

8. I.R. Philips, C. Dolphin, P. Clair, M.R. Hadley, A.J. Hutt, R. McCombie, R.L. Smith, and E.A. Shephard, 1995, The molecular biology of the flavin-containing monooxygenases of man, *Chemico-Biol. Interact.* **96:** 17–32.

9. N. Lomri, Q. Gu, and J.R. Cashman, 1992, Molecular cloning of the flavin-containing monooxygenase (Form II) form adult human liver, *Proc. Natl. Acad. Sci.* USA, **89:** 1685–1689. Correction, 1995, *Proc. Natl. Acad. Sci.* USA **92:** 9910.

10. A.J.M. Sadeque, K.E. Thummel, and A.E. Rettie, 1993, Purification of macaque liver flavin-containing monooxygenase: A form of the enzyme related immunochemically to an isozyme expressed selectively in adult human liver, *Biochim. Biophys. Acta,* **116:** 127–134.

11. J.R. Cashman, S.B. Park, Z-C. Yang, C.B. Washington, D.Y. Gomex, K.M. Giacomini, and C.M. Brett, 1993, Chemical, enzymatic, and human enantioselective S-oxidation of cimetidine, *Drug Metab. Dispo.* **21:** 587–597.

12. L.H. Overby, A.R. Buckpitt, M.P. Lawton, E. Atta-Asafo-Adjei, J. Schulze, and R.M. Philpot, 1955, Characterization of flavin-containing monooxygenase five (FMO5) from human and guinea pig: Evidence that the unique catalytic properties of FMO5 are not confined to the rabbit ortholog, *Arch. Biochem. Biophys.* **317:** 275–284.

13. S.A. Wrighton, M. Vandenbranden, J.C. Stevens, L.A. Shipley, B.J. Ring, A.E. Rettie, and J.R. Cashman, 1993, In vitro methods for assessing human hepatic drug metabolism: their use in drug development, *Drug Metabol. Rev.* **25:** 453–484.

14. N. Lomri, Z. Yang, and J.R. Cashman, 1993, Regio-and stereoselective oxygenations by adult human liver flavin-containing monooxygenase 3. Comparison with forms 1 and 2, *Chem. Res. Toxicol.* **6:** 800–807.

15. E. Atta-Asafo-Adjei, M.P. Lawton, and R.M. Philpot, 1993, Cloning, sequencing, distribution and expression of a mammalian flavin-containing monooxygenase from a third gene subfamily, *J. Biol Chem.* **268:** 9681–9689.

16. B. Blake, R.M. Philpot, P.E. Levi, and E. Hodgson, 1996, Xenobiotic biotransforming enzymes in the central nervous system: an isoform of flavin-containing monooxygenase (FMO4) is expressed in rabbit brain. *Chemico-biological Interactions,* **99:** 253–261.

17. K. Itagaki, G. Carver, and R.M. Philpot, 1996, Expression and characterization of a modified flavin-containing monooxygenase 4 (FMO4) from humans, *J. Biol. Chem.* **271:** 20102–20107.

18. D. M. Ziegler, , L.L. Poulsen and E.M. McKee ,1971, Interaction of primary amines with a mixed-function amine oxidase isolated from pig liver microsomes, *Xenobiotica* **1:** 523–531.

19. R.E. Tynes, P.J. Sabourin, E. Hodgson, and R.M. Philpot, 1986, Formation of hydrogen peroxide and N-hydroxylated amines catalyzed by pulmonary flavin-containing monooxygenases in the presence of primary alkylamines. *Arch. Biochem. Biophys.* **251:** 654–664.

20. L.L. Poulsen, K. Taylor, D.E. Williams, B.S.S. Masters, and D.M. Ziegler, 1986, Substrate specificity of the rabbit lung flavin-containing monooxygenase for amines: Oxidation products of primary alkylamines, *Mol. Pharmacol.* **30:** 680–685.

21. D.E. Williams, S.E. Hale, A.S. Muerhoff, and B.S.S. Masters, 1985, Rabbit lung flavin- containing monooxygenase, purification, characterization, and induction during pregnancy, *Mol. Pharm.* **28:** 381–390.

22. D.M. Ziegler, 1993, Recent studies on the structure and function of multisubstrate flavin-containing monooxygenases, *Annu. Rev. Pharmacol. Toxicol.* **33:** 179–199.

23. Kim, J. and P.N. Devreotes, 1994, Random chimeragenesis of G-protein-coupled receptors, *J. Biol. Chem.* **269:** 28724–28731.

6

CORRELATIONS BETWEEN IN VIVO AND IN VITRO STUDIES IN HUMAN DRUG METABOLISM

Michel Eichelbaum

Dr. Margarete Fischer-Bosch-Institut für Klinische Pharmakologie
Auerbachstr. 112
70376 Stuttgart, Germany

1. INTRODUCTION

Selection of the right dose is a major problem in drug therapy because following administration of the same dose of a drug to a patient population drug response varies substantially. A major factor responsible for variability in drug response is the activity of drug metabolising enzymes. The variability in the drug elimination process poses problems especially in the early phase of drug development. If a drug is mainly metabolised by an enzyme which exhibits a genetic polymorphism it is not unusual that a 50 fold difference in plasma concentrations will be observed if the same drug dosage is used. Thus during preclinical development of a drug and prior to its first administration to humans one would like to know the following facts about the drug. How is the drug eliminated from the body and which fraction of the dose is cleared by metabolism. If metabolism is the predominant route of elimination one can predict substantial variability in its clearance due to the well-known interindividual variation in the activity of drug metabolising enzymes. Less variability in clearance, however, will be observed if the drug is renally excreted because in subjects with normal kidney function only small interindividual differences in renal clearance occur. It is also of interest to know which enzymes (cytochrome P450s or the various phase II enzymes) catalyse the biotrans-formation of the drug. Furthermore we are interested to know the contribution of the different pathways to total metabolic clearance and what CYPs are involved in the formation of the different metabolites. Based on the knowledge of the enzymes involved in the biotransformation of a drug and knowing the expression of the enzymes in the liver we can make an educated guess about the contribution of liver to the metabolism of the drug under study. This is especially of interest in the case of drugs which are orally administered and where substantial first pass metabolism can occur. If a polymorphic enzyme is involved in the metabolism it is essential to know the

Molecular and Applied Aspects of Oxidative Drug Metabolizing Enzymes,
edited by Arinç *et al.* Kluwer Academic / Plenum Publishers, New York, 1999.

fraction of the dose cleared via this pathway. This will allow to predict the magnitude of phenotype related differences in clearance in patients. Quite often patients will receive in addition to the drug under study other drugs and therefore it is quite likely that metabolic drug interactions will occur. By knowing the enzymes involved in the biotransformation of a drug its potential to cause drug interactions or to be subject to a drug interactions can be anticipated.

Thus in vitro approaches which predict the human drug metabolism in vivo both in qualitative and quantitative terms would be extremely helpful in the clinical setting. Tissue slices, cell cultures, subcellular fractions and stable expressed enzymes of human origin have been studied in this respect. These in vitro methods have been used mainly to delineate metabolic pathways and to identify the enzymes involved in the formation of a metabolite. Most of the data generated so far are therefore restricted to qualitative aspects. Less effort has been expended to assess quantitative aspects, e.g. contribution of a given metabolite to the overall elimination of a drug in vivo and if these in vitro approaches can be used to predict in vivo clearance.

Within the context of this manuscript the focus will be on cytochrome P450 enzymes and will be restricted to the use of stable expressed enzymes, subcellular fractions, e.g., liver microsomes and hepatocytes.

2. STABLE EXPRESSED ENZYMES

Today almost all cytochrome P450 enzymes which are relevant for human drug metabolism are available as stable expressed enzymes. These enzymes are quite useful to identify the specific cytochrome P450 forms involved in the formation of a particular metabolite. Although inhibition studies with antibodies and drugs using liver microsomes are quite useful in identifying the enzymes catalysing the formation of a particular metabolite this approach has limitations. Some antibodies are not exclusively directed against one specific enzymes but recognise other enzymes from the same subfamily. The same holds true for chemical inhibitors. Furthermore inhibitory antibodies are not available for all cytochrome P450s. Antipyrine and verapamil will be used to demonstrate the advantages and limitations of this system.

Antipyrine has been used as a probe drug to assess a patients drug metabolising capacity. The drug is well suited for this purpose because after oral administration it is completely absorbed and its bioavailability is nearly 100%. Elimination proceeds almost exclusively through biotransformation, with a low clearance and is hence independent of liver blood flow. Protein binding is negligible. More than 95% of a given dose is excreted into urine as metabolites. The major metabolic routes are N-demethylation (formation of norantipyrine), 4-hydroxylation (formation of 4-hydroxyantipyrine) and 3-methylhydroxylation (formation of 3-hydroxymethylantipyrine) which together account for 50 to 80 % of the dose. It has been established that antipyrine metabolism proceeds through monooxygenases of the cytochrome P450 type. The approach that uses the total clearance of antipyrine to assess the drug metabolising capacity of a subject for other drugs has been rather disappointing because it was predictable only for few drugs. Attempts to improve the antipyrine test by measurement of the clearance to its three major metabolites did not significantly improve the predictive value of antipyrine as a probe drug, because each of the metabolites is formed by different enzymes as demonstrated recently.[1-7] Experiments with stable expressed enzyme showed the involvement of CYP1A2, CYP2C8, CYP2C9, and CYP2C18 in the formation of norantipyrine; CYP1A2, CYP2B6, CYP2C8, 2C9,

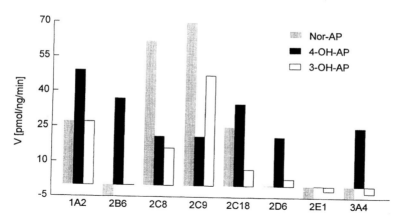

Figure 1. Formation of norantipyrine (Nor-AP), 4-hydroxy antipyrine (4-OH-AP) and 3-hydroxymethyl antipyrine (3-OH-AP) by stable expressed enzymes. Antipyrine at a concentration of 50 mM was incubated for 15 minutes in a 0.1 mol phosphate buffer at pH 7.4.[8]

2C18, CYP2D6 and CYP3A4 in the formation of 4-hydroxyantipyrine; and CYP1A2 and CYP2C9 in the formation of 3-hydroxymethyl antipyrine. Small amounts of 3-hydoxymethyl antipyrine were found in incubations with CYP2C18 and CYP2D6. Especially in the case of the formation of 4-hydroxyantipyrine the limitations become apparent. 4-hydroxyantipyrine was also formed in control microsomes prepared from yeast that contained the vector but not the coding sequence for a cytochrome to an extent that for some enzymes was almost as high as formation in microsomes from stables expressed enzymes (Figure 1). In contrast to stable expressed enzymes formation of 4-hydroxyantipyrine in microsomes was not inhibited by antibodies directed against the CYP2C subfamily. Furthermore sulphenazole which is an inhibitor of CYP2C9 did not inhibit the formation of 4-hydroxyantipyrine. The finding that the polymorphic enzyme CYP2D6 is involved in the formation of the 4-hydroxy metabolite could also be ruled out because antibodies directed against this cytochrome P450 enzyme did not inhibit its formation.[8,9] In addition in vivo data in panels of poor and extensive metabolisers of sparteine/debrisoquine could not demonstrate any difference in the clearance to the 3 major metabolites.[10] Therefore data derived with stable expressed enzymes should be supplemented by experiments with other in vitro system in order to reach meaningful conclusions for the in vivo situation.

Data from experiments with stable expressed enzymes can be used to estimate the contribution of a particular isozymes to the formation of a metabolite. By calculating the intrinsic clearance values for each expressed enzyme and by taking into account the proportion of a particular enzyme of the total cytochrome P450 content in the liver a semiquantitative assessment can be made.[11] In combination with inhibition experiments with antibodies and chemical inhibitors the contribution of the various cytochromes to the formation of a particular metabolite can be estimated. Using this approach it could be demonstrated that in the formation of norantipyrine CYP2C9 is the major enzyme whereas 4-hydroxylation is mainly catalysed by CYP3A4. In the case of 3-hydroxymethyl antipyrine both CYP1A2 and 2C9 contribute to its formation in similar proportions.[8]

Verapamil undergoes extensive first pass metabolism to a great number of metabolites. Most of the metabolites are devoid or have only weak calcium channel blocking activities. The biotransformation of verapamil involves N-dealkylation and O-dealkylation

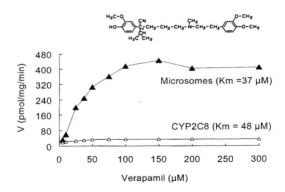

Figure 2. Formation of the O-demethylated metabolite D703 from verapamil by human liver microsomes and stable expressed CYP2C8 (for details: Busse et al., 1995[14]).

which are catalysed by different P450s. The N-demethylation to norverapamil and the N-dealkylated metabolites is mediated by CYP3A4 and in addition to a minor extent by CYP1A2 in the case of the N-dealkylated metabolites.[12] No formation of the O-demethylated metabolites was observed by stable expressed CYP3A4 and 1A2. The fact that formation of these metabolites by human liver microsomes was inhibited by anti-LKM2 antibodies which are directed against CYPs2C9 and 2C19 indicated the involvement of the enzymes of the 2C subfamily. Since the formation of the 2 O-demethylated metabolites D702 and D703 was inhibited only by 53 and 82%, respectively additional cytochromes of the 2C subfamily must be involved. The identification of these additional cytochromes of the CYP2C subfamily was accomplished with stable expressed enzymes. These experiments revealed that the formation of D703 was catalysed by CYP2C8, CYP2C9 and CYP2C18. In the case of D702 formation of this metabolite was observed only for CYP2C9 and CYP2C18. The ranking of the maximum turnover rate was CYP2C8 > CYP2C18 > CYP2C9. At first sight these data indicated CYP2C8 to be the major enzyme involved in the formation of D703 which would contradict results obtained in the inhibition studies with the LKM2 antibodies.

Taking into account, however, the relative abundance of the 3 enzymes in human liver which is about 10 fold hihger for CYP2C9 than for CYP2C8 or CYP2C18 despite its higher intrinsic clearance the contribution of CYP2C8 becomes relatively decreased as compared to CYP2C9 due to the much greater in vivo abundance of the latter (Figure 2).[13,14]

3. LIVER MICROSOMES

Human liver microsomes are the most widely employed in vitro system to study human metabolism. They allow to assess the overall metabolism of a drug. In addition by using samples from a number of subjects it provides an indication of the interindividual variability in metabolism in the population. The system is quite stable and liver microsomes can be stored at -80°C without loss of activity over a long period of time.

The formation of drug metabolites by human liver microsomes allows the to assess the relative contribution of individual enzymes and individual pathways to overall metabolism. By correlating the formation of a metabolite with the expression of the CYP content in the liver an indication which enzymes could be involved in the formation of the metabolites is obtained. The interindividual variation in enzyme expression in a liver bank results in variability of V_{max} and K_m. Intrinsic clearance derived from V_{max} and K_m values can be used to predict variabitity of hepatic in vivo clearance by using scaling factors and an appropriate liver perfusion model.[11,15] If several enzymes are involved in the formation of a metabolite the contribution of a specific cytochrome isoform to the total intrinsic clearance can be accessed by the use of specific antibodies and chemical inhibitors. In the case of a polymorphic enzyme the system can be used to see which metabolites are formed by this enzyme. Based on the intrinsic clearances for the various metabolites the contribution of the polymorphic enzyme to their formation can be estimated and a prediction can be made as far as phenotype related differences of in vivo clearance are concerned. The system will also provide useful informations on metabolic drug interactions. The validity of this approach will be demonstrated by using the example of the 5-HT$_3$-receptor antagonists tropisetron.

Tropisetron a highly potent and selective 5-HT$_3$-receptor antagonist, has been shown to be a potent antiemetic agent, superior to metoclopramide and comparable to the most effective combinations of antiemetic compounds used so far.[16-19]

Studies in healthy volunteers revealed pronounced interindividual variation in its pharmacokinetics.[20] Tropisetron is extensively biotransformed by hydroxylation of the indol ring in 5, 6, and 7 positions. The hydroxymetabolites are subsequently conjugated with glucuronid acid and sulfate. The major metabolic pathway seems to be 6-hydroxylation, the most abundant metabolites in urine were 6-hydroxytropisteron (31%) and its sulfate (13%) and 5-hydroxytropisteron as sulfate and glucuronide conjugates (12%). Oxidative N-demethylation and N-oxygenation at the tropinyl nitrogen also occur but in a minor extend. Only traces of N-demethyl and N-oxidmetabolites were found.[21]

The formation of 5- and 6-hydroxy metabolite was studied in the microsomal fraction of 7 EM and 1 PM livers for the sparteine/debrisoquine polymorphism. The formation of 5- and 6-hydroxytropisetron was biphasic with a high and a low affinity component in all 7 EM livers. In the PM liver the high affinity component was absent and only the low affinity component was observed. V_{max} of the high affinity component of the 5- (V_{max}: 1.88 ± 0.73 pmol/mg/min; K_m: 3.9 ± 2.1 µM) and 6- (V_{max}: 4.00 ± 1.77 pmol/mg/min; K_m: 4.66 ± 1.84 µM) hydroxy metabolites was correlated with the CYP2D6 amount (5-OH-ICS r=0.88, p <0.01; 6-OH-ICS r=0.98, p <0.01). In contrast no such a correlation was observed between the low affinity component and CYP2D6 content. There was no relationship between the V_{max} values of 5- and 6-hydroxytropisetron formation and the contents for the other P450 enzymes.

The influence of LKM1 antibodies on metabolite formation was studied in the microsomes of EM and PM livers. The degree of inhibition in the 5- and 6-OH formation by LKM1 in the EM livers was dependent on substrate concentration. At low substrate concentration of 10 µM the antibodies almost completely inhibited metabolite formation, whereas at tropisetron concentrations of 640 µM inhibition was incomplete. In the PM liver 5- and 6-OH formation was not affected by LKM1 (Figure 3). N-Demethyltropisetron formation was not inhibited by the antibodies in both EM and PM liver microsomes. Rate of formation of 5-OH and 6-OH-ICS in presence of LKM1 was correlated with the content of CYP3A4 (5-OH-ICS r=0.997; 6-OH-ICS r=0.844).

These data clearly indicate that at least two cytochrome P450 enzymes are involved in formation of 5- and 6-hydroxy tropisetrone. The high affinity component was identified as CYP2D6 and evidence for this assumption is based on the following observations. V_{max}

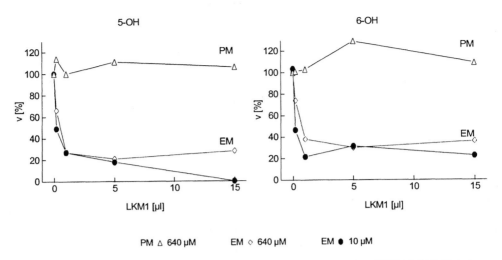

Figure 3. Inhibition of 5-OH and 6-OH tropisetron by antibodies directed against CYP2D6 (LKM1) in human liver microsomes from an EM and PM liver at 10 μM and 640 μM substrate concentration.

of this component was significantly correlated with the expression of CYP2D6 in the liver samples as derived from immunoblotting. The low affinity component was absent in the liver of PMs where no CYP2D6 is present. LKM1 antibodies which are directed against CYP2D6 completely inhibited formation of 5 and 6-OH tropisetron at low substrate concentrations. Accordingly, quinidine which is an inhibitor of CYP2D6 completely inhibited 5- and 6-hydroxy tropisetron formation at low substrate concentrations. Finally, 5- and 6-OH tropisetron were formed in yeast microsomes expressing CYP2D6.[22]

In contrast, the low affinity component involved in the formation of 5- and 6-OH tropisetron is not related to CYP2D6, since it could not be inhibited by antibodies directed against this enzyme and CYP2D6 inhibitors. Furthermore this component was also present in the PM liver. This low affinity component was identified as CYP3A4 which is again supported by several lines of evidence. The amounts of 5- and 6-OH tropisetron formed in presence of CYP2D6 blocking LKM1 antibodies correlated with the CYP3A4 content of the livers. The CYP3A4 inhibitor ketoconazole reduced 5- and 6-OH formation by 20 %. The degree of inhibition in 5- and 6-OH formation by ketoconazol corresponds to the percentage of metabolites formed in the presence of LKM1. Finally, small amounts of the 5- and 6-OH metabolites were formed in yeast microsomes containing CYP3A4.

The formation of the minor metabolite N-desmethyl tropisetron is catalyzed by CYP3A4 which is based on the following evidence. Incubation of tropisetron in presence of the microsomal fraction of yeast cells expressing CYP3A4 resulted in a high rate of formation of the N-desmethyl metabolite. Moreover, the CYP3A inhibitor ketoconazole reduced formation of this metabolite in a competitive manner. In contrast, CYP1A2 is not involved in the N-demethylation of tropisetron since inhibitors of this enzyme such as phenacetin and furaphylline did not impair the formation of N-desmethyltropisetron.

If one evaluates based on *in vitro* intrinsic clearance the contribution of CYP2D6 and CYP3A4 to the in vivo elimination of tropisetron several predictions can be made. CYP2D6 catalyzed 5 and 6-hydroxylation is the predominant route of elimination and hence large phenotype related differences in the formation of these metabolites and total plasma clearance have to be expected. With intrinsic clearance for the low affinity component being considerably smaller, formation of these two metabolites by CYP3A4 contributes only to a minor extent to the elimination and cannot compensate for the loss of CYP2D6 activity in PMs. Furthermore clearance to 6-hydroxy should be 2 to 3 times higher than to 5-hydroxytropisetron. Metabolism via CYP3A4 catalyzed N-demethylation is a minor pathway, since tropisetron concentrations achieved in vivo are far below the concentrations at which measurable amounts of N-desmethyltropisetron are formed in vitro. As in the case of CYP3A4 catalyzed hydroxylation this pathway cannot compensate for the loss of CYP2D6 in PMs. These in vitro prediction are in good agreement with the in vivo situation. EM subjects excrete 50 to 60% of the dose as 5- and 6-hydroxytropisetron whereas in PMs only trace amounts of these two metabolites are formed. The fraction of dose excreted in urine as 6-hydroxytropisetron is on average 3 times higher than the corresponding values for the 5-hydroxy metabolite. The N-demethyl metabolite accounts for only a minor part of the dose.[21]

These in vitro findings have several implications for the clinical use of tropisetron. Poor metabolizers will have considerably higher plasma concentrations than extensive metabolizers after administration of the same dose. With intoxication due to HT$_3$ receptor antagonists being unlikely given the rather wide therapeutic safety margin of these drugs it will be the subset of patients with extremely high CYP2D6 activity that may suffer from unexpectedly low efficacy of tropisetron. Recent data by Johansson et al (1993)[23] identified the molecular base of so called ultra rapid metabolizers (e.g. patients with extremely high CYP2D6 activity[24]) to be gene amplification. They reported three members of one family having 12 copies of the CYP2D6 gene. If tropisetron is given to a patient with this genetic disposition our in vitro data predict that regular doses will result in subtherapeutic plasma concentrations. Thus, this subset of patients will suffer from chemotherapy induced nausea despite administration of a potent 5-HT$_3$ antagonist. As a consequence dosage of tropisetron should be adjusted according to the individual CYP2D6 activity in particular in rapid metabolizers.

Beside interindividual variability in drug metabolism the question of drug interactions may play a pivotal role in net drug effect resulting from multiple drug administration during chemotherapy. In general our in vitro data predict that drugs which are substrates for CYP2D6 and CYP3A4 can inhibit tropisetron metabolism. Since CYP2D6 catalyzed 5- and 6-hydroxylation is the main route of elimination inhibition of this enzyme will have a major impact on tropisetron biotransformation. In view of their low K$_i$ values drugs such as quinidine[25] and propafenone[26] are expected to inhibit CYP2D6 completely. Coadministration of drugs known to be inducer of cytochrome P450 enzymes such as rifampicin should have no major impact on 5 and 6-hydroxylation and hence the overall disposition since CYP2D6 can not be induced.[27] However, a modest induction of tropisetron metabolism for the fraction of the dose which is metabolized via CYP3A4 can be expected. The same holds true for drugs which inhibit only CYP3A4. If drugs such as propafenone are administered which inhibit both CYP2D6 and CYP3A4 almost complete inhibtion of tropisetron metabolism is to be expected in both EMs and PMs whith the effects being much greater in the EM subset.

Since tropisetron is used in patients concomitantly with cytoctatic drugs the potential for drug interactions was investigated.

Of the cytostatic drugs studied, only vinblastine affected the 5 and 6-hydroxylation of tropisetron indicating that vinblastine is an inhibitor of CYP2D6. This is in agreement with data from Relling et al (1989) who reported that vinblastine was an inhibitor (K$_i$ 90µM) of the CYP2D6 catalyzed 1-hydroxylation of bufuralol. However, it is very unlikely that vinblastine will inhibit tropisetron metabolism in vivo, because the high vinblastine concentrations required (K_i 220 µM) will not be achieved. Whether or not vinblastine is itself metabolized by CYP2D6 is presently unknown. Provided this is the case tropisetron could inhibit vinblastine metabolism. The potential of tropisetron to cause interactions with drugs which are CYP2D6 and CYP3A4 substrates is rather small. Tropisetron concentrations required to inhibit these enzymes in vitro are at least an order of magnitude greater than the plasma concentration of tropisetron which are achieved in patients.

4. CELL CULTURE

In contrast to subcellular fractions and stable expressed enzymes hepatocyte cultures have the advantage that both phase 1 and phase 2 reactions can be studied thus providing a metabolic pattern which reflects the in vivo situation. Based on the rate of disappearance of the substrates in the incubation medium clearance can be estimated and by using appropriate scaling factors hepatic in vivo clearance can be predicted.[11,15] In addition aspects of regulation, e.g., induction can be investigated. The disadvantage of this system is its limited availability because it requires fresh liver tissue. Furthermore the P450 content decreases in primary hepatocyte cultures.

In Figure 4 the decline of verapamil concentrations in the incubation medium versus time in a primary hepatocyte culture is depicted. The half life is approx 8.5 hours which is very similar to the half live observed in vivo. Based on the number of hepatocyte present in the incubation medium and by assuming a hepatocyte number of 120×10^6 per gram of liver the clearance per gram of liver can be derived. By multiplying this clearance with total liver weight total clearance can be calculated. The clearances of 3–4 l/min obtained in these experiments were of the same order as those at 4–8 l/min observed in vivo. Incubation experiments carried out after preincubation of hepatocytes with rifampicin resulted in increased clearance. The magnitude of the induction phenomenon in vitro was very similar to the changes observed in healthy volunteers who had been pretreated with rifampicin (Figure 4).

If extrahepatic metabolism contributes substantially to overall drug elimination the use of liver as a tissue source results in poor prediction. This is especially the case for

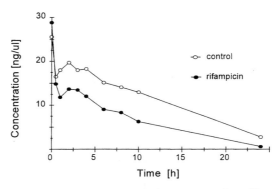

Figure 4. Decline of verapamil concentrations in primary hepatocytes culture. Hepatocytes (10^7 cells) obtained from partial hepatectomy were incubated with 50 µM verapamil. At various times aliquots of the incubation medium were analysed for verapamil and metabolite concentrations (unpublished results).

orally administered drugs with extensive presystemic metabolism by the gut wall. In such a situation in vivo predictions if based only on in vitro data derived from hepatic tissue will grossly underestimate in vivo clearance. The data may be even more misleading when enzyme induction occurs preferentially at extrahepatic sites. However, so far no in vitro approaches are available which allow to predict extrahepatic metabolism as precisely as it can be accomplished with hepatic metabolism.

REFERENCES

1. E.S. Vesell, 1991, The model drug approach in clinical pharmacology, *Clin. Pharmacol.Ther.* 50:239–248.
2. E.S. Vesell, 1979, The antipyrine test in clinical pharmacology: Conceptions and misconceptions, *Clin. Pharmacol. Ther.* 26: 275–86.
3. B.K. Park, 1982, Assessment of the drug metabolism capacity of the liver, *Br. J. Clin. Pharmacol.* 14:631–651.
4. M. Hartleb, 1991, Drugs and the liver part II: the role of the antipyrine test in drug metabolism studies, *Biopharm. Drug Dispos.* 12:559–70.
5. M. Eichelbaum, H.R. Ochs, G.M. Robertz, and A. Somogyi, 1982, Pharmacokinetics and metabolism of antipyrine (phenazone) after intravenous and oral administration, *Arzneimittelforschung* 32:575–78.
6. M.W.E. Teunissen, P. Spoelstra, C.W. Koch, B. Weeda, W.V. Duyn, and A.R. Janssens, 1984, Antipyrine clearance and metabolite formation in patients with alcoholic cirrhosis, *Br. J. Clin. Pharmacol.* 18:707–15.
7. M. Eichelbaum, M. Schomerus, N. Spannbrucker, and E. Zietz, 1976, The physiological disposition of 14C-antipyrine in man. *Naunyn Schmiedebergs Arch. Pharmacol.* 293 Suppl.:R63.
8. G. Engel, U. Hofmann, H. Heidemann, J. Cosme, and M. Eichelbaum, 1996, Antipyrine as a probe for human oxidative drug metabolism: Indentification of the cytochrome P450 enzymes catalyzing 4-hydroxyantipyrine, 3-hydroxymethylantipyrine, and norantipyrine formation, *Clin. Pharmacol. Ther.* 59:613–623.
9. J.E. Sharere and S.A. Wrighton, 1996, Identifation of the human cytochrome P450 involved in the in vitro oxidation of antipyrine, *Drug Metab. Dispos.* 24:487–494.
10. M. Eichelbaum, L. Bertilsson, and J. Säwe, 1983, Antipyrine metabolism in relation to the polymorphic oxidation of spartein and debrisoquine, *Br. J. Clin. Pharmacol.* 15:317–21.
11. T. Iwatsubo, N. Hirota, T. Ooie, H. Suzuki, N. Shimada, K. Chiba, T. Ishizaki, C.E. Green, C. Tyson, and Y. Sugiyama, 1997, Prediction of in vivo druf metabolism in the human liver from in vitro metabolism data, *Pharmacol. Ther.* 73:147–171.
12. H.K. Kroemer, J.-C. Gautier, P. Beaune, C. Henderson, C.R. Wolf, and M. Eichelbaum, 1993, Identification of P450 enzymes involved in metabolism of verapamil in humans, *Naunyn-Schmiedeberg's Arch. Pharmacol.* 348: 332–337.
13. H.K. Kroemer, H. Echizen, H. Heidemann, and M. Eichelbaum, 1992, Predictabilty of the in vivo metabolism of verapamil from in vitro data: contribution of individual metabolic pathways and stereoselective aspects, *J. Pharmacol. Exp. Ther.* 260: 1052–1057.
14. D. Busse, J. Cosme, P. Beaune, H.K. Kroemer, and M. Eichelbaum, 1995, Cytochromes of the P450 2C subfamily are the major enzymes involved in the O-demethylation of verapamil in humans, *Naunyn-Schmiedeberg's Arch. Pharmacol.* 353: 116–121.
15. J.B. Houston, 1994, Utility of in vitro drug metabolism data in predicting in vivo metabolic clearance, *Biochem. Pharmacol.* 47:1469–1479.
16. L. Dogliotti, R.A. Antonacci, E. Paze, C. Ortega, A. Berruti, and R. Faggiuolo, 1992, Three years' experience with tropisetron in the control of nausea and vomiting in cisplatin-treated patients, *Drugs* 43(3): 6–10.
17. U. Bruntsch, S. Drechsler, E. Hiller, W. Eiermann, A.H. Tulusan,M. Bühner, R. Hartenstein, H.J. Koenig, and W.M. Gallmeier, 1992, Prevention of chemotherapy-induced nausea and emesis in patients responding poorly to previous antiemetic therapy. Comparing tropisetron with optimised standard antiemetic therapy, *Drugs* 43(3): 23–26.
18. H. Bleiberg, S. Van Belle, R. Paridaens, G. De Wasch, L.Y. Dirix, and M. Tjean, 1992, Compassionate use of a 5-HT$_3$-receptor antagonist, tropisetron, in patients refractory to standard antiemetic treatment. *Drugs*, 43(3): 27–32, 1992.
19. B. Sorbe and A.M. Berglind, 1992, Tropisetron, a new 5-HT$_3$-receptor antagonist, in the prevention of radiation-induced nausea, vomiting and diarrhoea. *Drugs* 43(3): 33–39.
20. K.M. De Bruijn, 1992, Tropisetron: A review of the clinical experience, *Drugs* 43(3): 6–10.

21. V. Fischer, J.P. Baldeck, and F.L.S. Tse, 1992, Pharmacokinetics and metabolism of the 5-hydroxytryp-tamine antagonist tropisetron after single oral doses in humans, Drug Met. Disp. 20:603–607.

22. L. Firkusny, H.K. Kroemer, and M. Eichelbaum 1995, *In vitro* characterization of cytochrome P450 cata-lysed metabolism of the antiemetic tropiestron, *Biochem. Pharmacol.* 49: 1777–1784.

23. I. Johansson, E. Lundqvist, L. Bertilsson, M. L. Dahl, F. Sjöqvist, and M. Ingelman-Sundberg, 1993, Inher-ited amplification of an active gene in the cytochrome P450 CYP2D locus as a cause of ultrarapid metabo-lism of debrisoquine. *Proc. Natl. Acad. Sci. USA.* 90:11825–11829.

24. L. Bertilsson, M.L. Dahl, F. Sjöqvist, A. Åberg Wistedt, M. Humble, I. Johansson, E. Lundqvist and M. Ingelman-Sundberg, 1993, Molecular basis for rational megaprescribing in ultrarapid hydroxylators of de-brisquine, *Lancet* 341:63.

25. C. Funck-Bretano, H.K. Kroemer, H. Pavlou, R.L. Woosley, and D.M. Roden, 1989, Genetically-deter-mined interaction between propafenone and low dose quinidine: role of active metabolites in modulating net drug effect, *Br. J. Clin. Pharmacol.* 27:435–444.

26. H.K. Kroemer, G. Mikus, T. Kronbach, U.A. Meyer, and M. Eichelbaum, 1989, In vitro characterization of the human cytochrome P-450 involved in polymorphic oxidation of propafenone, *Clin. Pharmacol. Ther.* 45:28–33.

27. M. Eichelbaum, S. Mineshita, E.E. Ohnhaus, and C. Zekorn, 1986, The influence of enzyme induction on polymorphic sparteine oxidation, *Br. J. Clin. Pharmacol.* 22: 49–53.

28. M.V. Relling, W.E. Evans, R. Fonne-Pfister, and U.A. Meyer, 1989, Anticancer drugs as inhibitors of two polymorphic cytochrome P450 enzymes, debrisoquin and mephenytoin hydroxylase, in human liver mi-crosomes. *Cancer Res.* 49:68–71.

7

PHARMACOGENETICS

Polymorphisms in Xenobiotic Metabolism

Frank J. Gonzalez

Laboratory of Metabolism
National Cancer Institute
Bethesda, Maryland 20892

1. INTRODUCTION

1.1. Definitions and General Properties

A polymorphism is a monogenic trait that is caused by the presence in the population of more than one allele at the same locus (gene) that yields more than one phenotype in the organism. The frequency of the less common allele is usually > 1%. Polymorphisms usually do not cause any sickness or other problems that would decrease reproductive efficiency. This is why a polymorphism can be found in a population at such a high frequency (> 1%). In contrast, a genetic defect or mutation that causes a serious disease such as cystic fibrosis or phenylketonuria are found at extremely low frequencies in humans because sick people do not efficiently reproduce and transmit the trait. Only the heterozygotes carrying an allele for a genetic defect that is autosomal recessive, readily transmit the trait to offspring. An inborn error of metabolism is not a polymorphism.

Pharmacogenetic polymorphism is associated with metabolism of xenobiotics. The phenotype can be determined by analysis of enzyme activity but is not usually recognized without the administration of drugs associated with the metabolic deficiency.

1.2. History of Pharmacogenetics

Pharmacogenetics was established as a field in the 1950's with the discovery of enzyme variation in glucose 6-phosphate dehydrogenase, pseudocholinesterase variation and the N-acetyltransferase deficiency using the drug isoniazid. Vogel in Germany started use of the term "Pharmacogenetics" in 1957. Pharmacogenetics became popular with the discovery in Germany and England of the "debrisoquine/sparteine polymorphism" which is due to a defect in the cytochrome P450 CYP2D6. Debrisoquine was licensed for use in Europe as an antihypertensive drug but soon the physicians found a number of patients that had adverse reactions or toxicity's to the drug when it was administered at a standard

Molecular and Applied Aspects of Oxidative Drug Metabolizing Enzymes,
edited by Arınç *et al.* Kluwer Academic / Plenum Publishers, New York, 1999.

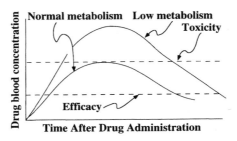

Figure 1. Schematic diagram of therapeutic index. The efficacy of a drug is optimal above an experimentally determined blood concentration. At higher concentrations, the drug can be toxic. This window is indicated by the dashed line. The level of drug is inversely proportional to its rate of metabolism and excretion. Low metabolism results in high drug levels that can potentially lead to adverse drug reactions.

dose. Other patients benefited from the drug. Patients that exhibited toxicity from the drug were found to lack the ability to convert debrisoquine to its 4-hydroxyl derivative.

1.3. Therapeutic Index and Metabolism of Drugs

This is an index that determines the safe concentrations of a drug in the blood that yields therapeutic efficacy. Drugs with a wide therapeutic index are relatively safe and their use is not complicated by differences in metabolism. Drugs with narrow therapeutic indexes can be dangerous since low rates of metabolism can cause them to reach the toxic threshold. Any drug that has a narrow therapeutic index and is rapidly metabolized, can be affected by a polymorphism in a drug metabolizing enzyme. Drugs with narrow therapeutic indexes include many cancer chemotherapy drugs and cardiovascular drugs.

2. MECHANISM OF POLYMORPHISMS

2.1. Functional vs. Non-Functional Polymorphisms

A functional polymorphism is a change in the DNA sequence of a gene that results in different levels of expression of protein or enzyme. For example, the introduction of a protein termination codon TGA from a codon for the amino acid cysteine (TGC) by mutation of the C to an A. A non-functional polymorphism can be considered a change in DNA sequence that does not result in different levels of expression of protein or enzyme. This could be a silent mutation in which the codon for an amino acid does not change due to the redundancy in the genetic code. An example would be a change from CTT to CTC, both of which encode a leucine. A change in DNA sequence within an intron could be detected but may also be considered non-functional since it does not influence gene expression.

It can never be certain that a DNA base change does not influence expression of the gene product. However, certain DNA changes, like introduction of a base change that produces a termination codon of a major deletion of a part of a gene's coding region are more likely to have functional significance than a silent mutation or a base change in the middle of an intron.

2.1.1. Definition of Allele. A gene can be defined as a specific unit of DNA coding sequence that is capable of producing a functional RNA, such as ribosomal and tRNA or

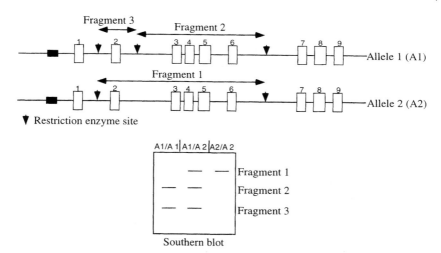

Figure 2. Determination of restriction fragment length polymorphisms (RFLP). A change in DNA sequence in the second intron of Allele 1 (A1) results in loss of a restriction enzyme site giving rise to Allele 2 (A2). This can be detected as an RFLP by Southern blot analysis. As shown, homozygotes for A1 can be distinguished from homozygotes for A2 and heterozygotes A1/A2 by the size of the restriction fragments. Fragments 2 and 3 that result from enzyme digestion, are diagnostic for allele A1 while the restriction enzyme-resistant fragment 1 is diagnostic for presence of allele A2.

protein. A particular gene can exist in a population in a number of different forms called alleles. Two different alleles can be distinguished by a single DNA base change. In an outbred population, like humans, every allele probably differs from every other allele in at least one base, particularly when one considers intron sequences that can accumulate changes without functional significance. More typically, alleles are only distinguished when they are functionally different, for example, have an amino acid difference that results in enzyme forms with different activities. However, this is not always true.

The definition of a gene mutation can also be discussed. Classically, a mutation is considered as a DNA base change that alters expression of a gene or gene product. However, there are many changes that occur that do not have any influence on functional activity. It as thus become common to use the term mutations when there in a functional alteration in a gene or gene product.

2.1.2. Definition of a Variant. It has become common in the field of genetics to use the term variant when describing an allele that gives rise to an enzyme with altered activity, but not deficient. For example, an amino acid change in a cytochrome P450 can result in an enzyme that has altered substrate specificity can be considered a "variant enzyme".

2.2. Methods for Detecting Polymorphisms or Gene Mutations

There are a number of methods that can be used to determine polymorphisms in genes. The earliest used was the restriction fragment length polymorphism or RFLP in which the base change in or around a gene causes formation or loss a recognition site for a restriction endonuclease. This method has largely been replaced by a number of other procedures that do not rely on the use of restriction enzymes and the analysis of DNA by Southern blotting electrophoresis. However, it is useful to discuss the RFLPs to illustrate

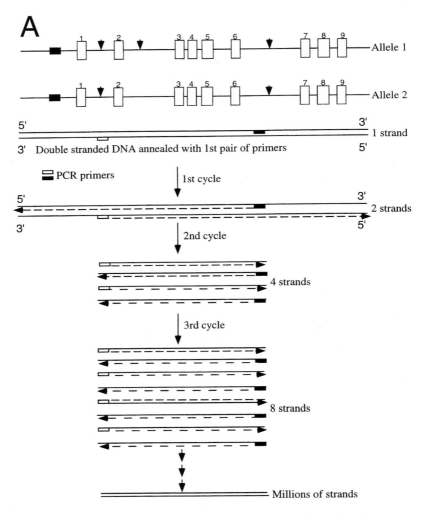

Figure 3. (A) PCR analysis for detection of specific mutations in genes. Genotyping a heterozygote, Allele 1 (A1) and Allele 2 (A2) by PCR. Each cycle of PCR: The three repetitive steps in PCR: 1) Denature DNA, 2) Anneal primers, 3) Extend new DNA (polymerize). This results in an exponential increase of 2^N where N = cycle number. Ten cycles = 1024 molecules of double stranded DNA from one starting molecule. Panel A shows the results of PCR and panel B shows how the PCR products can be analyzed for a specific mutation that changes a restriction enzyme recognition site.

differences between genes that give rise to polymorphisms and to introduce a more widely used method of polymerase chain reaction (PCR) for detection of gene mutations.

The Polymerase Chain Reaction (PCR) is the most commonly used method for detecting DNA base changes. This method allows the amplification of segments of a gene that can then be analyzed for the presence of base changes.

There are other techniques that are used to genotype for different alleles including ligation-mediated PCR and DNA microchips that are more easily automated for high throughput screening. The Affimetrics Co., is developing DNA microchips that will be used in determining gene mutations and polymorphisms. This requires three steps, 1) Am-

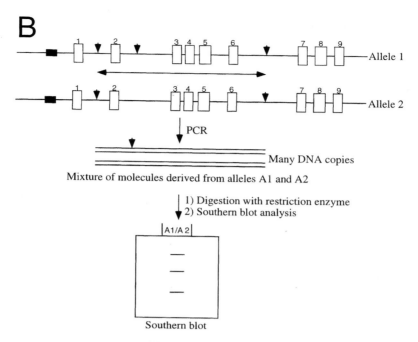

Figure 3. (*Continued*)

plify segment of DNA containing mutation. 2) Hybridize amplified DNA with a silicone microchip containing sequence of interest. 3) Determine whether the sequence is present by analyzing for hybrids using laser scanning.

2.3. Evolution of Xenobiotic-Metabolizing Enzymes

There are several indications suggesting that many xenobiotic-metabolizing enzymes are not essential for mammalian development and normal physiological homeostasis. These include the facts that 1) Numerous species differences exist for xenobiotic-metabolizing enzymes. For example, cats do not carry out glucuronidation and dogs do not have *N*-acetyltransferase activity.[1] Expression and catalytic activities of different forms of P450s and phase II enzymes can markedly vary among different species. A single drug will be metabolized differently or at different rates depending on the species. Even rats and mice differ in their xenobiotic-metabolizing enzymes. For example, mice are resistant to aflatoxin B1-induced liver cancer while rats are sensitive.[2] Mice have high level expression of a glutathione S-transferase (GST) that inactivates an epoxide of aflatoxin in their livers. Rats have low levels of this enzyme in liver. 2) Polymorphisms exist in xenobiotic-metabolizing enzymes. Polymorphisms are found in humans, rats, mice, rabbits and monkeys. These polymorphism have no impact on development or reproductive vitality. 3) P450-null (gene knockout) mice have no observable phenotype.[3,4]

Among the reasons for species differences and polymorphisms are 1) Overlapping substrate specificity. Loss of one enzyme is compensated for by another enzyme(s) that carries out same reaction on same substrates. 2) Xenobiotic-metabolizing enzymes are only required for metabolism of foreign chemicals and have only a minor role or no important role in metabolism of endogenous chemicals such as hormones.

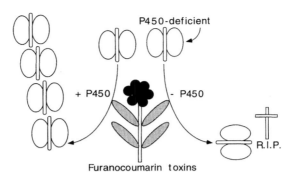

Figure 4. Schematic representation of the Monarch Butterfly P450 polymorphism. A strain of the Monarch butterfly lacks a P450 that metabolizes furanocoumarin toxins. The strain that does not contain the P450 is killed by toxin while the strain that contains P450 can survive on this plant as a sole food source. Other examples of plant-insect warfare exist involving P450s that metabolize toxins.

2.4. "Plant-Animal Warfare" Hypothesis

The "Plant-Animal Warfare" hypothesis was developed to account for the evolution of xenobiotic-metabolizing enzymes, in particular the cytochromes P450.[5] This theory is based on data from the evolutionary distances between P450 proteins and the "Molecular Drive" theory of evolution.[6] In this thesis, the xenobiotic-metabolizing enzymes evolved to metabolize and detoxify chemicals found in the diet, especially a diet of vegetation. Plants are known to produce a large number of chemicals that only appear to exist to kill plants. Since plants cannot run away and these chemical are their only defense against animal predators. In plant ecology, this has been most widely studies in the attack of insects on plants. Plant toxins that kill insects are called phytoalexins. Insects have enzymes that can detoxify phytoalexins allowing them to consume plants. This situation has been well documented with the Monarch butterfly[7,8] where strains exist that have a P450 that detoxifies a furanocoumarin toxin found in a dietary plant source. In much the same way, animals can have enzymes that allow them to consume plants that produce toxic chemicals.

Another system with agricultural significance has been studied that involves fungal-plant warfare in which a pea plant is preyed upon by the fungus *Nectria haematococca*.[9,10] In response to fungal infection, the pea produces an antimicrobial phytoalexin pisatin that is capable of killing the fungus. However, some strains of fungus contain a P450 called pisatin demethylase that detoxifies the toxin. It is noteworthy that the biosynthetic pathway for pisatin also involves P450 monooxygenases.[11] Thus, P450s participate in both production of toxins by plant and in degradation of toxins by parasites.

The evolution of xenobiotic enzymes could be driven by plants producing new chemicals that are toxic to animals and animal making new enzymes that can metabolize these toxins. In the Molecular Drive hypothesis,[6,12] an organism is constantly changing its genetic makeup by a process referred to as DNA metabolism in which DNA is altered by base changes, base slippage during replication, deletions, transpositions, and recombination/gene conversions to homogenize regions of the genome to produce novel genes. This process, also called genetic homogenization, would be totally random and would not due to selective pressures. An example of the evolution of a small gene family is shown in Figure 5.

Within a population of individuals, one individual may emerge with a gene or combination of genes that allows it a better advantage to survive in its environment and, as a result, it

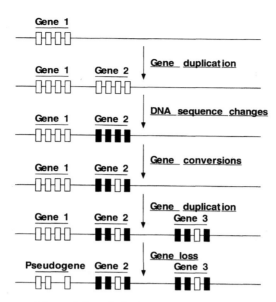

Figure 5. Schematic diagram of the evolution of a family of genes. Over many generations a family of genes can significantly change due to random DNA metabolism. This could lead to a duplication of gene 1 followed by nucleotide and amino acid codon changes that results in a gene encoding an enzyme with a different activity (gene 2). A gene conversion could occur in which an exon or several exons from a neighboring gene are transferred to the gene. This could also change activity. Another duplication of gene 2 could occur and the process could continue. One member of this small gene family could lose function through deletion of an exon.

will reproduce more efficiently and spread its useful genes. For example, it may now have a P450 that can detoxify a chemical produced by a dietary plant source thus allowing it to exploit a new source of food. Over the course of time, individual populations and different species would have their own unique "battery" of enzymes that are reflective of their history of dietary exposure to plants. This theory predicts that enzymes involved with detoxifying dietary toxins would have no role in development and physiological homeostasis. In this case evolution would give rise to species differences and polymorphisms. The latter would form during a transition period when a gene is lost or a new one is being formed in a population. In all cases of polymorphisms that have been described in rodent model systems and in humans, loss of a particular gene encoding a P450 or other xenobiotic-metabolizing enzyme does not have any detrimental effects except when a drug substrate or pro-toxin is administered.

The CYP2D subfamily can be used to illustrate species differences and polymorphisms.[13] This subfamily is most well studied because the human P450 CYP2D6 is responsible for the debrisoquine/sparteine drug oxidation polymorphism in which 10 % of Caucasians are not able to metabolize a large number of drugs due to the absence of the P450 CYP2D6. The number of genes in this subfamily exhibit differences; mice and rats each have five genes and humans have one active gene and two inactive genes. The regulation of these genes also show species differences. The mouse CYP2D genes show sex-specific expression in liver. This is not found in the rat or human. The mouse and rat genes appear to be regulated by different transcription factors that are distinct from the factor that controls expression of CYP2D6 in humans.[14,15] The activities of the CYP2D P450s differ; the rat CYP2D1 has substrate preferences that are similar to CYP2D6 but the among the mouse CYP2D P450s, none have been found that metabolize the drugs that are metabolized by CYP2D6.

3. POLYMORPHISMS OF XENOBIOTIC-METABOLIZING ENZYMES

3.1. Alcohol Metabolism

Major polymorphisms have been found that are associated with ethanol metabolism.[16-18] Some of these enzymes, especially the aldehyde dehydrogenases, also metabolize xenobiotics.

3.1.1. Alcohol Dehydrogenase Atypical Form (Atypical ADH).
There are three alleles of ADH_2, designated β_1, β_2, β_3. Active ADH is a dimer of two allelic products (or subunits). β_2 encodes a high activity variant enzyme subunit (called atypical form) having an Arg47His (the His leads to high activity) and individuals inheriting at least one β_2 allele have high activity. β_2/β_2 and β_2/β_1 or β_2/β_3 have high activity. β_2 is present at a high frequency in Asian populations (85% carry at least one allele) and at a low frequency of 5 to 20 % in Caucasians. The resulting phenotype is that Asians having a higher frequency of the β_2 allele and can metabolize ethanol more rapidly than Caucasians.

3.1.2. Aldehyde Dehydrogenase 2 (ALDH2).
A major allele for ALDH2 has a point mutation leading to Glu487Lys (change of amino acid from Glutamic acid to Lysine). This results in a dramatic change in charge at this amino acid position from - to +. The Lys form has low activity toward ethanol. About 50% of Asians are deficient in activity because they are homozygous for Glu487Lys allele.

The phenotype is expressed only during high ethanol consumption when acetaldehyde, produced by alcohol dehydrogenase (ADH), is slowly converted to acetic acid. This results in a flushing syndrome due to the build up of acetaldehyde which stimulates a dilation of facial blood vessels by release of catacholamines. Disulfiram, a drug used to treat alcoholics, accomplishes the same effect as having the polymorphism because it inhibits ALDH2, resulting in alcoholics becoming sick after ethanol consumption due to accumulation of acetaldehyde which is toxin at high blood levels. Acetaldehyde can also cause toxicity by damaging mitochondria.

3.1.3. Role of ADH and ALDH Is Alcohol Consumption.
Asian sensitivity to ethanol (flushing syndrome) is probably due to a high rate of production of acetaldehyde due to the ADH β_2 and low oxidation of acetaldehyde due to low activity of the ALDH2 Glu487Lys enzyme variant.

A higher frequency of the ALDH2 Glu487Lys enzyme is found in alcoholics and alcoholic liver disease.

3.2. Flavin-Containing Monooxygenase (Fish Odor Syndrome)

A form of flavin-containing monooxygenase is responsible for a disorder called trimethylaminuria (also called "fish odor syndrome") that causes halitosis due to the inability to metabolize trimethylamine (TMA) $(CH_3)_3$-NH to TMAO $(CH_3)_3$-N--O.[19,20] Their is a low olfactory threshold of 0.00037 ppm for the foul smelling TMA. TMA comes from dietary sources, especially fish that have very high levels. Precursors for TMA include choline, carnitine, ergothionine, betaine, lecithin and TMAO which can be reduced to TMA by bacteria in the intestines. A polymorphism was first found in chickens producing

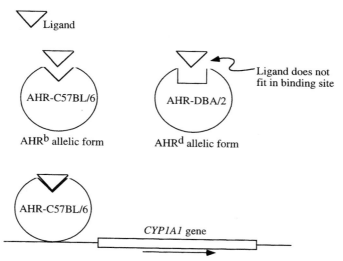

Figure 6. The AHR polymorphism. Two allelic forms exist in the laboratory mouse that differ in their affinity toward ligands. One form, frequently called the AHR^b allele, because it was first described in the C57BL/6 mouse, has a high ligand affinity while the other form, the AHR^d allele, first described in the DBA/2 mouse has a low affinity receptor. Both forms can bind to respond to high affinity ligands such a 2,3,7,8-tetrachlorodibenzo-p-dioxin.

eggs having a foul smell rendering them unacceptable to consumers. It a rare deficiency and only about 30 families have been found in which one or more family members cannot metabolize TMA and thus secrete high levels of the compound in their urine. A mutation in FMC causes trimethylamineuria. Symptoms can be controlled by dietary restrictions or by administration of drugs such as metronizole.[21] The biological significance, if any for this deficiency has been debated. It may allow predators to easily smell dead animals? There is also evidence that it is a type of anti-pheromone that is produced during menstruation due to down regulation of FMC.

3.3. Cytochromes P450

3.3.1. Aryl Hydrocarbon Receptor (AHR) Polymorphism. A polymorphism in the AHR in mouse causes altered response to xenobiotics such as polycyclic aromatic hydrocarbons and polyhalogenated biphenyl's.[22] The polymorphism in the AHR in mouse strains results in differences in inducibilities of P450s CYP1A1 and CYP1A2. Two alleles of the AHR exist in mice, one with a high affinity for ligands (AHR^b) that is found in C57BL/6 strain of mouse and another with a low affinity for ligand (AHR^d) found in DBA/2 strain of mouse. These receptors differ in a few amino acids.[23] Treatment of C57BL/6 mice with low concentrations of ligand will result in induction of the P450 gene due to high affinity of AHR^b for ligand. DBA/2 mice must be treated with at least ten-fold higher amounts of ligand to achieve induction, due to the low affinity of the AHR^d for ligand. While there are allelic variants in the human AHR receptor, none appear to be functionally significant based on ligand binding assays of recombinant receptor.[24,25]

3.3.2. Debrisoquine/Sparteine Drug Oxidation Polymorphism (CYP2D6). The human debrisoquine/sparteine drug oxidation polymorphism is due to mutations in the *CYP2D6*

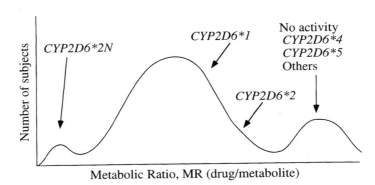

Figure 7. Frequency distribution histogram of debrisoquine metabolism in humans. A group of subjects would be administered debrisoquine and their rate of metabolism determined by urinary ratios of the parent drug debrisoquine/the metabolite 4-OH debrisoquine. This yields the Metabolic Ratio (MR). Each subject is then genotyped for the presence of *CYP2D6* alleles. The prevalence of the various alleles in a frequency distribution histogram is shown.

gene and is among the most important and most actively studied of the polymorphisms in xenobiotic-metabolizing enzymes.[13,26] The frequency of the polymorphism (homozygous for mutant *CYP2D6* alleles) ranges from about 10 % in Caucasians to < 1 % in Asians. CYP2D6 metabolizes a large number of drugs including cardiovascular agents, β-adrenergic blocking agents (Metoprolol), cardiovascular drugs (Encainide), tricyclic antidepressants (Amitriptyline) and pain medications (Codeine). Over 40 drugs or drug candidates are known to be metabolized by this P450. Those drugs exhibiting narrow therapeutic indexes such as the antiarrhythmic drugs can cause side effects in CYP2D6-deficient patients.[27] Over sixteen *CYP2D6* alleles have been found, 12 of which do not encode an active P450 enzymes.[28] Three alleles produce enzymes with decreased activity and one series of amplified *CYP2D6* genes cause higher levels of P450 expression and catalytic activities. Alleles are named by an * followed by a number. In some cases there are subgroups of an allele that differ in base changes that are silent and would not be expected to influence the enzyme activity of expression. *CYP2D6*1A* and *1B* are the alleles encoding the P450 with normal activity. A and B are subtypes. The *CYP2D6*4* allele group (with four subtypes) is the most commonly found of the alleles that do not encode an active enzyme due to a defect in splicing. The *CYP2D6*2N* is an amplification containing many copies of the *CYP2D6*2* gene producing increased activity. It is called *CYP2D6*2N* where N is the number of amplified copies which can range from 2 to 13 copies. *CYP2D6*5* is a deleted allele and is the second most abundant mutant allele.

3.3.3. S-Mephenytoin Polymorphism (CYP2C19). Mephenytoin, an anticonvulsant agent developed in the 1940s, is still widely used and very effective in control of seizures. It is metabolized by *N*-demethylation and 4'-hydroxylation which inactivate the drug. There are no complications in therapy that are due to differences in metabolism. Individuals that do not carry out metabolism do not have toxicity with mephenytoin due to a wide therapeutic window. Thus, the polymorphism is not relevant in *S*-mephenytoin therapy.

About 18 to 25 % of Asians and 2 to 5 % of Caucasians cannot carry out *S*-mephenytoin 4'-hydroxylation due to two mutant alleles of the *CYP2C19* gene.[29–31] *CYP2C19*1* is normal gene. *CYP2C19*2* has a G to A base change in exon 5 creating a splice site mutation and another mutant allele *CYP2C19*3* has a G to A base change in exon 4 creating a premature stop codon.

There are racial differences in frequency of the CYP2C19 polymorphism. In contrast to the CYP2D6 polymorphism which is more common in Caucasians, the CYP2C19 polymorphism is more frequent in Asians (up to 25 % are homozygous for mutant alleles) than Caucasians (less that 2% are deficient). The *CYP2C19*2* is about twice as abundant than *CYP2C19*3* in Japanese. Both mutant alleles are also found in Caucasians. *CYP2C19*2* is most common in Caucasians. *CYP2C19*3* is rare. The *CYP2C19*2* and *CYP2C19*3* have been found in Asians and Caucasians. Therefore, these mutations arose over 40,000 years ago, the estimated time of the Caucasian/Mongoloid fission (41,000 + 15,000 years ago).

3.3.4. Tolbutamide Hydroxylation (CYP2C9). Allelic variants with amino acid substitutions exist that have altered activity toward different substrates were found for CYP2C9. *CYP2C9*1* in the most common allele having Arg at 144 and Ile at 359. These variants have amino acid changes that affect catalytic activities toward selective CYP2C9 substrates. *CYP2C9*2* (Arg144Cys) has low warfarin 7-hydroxylase activity but normal tolbutamide hydroxylase activity.[32–34] *CYP2C9*3* (Ile359Leu) is deficient in activity toward certain substrates such as tolbutamide.[35] The allele frequencies are *CYP2C9*1* (0.79), *CYP2C9*2* (0.125) and *CYP2C9*3* (0.085).

3.3.5. Coumarin 7-Hydroxylase (CYP2A6). CYP2A6 is not known to metabolize many clinically-used drugs. It has high activity toward coumarin, an odorant found in plants and used in cosmetics. A polymorphism exists in CYP2A6 in which 1 % of Caucasians and 5 to 10 % Asians cannot metabolize coumarin and other substrates.[36,37] Two mutant alleles alleles have been identified. *CYP2A6*2* (Leu160His) encodes a P450 with no activity. *CYP2A6*3* has several amino acid differences Ser131Ala, Ala153Ser, Asp158Glu, Leu160Ile, Glu162Ser and is also catalytically inactive. The CYP2A6*3 variant allele*CYP2A6*3* was apparently formed by a gene conversion between *CYP2A6*1* and *CYP2A7*. The *CYP2A7* is probably not expressed and does not encode an active P450.

3.3.6. Non-Functional P450 Polymorphisms. There are polymorphisms in the *CYP1A1*[38] and *CYP2E1*[39] genes that are the result of base changes in introns or flanking DNA. These polymorphisms have not been shown to result in changes in P450 enzyme activities and therefore, they are considered non-functional polymorphisms until proven otherwise.

4. POLYMORPHISMS IN PHASE II ENZYMES

4.1. UDP-Glucuronosyltransferase (UGT)

The UGTs consist of two families of genes UGT1 and UGT2, with multiple different forms that can differ in their substrate specificity's, although overlapping specificity's are common.[40] The only characterized polymorphism in the UGTs is associated with the UGT1 locus.[41] The UGT1 is an unusual gene that is actually a single transcription unit in which four common exons are preceded by up to 13 different first exons.[42] Each exon has its own promoter and is spliced to the common exon 2 producing mRNAs encoding enzymes with different sequences and activities. The UGT1 locus encodes the bilirubin UGT and UGTs responsible for conjugating planer phenolic compounds.

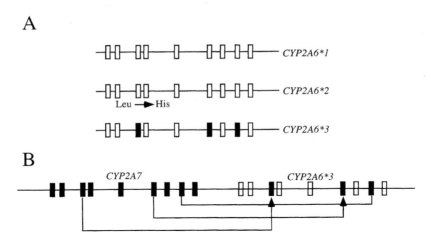

Figure 8. Panel A is a summary of the three *CYP2A6* alleles. Panel B is a schematic diagram of gene conversion between the *CYP2A7* gene and *CYP2A6* to generate *CYP2A6*3*.

The Gunn rat lacks the ability of glucuronidate bilirubin due to a mutation (base deletion) in the common coding region of the gene UGT1.[43] These rats become yellow due to high levels of unconjugated serum bilirubin. Since the common region of the *UGT1* gene is mutated resulting in mRNAs encoding truncated (prematurely terminated) protein, all of UGT1 enzymes are defective in the Gunn rat.

In humans, a rare genetic deficiency exist called the Crigler-Najjar (CN) syndrome type I. CN Type I has complete loss of UGT activity toward bilirubin.[42,44] Mutations have been found in exons 2, 3 and 4 (common exons for all UGT1 enzymes) and in the first unique exon in different patients. Giblet's disease and CN type II are more common disorders that are partial deficiencies in UGT1 resulting in high serum bilirubin that may result from heterozygotes or missense mutations (amino acid changes). Babies born with the severe CN type I have neurological abnormalities due to the accumulation of unconjugated bilirubin in the brain. They are treated with inducers of drug metabolizing enzymes (P450s) that will metabolize bilirubin by an alternative pathway. Phenobarbital and phenothiazines decrease serum bilirubin levels in the jaundiced (medical term for yellow skin and eyes) Gunn rat and in human CN type I patients.

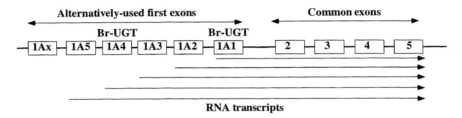

Figure 9. Structure and transcription of the UGT1 locus. The common exons 2 through 5 are preceded by several different first exons that each contains a unique promoter. The arrows represent the primary pre mRNAs that after splicing encode a single UGT enzyme.

Table 1. BChE alleles

Allele name	Allele frequencies	Phenotype
BCHE	0.98	Wild type
BCHE*70G	0.017	Low affinity/activity
BCHE*539T	0.10	Fluoride resistant
Others	0.083	Various

4.2. Sulfotransferases

Sulfotransferases are a family of enzymes involved in conjugation of drugs and endogenous steroids. Two sulfotransferase polymorphisms have been found.[45] A human polymorphism exist in a sulfotransferase called thermolabile phenol sulfotransferase (TL-PST) which is expressed in human platelets. The gene frequency estimates suggest that about 10% of the alleles encode a high activity form of the enzyme or an enzyme that is expressed at higher levels. The thermostable phenol sulfotransferase (TS-PST) is also believed to be polymorphic based on enzyme activity measurements with an allele frequency of 20% for the high activity form. TL-PST and TS-PST catalyze conjugation of acetaminophen. There are no adverse drug reactions associated with these polymorphisms. Low TS-PST has been correlated with diet-induced migraine headaches that are thought to be due to phenolic compounds in the diet that cannot be conjugated. Molecular mechanisms for these polymorphisms are not known but the data suggest that they are due to missense mutations that change activity or mutations that alter the regulation of the enzymes.

4.3. Catechol-*O*-Methyltransferase (COMT)

COMT carries out biotransformation of important neurotransmitters such as norepinephrine and related drugs such as L-DOPA a drug for Parkinson's disease.[46] The polymorphism may be related to neuropsychiatric conditions involving catacholamine neurotransmission such as mood disorders, schizophrenia, obsessive compulsive disorders, alchohol and drug abuse and attention deficit disorders. One mutant allele *COMT*2* encoding a missense Val158Met that causes lower activity and instability (higher rate of degradation) of the enzyme, as compared to the high activity allele *COMT*1*. The allele frequency for *COMT*2* is about 46%.

4.4. Thiopurine Methyltransferase (TPMT)

S-Methylation of drugs such as the antihypertensive agent captopril, anti-rheumatic fever drug D-penicillamine, antidiuretic spironolactone and anticancer drug 6-mercaptopurine are due to TPMT. A polymorphism has been described that affects toxicity and efficacy of thiopurine anti-cancer drugs mercaptopurine, azathioprine and thioguanine.[47] These drugs are actually prodrugs that must be converted to thiopurine nucleotides to have activity. TPMT competes with the hypoxanthine phosphoribosyl transferase (HPRT) and inactivates the drugs by *S*-methylation. TMPT-deficient patients exhibit high toxicity to the drugs to increase levels of thioguanine nucleotides (TGN). Two mutant alleles have been described; the *TPMT*2* allele encodes a missense mutation Ala80Pro that results in lower catalytic activity and the *TPMT*3* allele encodes a variant enzyme having Ala154Thr and Tyr240Cys. TPMT*3 encodes an enzymes that is more susceptible to degradation resulting in lower levels of the protein (about 100-fold lower than the normal enzyme from the *TPMT*1* allele).

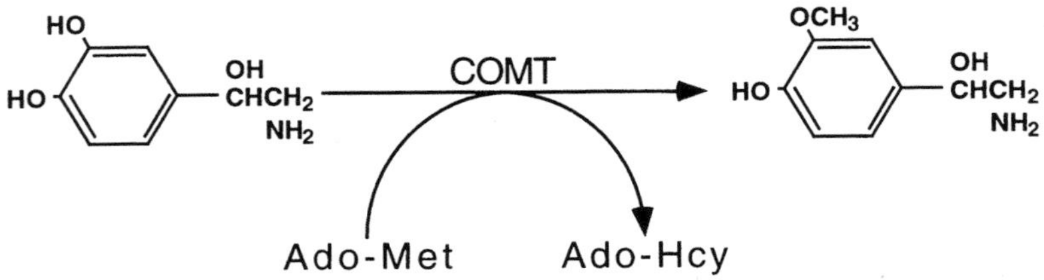

Figure 10. Reaction carried out by COMT.

4.5. Butyrylcholinesterase

In 1951 a few patients were found, who after being administered the muscle relaxant drug succinylcholine, exhibited prolonged muscle paralysis. The abnormal response was found to be due to a variant in butyrylcholinesterase (BChE) having a low affinity for succinylcholine. BChE is found in the serum and hydrolyzes succinylcholine to succinylmonocholine and choline, resulting in inactivation of the drug. Other drugs inactivated by this enzyme include procaine, isobucaine (local anesthetic), aspirin (analgesic), and mivacurium (muscle relaxant). A low affinity variant of BChE encoded by the *BCHE*70G* gene, is due to a mutation in codon 70 (Asp70Gly). Other minor alleles include a mutant allele that does not encode an active enzyme due to a base insertion that causes a protein reading frame shift (*BCHE*FS117*) and missense variants that are resistant to the inhibitor fluoride.

4.6. Paraoxonase/Arylesterase

Paraoxonase found in liver and serum and hydrolyzes organophosphates including paraoxon, a P450 metabolite of the commonly-used insecticide parathion. It also catalyzes the hydrolysis of aromatic carboxylic acid esters such as the naturally occurring phenyl acetate. A polymorphisms was found that is due to two alleles, designated A type (*PONA*) having low activity, and B type (*PONB*) having high activity.[48,49] PONA has a glutamine at position 192 and PONB has an arginine at position 192. The role of this polymorphism in toxicity to parathion and other organophosphates is not known (toxicity is due to inhibition of the enzyme acetylcholinesterase, a serine-type esterase that terminates the action of acetylcholine). Allele frequencies Caucasians 0.70 and 0.30 for *PONA* and *PONB*, respectively and Asians 0.30 and 0.70 for *PONA* and *PONB* type, respectively. Paraoxonase is found in serum in association with high density lipoproteins (HDL). Paraoxonase catalyzes the hydrolysis of lipid hydroperoxides. Attempts have been made to determine if this polymorphism is associated with susceptibility to coronary heart disease. However, these results are not conclusive.

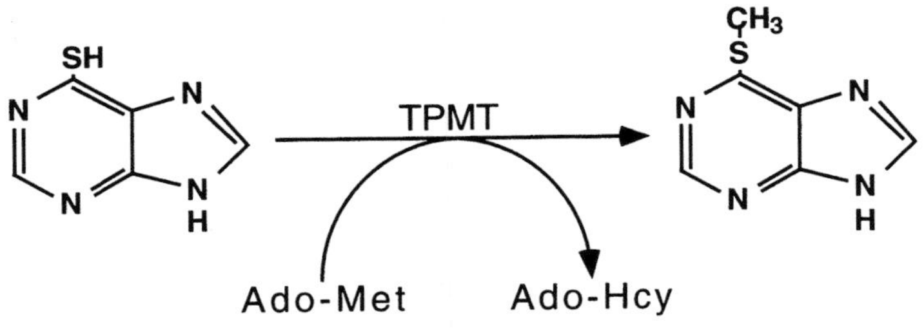

Figure 11. Reaction carried out by TPMT.

4.7. *N*-Acetyltransferases (NAT)

The NAT polymorphism was discovered in 1957 during clinical use of the anti-tuberculosis antibiotic isoniazid in which about 60 % of patients administered the drug excreted it in the urine unchanged and 25 % excreted inactive metabolites. Phenotyping test using urine metabolites were developed with the acetylation polymorphism. Other drugs under the influence of the acetylation polymorphism include sulfamethazine and other sulfa-type agents; procainamide, an antiarrhthmic; hydralzine, an antihypertensive; dapsone, and antileprotic; nitrazepan, a hypnotic; clonazepam, an antiepileptic; and caffeine, a stimulant.

There are two *NAT* genes, *NAT1* and *NAT2*. A common polymorphism is due to mutations in the *N*-acetyltransferase 2 gene *NAT2*.[50,51] The wild type (normal)*NAT2* gene is *NAT2*4* and the mutant *NAT2* alleles are *NAT2*5* through *NAT2*19*.[52] The major missense mutations are: *NAT2*5A, 5B and 5C* is Ile114Thr and is the most common in Caucasians (40 %) and is at low frequency in Asians (2 to 5%). Note that the three *NAT2*5* alleles differ in two other silent mutations. *NAT2*6A* is Arg197Gln and is common in Caucasians (35%) and Asians (30%). *NAT2*7A* is Gly286Glu and is the most common in Asians (15 %) and is found at only low frequency in Caucasians (2%). These three alleles together account for about 75% and 45% of all *NAT2* alleles in Caucasians and Asians, respectively. The frequencies of slow acetylators as determined by phenotyping is about 75% and 35% for Caucasians and Asians, respectively.

NAT2 is required for the metabolic activation of arylamine and heterocyclic amine carcinogens (Figure 12). These chemicals are activated by *N*-hydroxylation due to the activity of CYP1A2 followed by *O*-acetylation by sulfotransferase and NAT2.[53] The esters formed are unstable and through spontaneous intramolecular rearrangement, are converted to electrophilic derivatives that can bind to cellular macromolecules. NAT2 expression in liver can lead to inactivation of the arylamine-containing chemicals through *N*-acetylation. Thus, there is a competition between *N*-hydroxylation by CYP1A2 and *N*-acetylation by NAT2. Based on these findings, efforts have been made to determine if the NAT2 polymorphism is associated with cancer susceptibility in humans. High CYP1A2 and low NAT2 have been found to lead to increased risk for colon cancer, especially in individuals with high intake of heterocylic amine food mutagens due to consumption of over-cooked meats.[54]

A polymorphism in NAT1 has also been found in humans that is due to the *NAT1*10* allele, that accounts for about 20% of all alleles.[55] This allele has no mutations in the protein coding region but a mutation is in the 3' untranslated region of the mRNA that may affect abundance of the mRNA. The NAT1 mRNA encoded by the *NAT1*10* allele may be more abundant resulting in more protein and more activity.

4.8. Glutathione S-Transferases

Common polymorphisms have been found in the GSTM1 and GSTT1.
Both polymorphisms are due to a gene deletion. This is usually referred to as a null genotype.[56,57]
Allele frequencies:

GSTM1-null	0.74 (74% of all *GSTM1* alleles)
GSTT1-null	0.38 (38% of all *GSTT1* alleles)

There are two active *GSTM1* alleles that differ in a single amino acid Lys172Asp and called *GSTM1A* and *GSTM1B*, respectively. They do not appear to encode enzymes having different catalytic activities.

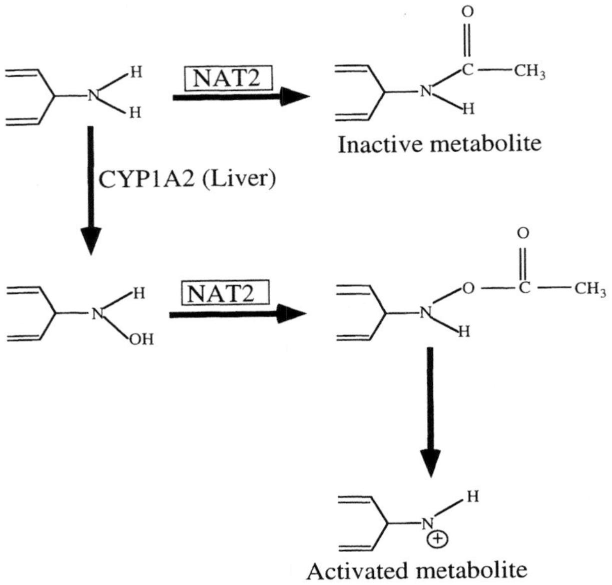

Figure 12. Mechanism for the metabolic activation of arylamine and heterocyclic amine procarcinogens. *N*-hydroxylation by CYP1A2 occurs in the liver and the *N*-OH metabolite can re-enter the bloodstream and be transferred to other tissues where it encounters the transferases. Upon conjugation with acetate, the active, ultimate carcinogenic metabolite is formed that can cause gene mutations.

A polymorphism for GSTM3 was found that may not be functionally significant.

The *GSTM3* gene has at least two alleles *GSTM3*A* and *GSTM3*B* that are identical across the protein coding region of the enzyme but differ at an upstream regulatory region that is though to control gene inducibility. This has not yet been proven. The *GSTM3*A* and *GSTM3*B* have allele frequencies of 0.80 and 0.20.

4.9. Dihydropyrimidine Dehydrogenase (DPD)

A genetic defect in the enzyme DPD was found to cause toxicity in patients administered fluoropyrimidine drugs such a 5-fluorouracil.[58,59] DPD is involved is a pyrimidine catabolic pathway that yields the amino acid β-alanine. Patients that have low DPD activity are susceptible to toxicity that can lead to paralysis and sometimes death. Individuals that are completely deficient in DPD due to two mutant alleles, have a condition called thymine-uraciluria. Babies born with this defect can have many abnormalities including developmental delay, convulsions/epilepsy, microcephaly.

A mutant allele designated *DPYD*2* was found that yields an inactive protein due to a splice mutation. Genotyping analysis confirmed that cancer patients that are heterozygous with one copy of the mutant allele are susceptible to toxicity. A patient homozygous for DPYD*2 was found that had no symptoms while a sibling was severely affected.[60]

4.10. Epoxide Hydrolase

Microsomal epoxide hydrolase is involved in hydration of epoxide of xenobiotics, including epoxides from by P450s. This enzyme is though to be critical in deactivating po-

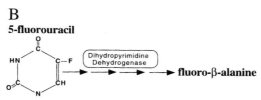

Figure 13. Metabolism of pyrimidines (panel A) and fluoropyrimidine drugs (panel B) by DPD.

tential electrophilic P450 metabolites that could damage cellular macromolecules and mutate DNA. A polymorphism that results in variation at amino acids 113 and 139 was found with the Tyr113His substitution thought to be the most important.[61,62] However, it is still uncertain whether the allelic products differ in enzymatic activity or level of expression.

4.11. NAD(P)H: quinone oxidoreductase (NQO1)

NQO1 catalyzes the two electron reduction of quinones, nitroaromatic compounds, and azo-dyes. The gene occurs as two allelic forms that differ at codon 187.[63,64] The Pro187Ser has no enzymatic activity and it is present at an allele frequency of about 30%. This allele was found to be over represented in patients with lung,[65] kidney, and bladder[66] cancers.

REFERENCES

1. Williams, R. T. 1967. Comparative patterns of drug metabolism. *Fed. Proc.* 26:1029–1039.
2. Eaton, D. L. and E. P. Gallagher. 1994. Mechanisms of aflatoxin carcinogenesis. *Annu. Rev. Pharmacol. Toxicol.* 34:135–172.
3. Buters, J. T., B. K. Tang, T. Pineau, H. V. Gelboin, S. Kimura, and F. J. Gonzalez. 1996. Role of CYP1A2 in caffeine pharmacokinetics and metabolism: studies using mice deficient in CYP1A2. *Pharmacogenetics.* 6:291–296.
4. Lee, S. S., J. T. Buters, T. Pineau, P. Fernandez-Salguero, and F. J. Gonzalez. 1996. Role of CYP2E1 in the hepatotoxicity of acetaminophen. *J. Biol. Chem.* 271:12063–12067.
5. Gonzalez, F. J. and D. W. Nebert. 1990. Evolution of the P450 gene superfamily: animal-plant 'warfare', molecular drive and human genetic differences in drug oxidation. *Trends. Genet.* 6:182–186.
6. Dover, G. A. and R. B. Flavell. 1984. Molecular coevolution: DNA divergence and the maintenance of function. *Cell* 38:622–623.
7. Ma, R., M. B. Cohen, M. R. Berenbaum, and M. A. Schuler. 1994. Black swallowtail (Papilio polyxenes) alleles encode cytochrome P450s that selectively metabolize linear furanocoumarins. *Arch. Biochem. Biophys.* 310:332–340.

8. Berenbaum, M. R. 1995. The chemistry of defense: theory and practice. *Proc. Natl. Acad. Sci. U. S. A.* 92:2–8.

9. Maloney, A. P. and H. D. VanEtten. 1994. A gene from the fungal plant pathogen Nectria haematococca that encodes the phytoalexin-detoxifying enzyme pisatin demethylase defines a new cytochrome P450 family. *Mol. Gen. Genet.* 243:506–514.

10. Reimmann, C. and H. D. VanEtten. 1994. Cloning and characterization of the PDA6–1 gene encoding a fungal cytochrome P-450 which detoxifies the phytoalexin pisatin from garden pea. *Gene* 146:221–226.

11. Paiva, N. L., Y. Sun, R. A. Dixon, H. D. VanEtten, and G. Hrazdina. 1994. Molecular cloning of isoflavone reductase from pea (Pisum sativum L.): evidence for a 3R-isoflavanone intermediate in (+)-pisatin biosynthesis. *Arch. Biochem. Biophys.* 312:501–510.

12. Dover, G. A., T. Strachan, E. S. Coen, and S. D. Brown. 1982. Molecular drive [letter]. *Science* 218:1069.

13. Gonzalez, F. J. 1996. The CYP2D subfamily. In Cytochromes P450: Metabolic and toxicological aspects. C. Ioannides, editor. CRC Press, London. 183–210.

14. Gonzalez, F. J. and Y. H. Lee. 1997. Mechanisms for constitutive expression of hepatic P450 genes. *FASEB J* 10:1112–1117.

15. Cairns, W., C. A. D. Smith, A. W. McLaren, and C. R. Wolf. 1996. Characterization of the human cytochrome P4502D6 promoter. A potential role for antagonistic interactions between members of the nuclear receptor family. *J. Biol. Chem.* 271:25269–25276.

16. Crabb, D. W. 1995. Ethanol oxidizing enzymes: roles in alcohol metabolism and alcoholic liver disease. *Prog. Liver. Dis.* 13:151–172.

17. Yoshida, A. 1994. Genetic polymorphisms of alcohol metabolizing enzymes related to alcohol sensitivity and alcoholic diseases. *Alcohol. Alcohol.* 29:693–696.

18. Agarwal, D. P. 1997. Molecular genetic aspects of alcohol metabolism and alcoholism. *Pharmacopsychiatry.* 30:79–84.

19. Phillips, I. R., C. T. Dolphin, P. Clair, M. R. Hadley, A. J. Hutt, R. R. McCombie, R. L. Smith, and E. A. Shephard. 1995. The molecular biology of the flavin-containing monooxygenases of man. *Chem. Biol. Interact.* 96:17–32.

20. Cashman, J. R., Y. A. Bi, J. Lin, R. Youil, M. Knight, S. Forrest, and E. Treacy. 1997. Human flavin-containing monooxygenase form 3: cDNA expression of the enzymes containing amino acid substitutions observed in individuals with trimethylaminuria. *Chem. Res. Toxicol.* 10:837–841.

21. Treacy, E., D. Johnson, J. J. Pitt, and D. M. Danks. 1995. Trimethylaminuria, fish odour syndrome: a new method of detection and response to treatment with metronidazole. *J. Inherit. Metab. Dis.* 18:306–312.

22. Schmidt, J. V. and C. A. Bradfield. 1996. Ah receptor signaling pathways. *Ann. Rev. Cell Dev. Biol.* 12:55–89.

23. Chang, C., D. R. Smith, V. S. Prasad, C. L. Sidman, D. W. Nebert, and A. Puga. 1993. Ten nucleotide differences, five of which cause amino acid changes, are associated with the Ah receptor locus polymorphism of C57BL/6 and DBA/2 mice. *Pharmacogenetics.* 3:312–321.

24. Hayashi, S., J. Watanabe, K. Nakachi, H. Eguchi, O. Gotoh, and K. Kawajiri. 1994. Interindividual difference in expression of human Ah receptor and related P450 genes. *Carcinogenesis* 15:801–806.

25. Perdew, G. H. and C. E. Hollenback. 1995. Evidence for two functionally distinct forms of the human Ah receptor. *J. Biochem. Toxicol.* 10:95–102.

26. Gonzalez, F. J. and J. R. Idle. 1994. Pharmacogenetic phenotyping and genotyping. Present status and future potential. *Clin. Pharmacokinet.* 26:59–70.

27. Buchert, E. and R. L. Woosley. 1992. Clinical implications of variable antiarrhythmic drug metabolism. *Pharmacogenetics.* 2:2–11.

28. Daly, A. K., J. Brockmoller, F. Broly, M. Eichelbaum, W. E. Evans, F. J. Gonzalez, J. D. Huang, J. R. Idle, M. Ingelman-Sundberg, T. Ishizaki, E. Jacqz-Aigrain, U. A. Meyer, D. W. Nebert, V. M. Steen, C. R. Wolf, and U. M. Zanger. 1996. Nomenclature for human CYP2D6 alleles. *Pharmacogenetics.* 6:193–201.

29. Goldstein, J. A., T. Ishizaki, K. Chiba, S. M. de Morais, D. Bell, P. M. Krahn, and D. A. Evans. 1997. Frequencies of the defective CYP2C19 alleles responsible for the mephenytoin poor metabolizer phenotype in various Oriental, Caucasian, Saudi Arabian and American black populations. *Pharmacogenetics.* 7:59–64.

30. Xiao, Z. S., J. A. Goldstein, H. G. Xie, J. Blaisdell, W. Wang, C. H. Jiang, F. X. Yan, N. He, S. L. Huang, Z. H. Xu, and H. H. Zhou. 1997. Differences in the incidence of the CYP2C19 polymorphism affecting the S-mephenytoin phenotype in Chinese Han and Bai populations and identification of a new rare CYP2C19 mutant allele. *J. Pharmacol. Exp. Ther.* 281:604–609.

31. Goldstein, J. A. and S. M. de Morais. 1994. Biochemistry and molecular biology of the human CYP2C subfamily. *Pharmacogenetics.* 4:285–299.

32. Furuya, H., P. Fernandez-Salguero, W. Gregory, H. Taber, A. Steward, F. J. Gonzalez, and J. R. Idle. 1995. Genetic polymorphism of CYP2C9 and its effect on warfarin maintenance dose requirement in patients undergoing anticoagulation therapy. *Pharmacogenetics.* 5:389–392.

33. Rettie, A. E., L. C. Wienkers, F. J. Gonzalez, W. F. Trager, and K. R. Korzekwa. 1994. Impaired (S)-warfarin metabolism catalysed by the R144C allelic variant of CYP2C9. *Pharmacogenetics.* 4:39–42.

34. Crespi, C. L. and V. P. Miller. 1997. The R144C change in the CYP2C9*2 allele alters interaction of the cytochrome P450 with NADPH:cytochrome P450 oxidoreductase. *Pharmacogenetics.* 7:203–210.

35. Sullivan-Klose, T. H., B. I. Ghanayem, D. A. Bell, Z. Y. Zhang, L. S. Kaminsky, G. M. Shenfield, J. O. Miners, D. J. Birkett, and J. A. Goldstein. 1996. The role of the CYP2C9-Leu359 allelic variant in the tolbutamide polymorphism. *Pharmacogenetics.* 6:341–349.

36. Fernandez-Salguero, P. and F. J. Gonzalez. 1995. The CYP2A gene subfamily: species differences, regulation, catalytic activities and role in chemical carcinogenesis. *Pharmacogenetics.* 5 Spec No:S123-S128.

37. Fernandez-Salguero, P., S. M. Hoffman, S. Cholerton, H. Mohrenweiser, H. Raunio, A. Rautio, O. Pelkonen, J. D. Huang, W. E. Evans, and J. R. Idle. 1995. A genetic polymorphism in coumarin 7-hydroxylation: sequence of the human CYP2A genes and identification of variant CYP2A6 alleles. *Am. J. Hum. Genet.* 57:651–660.

38. Kawajiri, K., J. Watanabe, and S. Hayashi. 1996. Identification of allelic variants of the human CYP1A1 gene. *Methods Enzymol.* 272:226–232.

39. Hu, Y., M. Oscarson, I. Johansson, Q. Y. Yue, M. L. Dahl, M. Tabone, S. Arinco, E. Albano, and M. Ingelman-Sundberg. 1997. Genetic polymorphism of human CYP2E1: characterization of two variant alleles. *Mol. Pharmacol.* 51:370–376.

40. Mackenzie, P. I., I. S. Owens, B. Burchell, K. W. Bock, A. Bairoch, A. Belanger, S. Fournel-Gigleux, M. Green, D. W. Hum, T. Iyanagi, D. Lancet, P. Louisot, J. Magdalou, J. R. Ritter, H. Schachter, T. R. Tephly, K. F. Tipton, and D. W. Nebert. 1997. The UDP glycosyltransferase gene superfamily: recommended nomenclature update based on evolutionary divergence. *Pharmacogenetics.* 7:255–269.

41. Owens, I. S. and J. K. Ritter. 1995. Gene structure at the human UGT1 locus creates diversity in isozyme structure, substrate specificity, and regulation. *Prog. Nucleic Acid. Res. Mol. Biol.* 51:305–338.

42. Owens, I. S., J. K. Ritter, M. T. Yeatman, and F. Chen. 1996. The novel UGT1 gene complex links bilirubin, xenobiotics, and therapeutic drug metabolism by encoding UDP-glucuronosyltransferase isozymes with a common carboxyl terminus. *J. Pharmacokinet. Biopharm.* 24:491–508.

43. Owens, I. S. and J. K. Ritter. 1992. The novel bilirubin/phenol UDP-glucuronosyltransferase UGT1 gene locus: implications for multiple nonhemolytic familial hyperbilirubinemia phenotypes. *Pharmacogenetics.* 2:93–108.

44. Jansen, P. L., P. J. Bosma, and J. R. Chowdhury. 1995. Molecular biology of bilirubin metabolism. *Prog. Liver. Dis.* 13:125–150.

45. Weinshilboum, R. M., D. M. Otterness, I. A. Aksoy, T. C. Wood, C. Her, and R. B. Raftogianis. 1997. Sulfation and sulfotransferases 1: Sulfotransferase molecular biology: cDNAs and genes. *FASEB. J.* 11:3–14.

46. Lachman, H. M., D. F. Papolos, T. Saito, Y. M. Yu, C. L. Szumlanski, and R. M. Weinshilboum. 1996. Human catechol-O-methyltransferase pharmacogenetics: description of a functional polymorphism and its potential application to neuropsychiatric disorders. *Pharmacogenetics.* 6:243–250.

47. Krynetski, E. Y., H. L. Tai, C. R. Yates, M. Y. Fessing, T. Loennechen, J. D. Schuetz, M. V. Relling, and W. E. Evans. 1996. Genetic polymorphism of thiopurine S-methyltransferase: clinical importance and molecular mechanisms. *Pharmacogenetics.* 6:279–290.

48. Mackness, M. I., B. Mackness, P. N. Durrington, P. W. Connelly, and R. A. Hegele. 1996. Paraoxonase: biochemistry, genetics and relationship to plasma lipoproteins. *Curr. Opin. Lipidol.* 7:69–76.

49. Renczes, G., K. Rona, I. Szabo, and A. Somogyi. 1997. [Polymorphism and clinical significance of human paraoxonase enzyme]. *Orv. Hetil.* 138:2057–2059.

50. Spielberg, S. P. 1996. N-acetyltransferases: pharmacogenetics and clinical consequences of polymorphic drug metabolism. *J. Pharmacokinet. Biopharm.* 24:509–519.

51. Grant, D. M., N. C. Hughes, S. A. Janezic, G. H. Goodfellow, H. J. Chen, A. Gaedigk, V. L. Yu, and R. Grewal. 1997. Human acetyltransferase polymorphisms. *Mutat. Res.* 376:61–70.

52. Vatsis, K. P., W. W. Weber, D. A. Bell, J. M. Dupret, D. A. Evans, D. M. Grant, D. W. Hein, H. J. Lin, U. A. Meyer, and M. V. Relling. 1995. Nomenclature for N-acetyltransferases. *Pharmacogenetics.* 5:1–17.

53. Badawi, A. F., S. J. Stern, N. P. Lang, and F. F. Kadlubar. 1996. Cytochrome P-450 and acetyltransferase expression as biomarkers of carcinogen-DNA adduct levels and human cancer susceptibility. *Prog. Clin. Biol. Res.* 395:109–140.

54. Lang, N. P., M. A. Butler, J. Massengill, M. Lawson, R. C. Stotts, M. Hauer-Jensen, and F. F. Kadlubar. 1994. Rapid metabolic phenotypes for acetyltransferase and cytochrome P4501A2 and putative exposure to

food-borne heterocyclic amines increase the risk for colorectal cancer or polyps. *Cancer Epidemiol. Biomarkers. Prev.* 3:675–682.

55. Doll, M. A., W. Jiang, A. C. Deitz, T. D. Rustan, and D. W. Hein. 1997. Identification of a novel allele at the human NAT1 acetyltransferase locus. *Biochem. Biophys. Res. Commun.* 233:584–591.

56. Daly, A. K. 1995. Molecular basis of polymorphic drug metabolism. *J. Mol. Med.* 73:539–553.

57. Hayes, J. D. and D. J. Pulford. 1995. The glutathione S-transferase supergene family: regulation of GST and the contribution of the isoenzymes to cancer chemoprotection and drug resistance. *Crit. Rev. Biochem. Mol. Biol.* 30:445–600.

58. Wei, X., H. L. McLeod, J. McMurrough, F. J. Gonzalez, and P. Fernandez-Salguero. 1996. Molecular basis of the human dihydropyrimidine dehydrogenase deficiency and 5-fluorouracil toxicity. *J. Clin. Invest.* 98:610–615.

59. Gonzalez, F. J. and P. Fernandez-Salguero. 1995. Diagnostic analysis, clinical importance and molecular basis of dihydropyrimidine dehydrogenase deficiency [see comments]. *Trends. Pharmacol. Sci.* 16:325–327.

60. Fernandez-Salguero, P. M., A. Sapone, X. Wei, J. R. Holt, S. Jones, J. R. Idle, and F. J. Gonzalez. 1997. Lack of correlation between phenotype and genotype for the polymorphically expressed dihydropyrimidine dehydrogenase in a family of Pakistani origin. *Pharmacogenetics.* 7:161–163.

61. Hassett, C., L. Aicher, J. S. Sidhu, and C. J. Omiecinski. 1994. Human microsomal epoxide hydrolase: genetic polymorphism and functional expression in vitro of amino acid variants [published erratum appears in Hum Mol Genet 1994 Jul;3(7):1214]. *Hum. Mol. Genet.* 3:421–428.

62. Hassett, C., J. Lin, C. L. Carty, E. M. Laurenzana, and C. J. Omiecinski. 1997. Human hepatic microsomal epoxide hydrolase: comparative analysis of polymorphic expression. *Arch. Biochem. Biophys.* 337:275–283.

63. Kelsey, K. T., D. Ross, R. D. Traver, D. C. Christiani, Z. F. Zuo, M. R. Spitz, M. Wang, X. Xu, B. K. Lee, B. S. Schwartz, and J. K. Wiencke. 1997. Ethnic variation in the prevalence of a common NAD(P)H quinone oxidoreductase polymorphism and its implications for anti-cancer chemotherapy. *Br. J. Cancer* 76:852–854.

64. Traver, R. D., D. Siegel, H. D. Beall, R. M. Phillips, N. W. Gibson, W. A. Franklin, and D. Ross. 1997. Characterization of a polymorphism in NAD(P)H: quinone oxidoreductase (DT-diaphorase). *Br. J. Cancer* 75:69–75.

65. Rosvold, E. A., K. A. McGlynn, E. D. Lustbader, and K. H. Buetow. 1995. Identification of an NAD(P)H:quinone oxidoreductase polymorphism and its association with lung cancer and smoking. *Pharmacogenetics.* 5:199–206.

66. Schulz, W. A., A. Krummeck, I. Rosinger, P. Eickelmann, C. Neuhaus, T. Ebert, B. J. Schmitz-Drager, and H. Sies. 1997. Increased frequency of a null-allele for NAD(P)H: quinone oxidoreductase in patients with urological malignancies. *Pharmacogenetics.* 7:235–239.

8

CLINICAL ASPECTS OF POLYMORPHIC DRUG METABOLISM IN HUMANS

Michel Eichelbaum

Dr. Margarete Fischer-Bosch-Institut für Klinische Pharmakologie
Auerbachstr. 112
70376 Stuttgart, Germany

1. INTRODUCTION

Although it was already postulated at the beginning of this century by Garrod,[1] the founder of modern biochemical genetics, that hereditary determined differences in certain biochemical processes could be the cause for adverse reactions after the ingestion of drugs and food it was not until the late 1950s when it was proven that his hypothesis was correct.[2]

At that time it was demonstrated that haemolysis observed after the consumption of certain drugs and food was due to a genetic determined enzyme deficiency, e.g. glucose-6-phosphate dehydrogenase. Since that time it has been realised that genetic variability of enzymes involved in the action and disposition of drugs significantly contributes to variability in drug response and adverse drug reactions. The study of genetically determined variations in drug response is referred to as pharmacogenetics, a term coined by Vogel in 1957.[2]

Pharmacogenetics deals with clinically significant hereditary variations in the response to drugs due to altered drug disposition and abnormal drug reaction in affected persons. In general, two types of pharmacogenetic conditions can be differentiated.

1. Genetic conditions transmitted as single factors altering the way drugs act on the body (e.g., drug and food induced haemolysis due to glucose-6-phosphate dehydrogenase deficiency).
2. Genetic condition transmitted as single factors altering the way the body acts on drugs (altered drug metabolism, e.g., polymorphic N-acetylation and cytochrome P450 catalyzed drug oxidation reactions; atypical pseudo-cholinesterase).

Genetic polymorphisms are a common biological phenomenon.

The term genetic polymorphism defines a monogenic trait that exists in the population in at least two phenotypes and presumably at least two genotypes, neither of which is

Molecular and Applied Aspects of Oxidative Drug Metabolizing Enzymes,
edited by Arinç *et al*. Kluwer Academic / Plenum Publishers, New York, 1999.

rare, i.e., the rarest phenotype still occurs at the frequency of greater 1%. Thus rare genetic defects are those where the rare phenotype occurs with a frequency of less than 1% in the population.

2. GENETIC POLYMORPHISMS CAUSING ALTERED DRUG ACTION

Genetic variations due to altered drug response were the first examples demonstrating the consequences of a genetic polymorphism of an enzyme for food and drug toxicity. Because it is the most common inherited enzymopathy affecting at least 500 million people worldwide, glucose-6-phosphate dehydrogenase deficiency will be discussed.

2.1. Glucose-6-Phosphate Dehydrogenase (G-6-PD) Deficiency

The first description of this pharmacogenetic trait by Lucretius Caro is 2000 years old. It refers to the haemolysis which develops after the consumption of fava beans. Quod aliis cibus est, aliis fuat acre venenum (What is one man's food is another man's poison).

Glucose-6-phosphate dehydrogenase is a cytoplasmic enzyme that is distributed in all cells. G-6-PD catalyses the first step in the hexose monophosphate pathway, and it produces NADPH required for the preservation and regeneration of the reduced form of glutathione. It is also essential for the stability of catalase. Since catalase and glutathione are essential for the detoxication of hydrogen peroxide, the defense of cells against this compound depends ultimately and heavily on G-6-PD. This is especially true in red cells, which are exquisitely sensitive to oxidative damage and in which other NADPH-producing enzymes are lacking

G-6-PD deficiency is the most common known enzymopathy. It affects approximately 400 million people worldwide. The highest prevalence rates with gene frequency in the range of 5 to 50 percent are found in tropical Africa, in the Middle East, in tropical and subtropical Asia, in some areas of the Mediterranean and in Papua New Guinea.

The gene encoding G-6-PD maps to the telomeric region of the long arm of the X chromosome. It consists of 13 exons and 12 introns spread over a region of 100 kilobases. The normal enzyme is designated as G-6-PD B. It represents the most common type of enzyme encountered in all the population groups that have been studied. So far more than 350 variants have been identified. Among persons of African descent a mutant enzyme with normal activity is prevalent. It is known as A+ and migrates electrophoretically more rapidly than the B enzyme. A single amino acid substitution has been identified at nucleotide position 376. The most common form associated with severe enzyme deficiency in Africa is the A- form which is due to an additional mutation at nucleotide 202 that accounts for its in vivo instability. Among Caucasians population the most common variant enzyme was originally designated as G-6-PD Mediterranean. Many similar but distinct variants exist in the Mediterranean region. It has the same electrophoretically mobility as the B enzyme. Although it is formed in normal amounts there is no activity present because this enzyme has a decreased stability in vivo. This suggests that the mutant protein is unusually sensitive to proteolysis in the environment of the erythrocyte.[3]

The high frequency of G-6-PD-deficient genes in many populations implies that G-6-PD deficiency confers some sort of selective advantage. The geographical correlation of its distribution with the historical endemicity of malaria suggest that this disorder has risen in frequency through natural selection by malaria. However, attempts to confirm that

G-6-PD-deficiency is protective in case control studies of malaria have yielded conflicting results. So far it has been unclear whether both male hemizygotes and female heterozygotes are protected or as frequently suggested only females. In a recent study involving two large case control studies of over 2000 African children the common African form A- was shown to be associated with a 46 to 58 percent reduction in risk of severe malaria for both female heterozygotes and male hemizygotes. These study also could demonstrate that a counterbalancing selective disadvantage associated with this enzyme deficiency has retarded its rise in frequency in malaria endemic regions.[4]

Most G-6-PD deficient persons never suffer any clinical manifestation from this common genetic trait. The major clinical consequences is haemolytic aenemia. In general, haemolysis is associated with stress, most notably drug administration, infection, the newborn period and exposure to fava beans. Drug induced haemolysis is usually more severe in persons with the Mediterranean type of deficiency than in those with the milder A-African or Canton type deficiency. Favism is potentially one of the gravest clinical consequences of G-6-PD deficiency. It occurs much more commonly in children than in adults. Before transfusion became available most of the afflicted children died.

3. GENETIC POLYMORPHISMS CAUSING ALTERED DRUG METABOLISM

For almost all enzymes involved in the biotransformation of xenobiotics and drugs genetic polymorphisms or rare genetic defects have been described.

These genetic polymorphisms of drug metabolising enzymes give rise to distinct subgroups within the population which differ in their ability to metabolise drugs which are substrates for the affected enzymes. Since the activity of drug metabolising enzymes is a major deteminant of both the intensity and duration of a drug action genetic variations in the drug elimination process explain why some patients do not obtain the expected drug effects or show exaggerated drug response and serious toxicity when taking the standard and safe dose of a drug.

Among the drugs metabolising enzymes cytochrome P450 enzymes are of paramount importance for the metabolism of many xenobiotics. Most of the drugs in therapeutic use are substrates for these enzymes which oxidise both endogenous and exogenous compounds. Cytochrome P450 enzymes are part of a supergene family and nearly 500 genes have been identified. In mammals 14 gene families and 26 subfamilies are known. In man 20 subfamilies have been identified so far. The cytochrome P450 which are of importance for human drug metabolism belong to the families 1A, 2B, 2C, 2D, 2E and 3A.[5]

3.1. Cytochrome P450 2D6 (Debrisoquine/Sparteine) Polymorphism

This first example of a genetic polymorphism of the cytochrome P450 enzymes was discovered independently at St Mary's Hospital Medical School in London and at the Department of Internal Medicine of the University of Bonn by the serendipitous observation that some volunteers participating in pharmacokinetics studies of debrisoquine and sparteine, respectively experienced side-effects after administration of so called standard doses.[6,7] In the ensuing studies carried out to elucidate the mechanisms responsible for it became apparent that these volunteers showed an odd metabolic behaviour because they could not metabolise the drugs. It was then demonstrated that the metabolism of the two drugs was under monogenic control and that subjects with impaired metabolism were homozygous for a recessive allele. Based on the metabolic handling the of the two probe drugs the population could be divided into two distinct groups, the extensive metaboliser (EM) and the poor metaboliser (PM) phenotype.[8,9] In extensive population studies it was then shown that between 5 to 10 % of individuals of white populations were of the PM phenotype. In contrast the prevalence of the PM phenotype in oriental populations was much lower with a frequency of 0.5 to 1%.

3.2. Molecular Mechanism

The CYP 2D6 gene locus in humans is on the long arm of chromosome 22 (q13.1)[10,11] and consists of three highly homologous genes, CYP2D8P, CYP2D7P and CYP2D6. They are located in this orientation (5′ to 3′) on a contiguous region of about 45 kb.[12–14] The CYP2D6 gene consists of 9 exons and 8 introns. The CYP2D8P is a pseudo-gene that contains multiple deletions and insertions. The CYP2D7P gene is very similar to CYP2D6 and its coding sequence has only a single inactivating mutation in the first exon. Since in RNA from 8 human livers no specific mRNA product was detected the data indicate that CYP2D7P is also a pseudogene.

The CYP2D6 gene is highly polymorphic and more than 20 different mutations have been identified leading either to a complete loss of function, decreased activity or increased enzyme function. The most common mutant allele is the CYP2D6*4 where several mutations occur. The mutation associated with the loss of function is the G to A substitution at the splice site of exon 4 leading to a shift of splice site and, as a consequence, no protein is synthesized. More than 17 nonfunctional alleles have been described. In addition several mutations are associated with a reduced metabolic function. The so called ultrarapid metaboliser phenotype is due to a gene duplication which leads to two functional copies of CYP2D6*2. As a consequence of this gene duplication more enzyme is expressed (Table 1).[15–18]

As with the other polymorphisms significant ethnic differences exist in the frequency of the poor metaboliser phenotype, which is rare in most populations with the exception of Caucasians where 5 to 10% are carriers of nonfunctional alleles. The most common nonfunctional allele CYP2D6* 4 has a very low prevalence in the Oriental population and thus explains the lower frequency of the poor metaboliser phenotype in this population.[19–21] Compared with Caucasians, a considerably higher incidence of gene duplication and amplification alleles was reported for Ethiopians.

3.3. Clinical Aspects

The clinically relevance of the debrisoquine/sparteine polymorphism will depend on the nature of the drug, the relative importance of the affected metabolic pathway to the overall elimination of the compound, the therapeutic window of the drug and whether the parent drug or a metabolite is the active therapeutic moiety. Until now more than 40 drugs have been identified whose metabolism is in part or to a major extent catalysed by CYP2D6. This includes such important classes of drugs such as amphetamines, antiarrhythmics, antidepressants, ß-receptor antagonists, neuroleptics, hypoglycaemics, HT3-receptor antagonists and weak opioids (Table 2).

If the parent drug is solely responsible for therapeutic response it can be anticipated that following standard doses of the drug will be associated with exaggerated response, side effects and toxicity in the poor metaboliser patient. In a situation where both the parent compound and a metabolite contribute to the therapeutic efficacy the consequences of

Table 1. Mutations of CYP2D6 (from Zanger and Meyer, 1997)[89]

Allele–systematic name	Allele–trivial name	Protein	Selected nucleotide changes / key mutations	Xba I Haplo-type (kb)	Effect	Enzyme activity
CYP2D6*1A	Wild-type	CYP2D6 1	None	29		Normal
CYP2D6*1B			$G_{3916}A$	29		Normal
CYP2D6*2	CYP2D6L	CYP2D6 2	$C_{2938}T, G_{4268}C$	29	$R_{296}C, S_{486}T$	Normal (?)
CYP2D6*3	CYP2D6A		A_{2637} deletion	29	Frameshift	None
CYP2D6*4A	CYP2D6B		$G_{1934}A$	44/29/16+9	Splicing defect	None
CYP2D6*4B	CYP2D6B		$G_{1934}A$	29	Splicing defect	None
CYP2D6*4C	K29-1		$G_{1934}A$	44,29	Splicing defect	None
CYP2D6*4D			$G_{1934}A$		Splicing defect	None
CYP2D6*5	CYP2D6D		CYP2D6 deleted	11.5 or 13	CYP2D6 deleted	None
CYP2D6*6A	CYP2D6T		T_{1795} deleted	29	Frameshift	None
CYP2D6*6B	CYP2D6T		T_{1795} deleted	29	Frameshift	None
CYP2D6*7	CYP2D6E	CYP2D6 7	$A_{3023}C$	29	$H_{324}P$	None
CYP2D6*8	CYP2D6G		$G_{1846}T$		Stop codon	None
CYP2D6*11	CYP2D6F		$G_{971}C$		Splicing defect	None
CYP2D6*12		CYP2D6 12	$G_{212}A$	29	$G_{42}R, R_{296}C, S_{486}T$	None
CYP2D6*13			CYP2D7P/CYP2D6 hybrid	9	Frameshift	None
CYP2D6*14		CYP2D6 14	$G_{1846}A$	29	$P_{34}S, G_{169}R, R_{296}C, S_{486}T$	None
CYP2D6*15			T_{226} insertion	29	Frameshift	None
CYP2D6*16	CYP2D6D2		CYP2D7P/CYP2D6 hybrid	11	Frameshift	None
CYP2D6*18	CYP2D6J9	CYP2D6 18	TCACCCGTG insertion at 4213	29	Insertion of VPT at 468	None
CYP2D6*9	CYP2D6C	CYP2D6 9	A_{2701}-A_{2703} or G_{2702}-A_{2704} deleted	29	K_{281} deleted	Decreased
CYP2D6*10A	CYP2D6J	CYP2D6 10A	$C_{188}T$	44,29	$P_{34}S, S_{486}T$	Decreased
CYP2D6*10B	CYP2D6Ch1	CYP2D6 10A	$C_{188}T$	44,29	$P_{34}S, S_{486}T$	Decreased
CYP2D6*10C	CYP2D6Ch2	CYP2D6 10C	$C_{188}T$	44,29	$P_{34}S, S_{486}T$	Decreased
CYP2D6*17	CYP2D6Z	CYP2D6 17	$C_{1111}T, G_{1726}C, C_{2938}T, G_{4268}C$	29	$T_{107}I, R_{296}S, S_{486}T$	Decreased
CYP2D6*2XN	CYP2D6		$C_{2938}T, G_{4268}C$	42-175	N active genes, $R_{296}C, S_{486}T$	Increased
(N=2, 3, 4, 5 or 13)						

Table 2. Drugs which are substrates for CYP2D6 (from Fromm et. al., 1997)[88]

Substrate	Pathway	Substrate	Pathway
Alprenolol	Aromatic hydroxylation	Maprotiline	
Amiflamine	N-Demethylation	Methoxyamphetamine	O-Dealkylation
Amitriptyline	Benzylic hydroxylation	Methoxyamphetamine	Aromatic hydroxylation and N-demethylation
Aprindine	Aromatic hydroxylation	Methylenedioxymeth- amphetamine ('ecstasy')	Demethylenation
Brofaromine	O-Demethylation	Metoprolol	Aliphatic hydroxylation and O-dealkylation
Bufuralol	Aliphatic and aromatic hydroxylation	Mexiletine	Hydroxylation
Bunitrolol	Hydroxylation	Mianserin	Hydroxylation
Bupranolol	Hydroxylation	Minaprine	Hydroxylation
Cinnarizine	Aromatic hydroxylation	Norcodeine	O-Demethylation
Clomipramine	Aromatic hydroxylation	Nortriptyline	Benzylic hydroxylation
Codeine	O-Demethylation	N-Propylajmaline	Benzylic hydroxylation
Debrisoquine	Hydroxylation	Ondansetron	Hydroxylation
Desipramine	Aromatic hydroxylation	Oxycodone	O-Demethylation
Desmethylcitalopram	Demethylation	Paroxetine	Demethylenation
Dexfenfluramine		Perhexiline	Aliphatic hydroxylation
Dextromethorphan	O-Demethylation	Perphenazine	
Dihydrocodeine	O-Demethylation	Phenformine	Aromatic hydroxylation
Dolasetron	Hydroxylation	Promethazine	Aromatic hydroxylation
Encainide	O-Demethylation	Propafenone	Aromatic hydroxylation
Ethylmorphine	O-Deethylation	Propranolol	Aromatic hydroxylation
Flecainide	O-Dealkylation	Risperidone	Hydroxylation
Flunarizine	Aromatic hydroxylation	Sparteine	Hydroxylation
Fluvoxamine		Thioridazine	Side chain sulfoxidation
Guanoxan	Aromatic hydroxylation	Timolol	O-Dealkylation
Haloperidol		Tomoxetine	
Hydrocodone	O-Demethylation	Tropisetron	Aromatic hydroxylation
Indoramin	Aromatic hydroxylation	Venlafaxine	O-Demethylation
Imipramine	Aromatic hydroxylation	Zuclopenthixol	

polymorphic drug oxidation will depend on the potency of the metabolite relative to the parent compound and also the quality of action of the metabolite. However, in the case where the parent compound is a prodrug and the metabolite formed from this prodrug is solely responsible for therapeutic efficacy poor metaboliser patient will not have the beneficial therapeutic effect because no active metabolite can be formed. In the following examples for these various situations will be discussed.

3.4. Cardiovascular Drugs

3.4.1. ß-Adrenoreceptor Antagonists. The metabolism of several ß-adrenoreceptor antagonists is catalysed at least in part by CYP2D6. The examples of metoprolol and propanolol illustrate the consequences of polymorphic drug oxidation on the pharmacokinetics and pharmacodynamics depending on the fraction of the dose which is that catalysed by CYP2D6. Whereas in the case of metoprolol which is to a large extent metabolised by CYP2D6 pronounced differences in the plasma concentration and the degree and duration of ß-blockade have been observed between extensive and poor metabolisers no differences in the pharmacokinetics of propanolol were seen in the two phenotypes.

In the case of bufuralol side-effect such as profuse sweating, nausea and vomiting which occurred 2 to 6 hours after drug intake was only observed in the volunteers with the poor metaboliser phenotype.

3.4.2. Antiarrhythmics.

Most antiarrhythmics drugs of Class 1 are substrates for CYP2D6. The extent to which each antiarrhythmic is metabolised by this polymorphic enzyme varies among the drugs. For some drugs only a minor fraction of the dose is cleared via this enzyme and therefore the impact on the pharmacokinetics and pharmacodynamics between the extensive and the poor metaboliser phenotype will be negligible. On the other hand in the case of drugs where a substantial fraction of the dose is cleared via the polymorphic enzyme profound differences do exist between the two phenotypes resulting in a 10 to 15 fold difference in plasma concentrations following administration of the same dose. Because the therapeutic range of antiarrhythmics is usually narrow side effects and toxicity will occur predominantly in the poor metaboliser patient. Since in high drug concentrations have been identified as a predisposing factor for proarrhythmic effects of these drugs in addition to an impaired left ventricular function poor metaboliser patience might be at a higher risk to develop these life threatening side effects.

The place of antiarrhythmics in therapy has considerably changed since the publictation of the results of the Cardiac Arrhythmia Suppression Trial (CAST). In the view of an increased mortality with treatment of encainide and flecainide in comparison to placebo in this study there have been speculations, whether CYP2D6 phenotype might have influenced therapeutic outcome. Although this question could not be clearly answered, data indicate no excess of drug effect (determined by QRS or QT intervals) with the patients who died during therapy.[22–24] Moreover, the continuous increase in mortality during the course of the study argues against a major role of reduced drug metabolism in poor metabolizers, since higher drug concentrations and an increased risk of toxicity would be expected primarily in the early phase of the trial.

Propafenone is extensively metabolized in man and formation of the active metabolite 5-hydroxypropafenone is catalyzed by CYP2D6[25–27] resulting in higher drug concentrations of the parent compound in the poor metabolizers. These higher drug concentrations of propafenone in poor metabolizers were associated with a considerably higher incidence of central nervous side effecs in poor metabolizers (14 % vs 67 %).[25] Three other issues have to be considered when patients are treated with propafenone. First, in extensive metabolizers nonlinear pharmacokinetics of propafenone have been observed with increased dosage, which has been attributed to saturation of CYP2D6-mediated first-pass metabolism.[25] Second, like several other antiarrhythmics propafenone is administered as a racemate with its enantiomers differing in their pharmacokinetic handling and pharmacodynamic profile.[28] Whereas both enantiomers of propafenone are equipotent sodium channel blockers, only the S-enantiomer exerts a modest degree of ß-blockade. Indeed, ß-blockade after administration of propafenone was observed in both phenotypes, but it was more pronounced in the poor metabolizers.[29] Finally, in vitro and in vivo data clearly showed that due to an enantiomer/enantiomer interaction first-pass metabolism of propafenone is altered thereby modifying ß-blocking effects of racemic propafenone.[30,31]

Keeping these findings in mind one could speculate that the frequency of adverse events, which was found in a randomized, placebo-controlled trial of propafenone in the prophylaxis of paroxysmal supraventricular tachycardia and paroxysmal atrial fibrillation, might primarily be determined by the 8% of poor metabolizers and possibly by the intermediate metabolizers in such a study population.[32] Unfortunately, neither phenotype nor propafenone plasma concentrations were determined within that trial to prove that hypothesis.

Metabolism and effect of another sodium channel blocker, encainide, are also affected by CYP2D6 phenotype. Metabolism of encainide is grossly impaired in poor metabolizers resulting in bioavailability of 83% in contrast to 30% in extensive metabolizeres after chronic administration.[33] Since sodium channel blocking activity after administration of encainide resides mainly in its metabolite O-desmethyl encainide[34] and since formation of O-desmethyl encainide and 3-methoxy-O-desmethyl encainide is catalyzed by CYP2D6,[33-36] QRS interval prolongation is more pronounced in extensive metabolizers in comparison to poor metabolizers.[33,37] Again, consequences of these differences for therapeutic effects and side effects remain unclear, although some data indicate an increased risk of proarrhythmic events in extensive metabolizers with equal therapeutic efficacy in both phenotypes.[38]

The major antiarrhythmic activity of flecainide resides in the equipotent S- and R-enantiomers of the parent compound. Although 50% of an administered dose of flecainide is excreted in urine unchanged, oral clearance was considerably smaller in poor metabolizeres of sparteine when compared to extensive metabolizers after administration of a single dose.[39-41] During steady-state conditions these differences were statistically not significant.[42] Mean steady-state concentrations and QRS-interval prolongation were not different between both phenotypes, but nonlinear pharmacokinetics and a shortened half-life were observed in the extensive metabolizers. Poor metabolizers, however, with an additional renal failure, are likely to be of greater risk attaining higher flecainide concentrations and hence adverse reactions, e.g. proarrhythmic events.[43, 44]

3.5. Neuroactive Drugs

Since CYP2D6 plays an important role in metabolism of psychotropic drugs, a European Consensus conference recommended that routine phenotyping should be conducted in psychiatric patients.[45,46] Although large prospective studies have not finally proven the usefulness of CYP2D6 phenotyping before starting administration of psychotropic drugs, some data indicate an increased therapeutic efficacy after routine phenotyping e.g., in elderly patients.[47] A recently published report of an expert group in the field of antidepressants and drug-metabolizing enzymes[48] also states the importance of therapeutic drug monitoring, phenotyping and genotyping for prevention of problems with both efficacy and tolerability. There is clear evidence that patients with the poor metabolizer status are of increased risk for high serum concentrations of many antidepressants.[49]

Two recent prospective studies have explored in patients with depressions, who were treated with antidepressant drugs, whether the CYP2D6 phenotype/genotype is related to clinical outcome and the occurrence of adverse effects (Chen et al.,[50] 1996; Spina et al.,[51] 1997). Both studies could demonstrate that PM patients have a higher risk for adverse drug effects. Furthermore, in the study by Spina[51] et al., where patients received the same dose of desimipramine, a close correlation between the dextromethorphan metabolic ratio (MR) and desimipramine plasma levels was observed. The patients complaining of severe adverse effects had the highest plasma concentrations and were PMs. After dose reduction these side effects disappeared (Spina et al.,[51] 1997). Thus, in order to avoid concentration dependent side effects in PMs, phenotyping/genotyping prior to institution of drug treatment could identify the patients at risk.

3.5.1. Antidepressants. The metabolism of allmost all tricyclic and tetracyclic anti-depressants cosegregates with the sparteine/debrisoquine polymorphism[46] resulting in marked alterations of first-pass metabolism (e.g. for desipramine, imipramine). Very early a broad interindividual variability in plasma concentrations of antidepressants has been recognized (e.g. for desipramine[52]). It is now generally accepted that therapeutic effects of tricyclic antidepressants are depend on concentration of active parent compound and/or active metabolites. Serious side effects usually occur when drug levels are well above the therapeutic range.[53] The CYP2D6 polymorphism is a major cause for the observed interin-dividual variability. Hydroxylation of the secondary amine tricyclics nortriptyline and desipramine and of the tertiary amine tricyclics amitriptyline, imipramine and clomi-pramine is catalyzed by CYP2D6 (Table 1). N-demethylation of amitriptyline, imipramine and clomipramine, however, to its metabolites nortriptyline, desipramine and demethyl-clomipramine, respectively, is not affected by this genetic polymorphism.[46,54,55]

Formation of 2-hydroxydesipramine is an important pathway of desipramine. Stud-ies in healthy volunteers indicate a clear correlation between the of 2-hydroxylation of desipramine and 4-hydroxylation of debrisoquine.[56–58] Moreover, plasma levels of despi-ramine can be predicted by the metabolic ratio of debrisoquine.[59] Marked differences in bioavailability of desipramine have been observed between extensive and poor metabo-lizers after administration of a single dose (44 % vs 14 %).[60] Cardiotoxicity has been asso-ciated in a single case report with elevated desipramine levels and a CYP2D6 poor metabolizer status.[61]

Bertilsson et al.[62] pointed out that poor metabolizers developing high nortriptyline concentration at standard doses are not the only group of patients, to which special atten-tion should be paid on. Ultrarapid metabolizers of debrisoquine will require unusual high doses of nortriptyline in order to achieve therapeutic drug levels.[62] As the molecular mechanism responsible gene amplification has been identified.

In addition to the hydroxylation of the secondary amine tricyclic antidepressants CYP2D6 catalyzes hydroxylation of the tertiary amine tricyclic antidepressants amitrip-tyline, imipramine and clomipramine (Table 1). First-pass metabolism of imipramine is impaired in poor metabolizers of sparteine (F = 29%) in comparison to extensive metabo-lizers after administration of a single dose (61 %).[60] Apparent oral clearance of clomi-pramine is reduced in poor metabolizers due to impaired 2- and 8-hydroxylation of clomipramine and 8-hydroxylation of demethylclomipramine.[55,63] Differences in pharma-cokinetics after administration of a single dose, however, can not always be extrapolated to steady-state conditions. Dose-dependent pharmacokinetics of imipramine and desipramine have been reported and saturation of CYP2D6-catalyzed imipramine- and desipramine-2-hydroxylation have been suggested (for review[53]).

Selective serotonin reuptake inhibitors interfere with CYP2D6 by two ways. First, some selective serotonin reuptake inhibitors are metabolized by CYP2D6 and their dispo-sition depends on the CYP2D6 phenotype. Second, therapeutically relevant drug interac-tions occur due to inhibition of CYP2D6 by selective serotonin reuptake inhibitors, which are not neccessarily caused by the selective serotonin reuptake inhibitors, which are also metabolized by CYP2D6 (similar to quinidine).

3.5.2. Neuroleptics. Panel and patient studies indicate that the metabolism of neuroleptics cosegregates with CYP2D6, since metabolism of the neuroleptics per-phenazine, zuclopenthixol, thioridazine, risperidone and haloperidol is impaired in poor metabolizers. The clinical relevance was confirmed by Jerling et al.,[64] who reported that CYP2D6 genotype predicts the oral clearance of perphenazine in patients. Furthermore,

Linnet et al.[65] detected differences in dose-corrected perphenazine steady-state concentrations in psychiatric patients with poor and extensive metabolizer genotype.

For the most widley used neuroleptic haloperidol in healthy poor metabolizers after a single dose[66] terminal half-life was prolonged and apparent oral clearance was increased twofold. This findings could be confirmal in 8 patients with schizophrenia were the highest haloperidol plasma concentrations were observed in the only poor metabolizer patient.[67] In this poor metabolizer high plasma concentrations of haloperidol were associated with a high dopamine D_2 receptor occupancy during the study (determined by positon emission tomography using [^{11}C] raclopride as radioligand) and with the highest parkinsonian total score throughout the study. The work by LLerena et al.[66] indicates that higher haloperidol plasma concentrations in extensive metabolizers might be associated with an increased risk of developing neurological side effects in comparison to poor metabolizers.

3.6. Opioids

Codeine has been used since its isolation in 1832 as an analgesic, antitussive and antidiarrhoeal agent. Identification of codeine metabolites and the enzymes catalyzing ist biotransformation explain the large interindividual variability in ist effects. Codeine O-demethylation to yield morphine proceeds via CYP2D6. Urinary excretion of morphine, morphine-3-glucuronide and morphine-6-glucuronide is negligble in poor metabolizers compared to extensive metabolizers who form 5–15% jof these metabolites after codeine administration.[68] If the analgesic activity of codeine would reside with the parent drug, this fact would have no clinical consequences. Several studies, however, indicate that the analgesic effects of codeine is mediated by its metabolites morphine and morphine-6-glucuronide.[69] In agreement with this assumpation are studies in healthy volunteers where analgesia could be observed in extensive metabolizers but not in poor metabolizers in after codeine administration.[70,71] These results were recently confirmed by Caraco et al.,[72] who reported greater respiratory, psychomotor and pupillary effects of codeine in extensive metabolizers in comparison to poor metabolizers. The authors conclude that poor metabolizers would not be expected to benefit from codeine therapy. Indeed, there is one study on the relief of postoperative pain with codeine, in which nine out of ten extensive metabolizers were effectively treated with codeine, but not the only poor metabolizer.[73]

4. CYTOCHROME P450 2C19 (S-MEPHENYTOIN POLYMORPHISM

The S-mephenytoin polymorphism was discovered almost 20 years ago by Küpfer and his colleagues in Bern. They observed a subject during a pharmacokinetic study with mephenytoin who showed a marked impairment to metabolise this drug to the major metabolite 4-hydroxymephenytoin. In extensive metabolisers this racemic drug is eliminated by 4-hydroxylation of the S-enantiomer, while R-mephenytoin is N-demethylated. In poor metabolisers of mephenytoin this stereoselectivity is lost. Phenotype assignment is nowadays based on the ratio of the S to R-enantiomer, poor metabolisers having a ratio of approximately 1 and extensive metabolisers below this value. The defect is inherited as an autosomal recessive trait. The frequency of the PM shows substantial racial differences with 2 to 3% of PM occuring in Caucasians and up to 23% in Oriental populations.[74–77]

4.1. Molecular Mechanism

It was not until recently that CYP2C19 was identified as the enzyme that catalysed the metabolism of S-mephenytoin.[78–80] The CYP2C subfamily consists of a cluster of genes at chromosome 10q24. The large scale structure and the exact gene number of the CYP2C cluster was recently determined with YAC clones bearing the CYP2C genes. A 2.4 Mb contiguous map was constructed. This map links the CYP2C cluster to an adjacent gene, the serum retinol binding protein gene (RBP4). The cluster spans approximately 500 kb and compromises four CYP2C genes on proximal 10q4. The CYP2C cluster is telometric to RBP4, giving the gene order and orientation from centromer to RBP4–2C18–2C19–2C9–2C8 to telomer.[81–84]

The principal molecular defect is a single base pair G to A mutation in exon 5 corresponding to base pair 681 of the cDNA. This mutation creates an aberrant splice site, which is apparently used exclusively as a splice site. It accounts for 75–83% of all defective alleles in Orientals and Caucasians. In addition, a G to A transition has been identified that introduces a premature stop codon. These two mutant alleles seem to account for almost all mutant alleles.[85,86]

4.2. Clinical Aspects

As far as the clinical consequences are concerned, there are less data available than for CYP2D6.

Table 3 lists the drugs which are metabolised by CYP2C19. In the case of diazepam differences in the oral clearance have been observed between the two phenotypes. Moreover, the disposition of the metabolite desmethyldiazepam is also different between the phenotypes. Several antidepressants appear to be substrates of CYP2C19 and CYP2D6. Whereas hydroxylation of imipramine cosegregates with the CYP2D6 polymorphism the N-demethylation seems to involve CYP2C19. In addition to imipramine, the biotransformation of clomipramine is also affected by both the CYP2D6 and CYP2C19 polymorphism.

An interesting aspect of this polymorphism is the pronounced difference in the frequency of the mutant alleles among people of different racial origin. This has consequences and implications for the use of drugs in different countries. In the case of the proton pump inhibitor omeprazole pronounced differences in the pharmacokinetics between extensive and poor metabolisers have been demonstrated. In addition to this pheno-

Table 3. Drugs which are substrates for CYP2C19
(from Fromm et. al., 1997)[88]

Substrate	Pathway
Carisoprodol	N-Dealkylation
Citalopram	Demethylation
Clomipramine	N-Demethylation
Diazepam	N-Demethylation
Hexobarbital	Hydroxylation
Imipramine	N-Demethylation
S-Mephenytoin	4'-Hydroxylation
Mephobarbital	Hydroxylation
Moclobemide	Hydroxylation
Omeprazole	Hydroxylation
Proguanil	Formation of cycloguanil
Propranolol	Side chain oxidation

type related differences in metabolism among homozygous metabolisers differences exist. If the same dose of omeprazole which has been studied in a population will a low frequency for the mutant allele is used in a population with a 5 to 10 higher frequency not only the number of poor metabolisers but also heterozygous extensive metaboliser will be much higher. As at consequence much higher drug concentrations are achieved in this population. These concentration might be in excess of the concentrations required for therapeutic efficacy and hence side effects might occur with a much higher frequency in this population. In the case of omeprazole the gastrin levels are certainly much higher.[86]

In addition to these drugs it has been shown recently that the metabolism of the proton pump inhibitor pantropazole and lansoprazole is controlled by the CYP2C19.

5. CONCLUSIONS

There can be no doubt that the discovery of genetic polymorphisms in drug metabolism has contributed a great deal to our understanding of the influence of genetic factors on the disposition and effects of drugs and has enabled us to elucidate the mechanisms responsible for certain adverse drug reactions. It is hoped that this knowledge will find its way into clinical practice in order to achieve our ultimate goal of making drug therapy more effective and safer for the benefit of the patient.

REFERENCES

1. A.E. Garrod, 1992, The incidence of alcaptonuria: A study in chemical individuality, *Lancet* **339**:1616–1620.
2. F. Vogel and A.G. Motulsky, 1986, *Human genetics. Problem and approches*, 2nd ed., Springer-Verlag, Berlin.
3. L. Luzzatto and A. Mehta, 1989, Glucose-6-phaophate dehydrogenase deficiency in: *The metabolic basis of inherited diseases*, (C.R. Scriver, A.L. Baudet, W.S. Sly, D. Vale, eds), Ch. 91, McGraw-Hill, New York.
4. C. Ruwende, S.C. Khoo, R.W. Snow, and S.N.R. Yates, D. Kwiatkowski, S. Gupta, P. Warn, C.E.M. Allsopp, S.C. Gilbert, N. Peschu, C.I. Newbold, B.M. Greenwood, K. Marsh, and A.V.S. Hill, 1995, Natural selection of hemi- and heterozygotes for G-6-PD deficiency in Africa by resistance to severe malaria, *Nature* **376**:246–249.
5. D.R. Nelson, L. Koymans, T. Kamataki, J.J. Stegeman, R. Feyereisen, D.J. Waxman, M.R.Waterman, O. Gotoh, M.J. Coon, R.W. Estabrook, I.C. Gunsalus, and D.W. Nebert, 1996, P450 superfamily: update on new sequences, gene mapping, accession number and nomenclature. *Pharmacogenetics* **6**:1–42.
6. A. Mahgoub, J.R. Idle, L.G. Dring, R. Lancester, and R.L. Smith, 1977, Polymorphic hydoxylation of debrisoquine in man, *Lancet* **2**:584–586.
7. M. Eichelbaum, N. Spannbrucker, B. Steinicke, and H.J. Dengler, 1979, Defective N-oxidation of sparteine in man: a new pharmacogenetic defect, *Eur. J. Clin. Pharmacol.* **16**:183–187.
8. M. Eichelbaum and A.S. Gross, 1990, The genetic polymorphism of debrisoquine/sparteine metabolism-clinical aspects, *Pharmacol. Ther.* **46**:377–394.
9. U.A. Meyer, R.C. Skoda, and U.M. Zanger, 1990, The genetic polymorphism of debrisoquine/sparteine metabolism-molecular mechanisms, *Pharmacol. Ther.* **46**:297–308.
10. M. Eichelbaum, M.P. Baur, H.J. Dengler, B.O. Osikowska Evers, G. Tieves, and C. Zekorn, and C. Rittner, 1987, Chromosomal assignment of human cytochrome P450 (debrisoquine/sparteine type) to chromosome 22, *Br. J. Clin. Pharmacol.* **23**:455–458.
11. F.J. Gonzales, F. Vilbois, J.P. Hardwick, O.W. McBride, D.W. Nebert, and H.V. Gelboin, U.A. Meyer, 1988, Human debrisoquine 4-hydroxylase (P450IID1): cDNA and deduced amino acid sequence and assignment of the CYP2D locus to chromosome 22, *Genomics* **2**:174–179.
12. S. Kimura, M. Umeno, R.C. Skoda, U.A. Meyer, and F.J. Gonzales, 1989, The human debrisoquine 4-hydroxylase (CYP2D) locus: sequence and identification of the polymorphic CYP2D6 gene, a related gene, and a pseudogene, *Am. J. Hum. Genet.* **45**:889–904.

13. A. Gaedigk, M. Blum, R. Gaedigk, M. Eichelbaum, and U.A. Meyer, 1991, Deletion of the entire cytochrome P450 CYP2D6 gene as a cause of impaired drug metabolism in poor metabolizers of the debrisoquine/sparteine polymorphism, *Am. J. Hum. Genet.* **48**:943–950.

14. M.H. Heim and U.A. Meyer, 1992, Evolution of a highly polymorphic human cytochrome P450 gene cluster: CYP2D6, *Genomics* **14**:49–58.

15. L. Bertilsson, M.L. Dahl, F. Sjöqvist, A. Åberg Wistedt, M. Humble, I. Johansson, E. Lundqvist and M. Ingelman-Sundberg, 1993, Molecular basis for rational megaprescribing in ultrarapid hydroxylators of debrisquine, *Lancet* **341**:63.

16. M.L. Dahl, I. Johansson, L. Bertilsson, M. Ingelman-Sundberg, and F. Sjoqvist, 1995, Ultrarapid hydroxylation of debrisoquine in a Swedish population. Analysis of the molecular genetic basis, *J. Pharamacol. Exp. Ther.* **274**:516–520

17. J.A. Agundez, M.C. Ledesma, J.M. Ladero, and J. Benitez, 1995, Prevalence of CYP2D6 gene duplication and its repercussion on the oxidative phenotype in a white population, *Clin. Pharmacol. Ther.* **57**:265–269.

18. E. Aklillu, I. Perssson, L. Bertilsson, F. Rodriques, and M. Ingelmann-Sundberg, 1996, Frequent distribution in a black African population carrying duplicated and multiduplicated function CYP2D6 alleles detection, *J. Pharmacol. Exp. Ther.* **278**:441–446

19. I. Johansson, M. Oscarson, Q.Y. Yue, L. Bertilsson, F. Sjoqvist, and M. Ingelman-Sunberg, 1994, Genetic analysis of the Chinese cytochrome P4502D locus: characterization of variant CYP2D6 genes present for debrisoquine hydroxylation, *Mol. Pharmacol.* **46**:452–459.

20. H. Yokota, S. Tamura, H. Furuya, S. Kimura, M. Watanabe, I. Kanazawa I. Kondo, and F.J. Gonzales, 1993, Evidence for a new variant CYP2D6 allele CYP2D6J in a Japanese population associated with lower in vivo rates of sparteine metabolism, *Pharmacogenetics* **3**:256–263.

21. M.L. Dahl, Q.Y. Yue, H.K. Roh, I. Johansson, J. Säwe, F. Sjöqvist and L. Bertilsson, 1995, Genetic analysis of the CYP2D locus in relation to debrisoquine hydroxylation capacity in Korean, Japanese and Chinese subjects, *Pharmacogenetics* **5**:159–164.

22. D.S. Echt, P.R. Liebson, L.B. Mitchell, R.W. Peters, D. Obias-Manno, A.H. Barker, D. Arensberg, A. Baker, L. Friedman, H.L. Greene, M.L. Huther, D.W. Richardson, and the CAST investigators, 1991, Mortality and morbidity in patients receiving encainide, flecainide, or placebo. The Cardiac Arrhythmia Suppression Trial, *N. Engl. J. Med.* **324**:781–788.

23. The Cardiac Arrhythmia Suppression Trial (CAST) Investigators, 1989, Preliminary report: effect of encainide and flecainide on mortality in a randomized trial of arrhythmia suppression after myocardial infarction, *N. Engl. J. Med.* **321**:406–412.

24. E. Buchert and R.L. Woosley, 1992, Clinical implications of variable antiarrhythmic drug metabolism, *Pharmacogenetics* **2**:2–11.

25. L.A. Siddoway, K.A. Thompson, C.B. McAllister, T. Wang, G.R. Wilkinson, D.M. Roden, R.L. Woosley, 1987, Polymorphism of propafenone metabolism and disposition in man: clinical and pharmacokinetic consequences, *Circulation* **75**:785–791.

26. H.K. Kroemer, G. Mikus, T. Kronbach, U.A. Meyer, and M. Eichelbaum, 1989, In vitro characterization of the human cytochrome P-450 involved in polymorphic oxidation of propafenone. *Clin. Pharmacol. Ther.* **45**:28–33.

27. C. Funck-Brentano, H.K. Kroemer, J.T. Lee, and D.M. Roden, 1990, Propafenone. *N. Engl. J. Med.* **322**:518–525.

28. H.K. Kroemer, C. Funck-Brentano, D.J. Silberstein, A.J. Wood, M. Eichelbaum, R.L. Woosley, and D.M. Roden, 1989, Stereoselective disposition and pharmacologic activity of propafenone enantiomers, *Circulation* **79**:1068–1076.

29. J.T. Lee, H.K. Kroemer, D.J. Silberstein, C. Funck-Brentano, M.D. Lineberry, A.J. Wood, D.M. Roden, and R.L. Woosley, 1990, The role of genetically determined polymorphic drug metabolism in the beta-blockade produced by propafenone, *N. Engl. J. Med.* **322**:1764–1768.

30. H.K. Kroemer, C. Fischer, C.O. Meese, M. Eichelbaum, 1991, Enantiomer/enantiomer interaction of (S)- and (R)-propafenone for cytochrome P450IID6-catalyzed 5-hydroxy-lation: in vitro evaluation of the mechanism, *Mol. Pharmacol.* **40**:135–142.

31. H.K. Kroemer, M.F. Fromm, K. Bühl, H. Terefe, G. Blaschke, and M. Eichelbaum, 1994, An enantiomer-enantiomer interaction of (S)- and (R)-propafenone modifies the effect of racemic drug therapy, *Circulation* **89**:2396–2400.

32. UK Propafenone PSVT Study Group, 1995, A randomized, placebo-controlled trial of propafenone in the prophylaxis of paroxysmal supraventricular tachycardia and paroxysmal atrial fibrillation, *Circulation* **92**:2550–2557.

33. T. Wang, D.M. Roden, H.T. Wolfenden, R.L. Woosley, A.J. Wood, and G.R. Wilkinson, 1984, Influence of genetic polymorphism on the metabolism and disposition of encainide in man, *J. Pharmacol. Exp. Ther.* **228**:605–611.

34. D.M. Roden, H.J. Duff, D. Altenbern, and R.L. Woosley, 1982, Antiarrhythmic activity of the O-demethyl metabolite of encainide, *J. Pharmacol. Exp. Ther.* **221**:552–557.

35. R.L. Woosley, D.M. Roden, G.H. Dai, T. Wang, D. Altenbern, J. Oates, and G.R. Wilkinson, 1986, Co-inheritance of the polymorphic metabolism of encainide and debrisoquin, *Clin. Pharmacol. Ther.* **39**:282–287.

36. R.L. Woosley, A.J. Wood, and D.M. Roden, 1988, Drug therapy. Encainide, *N. Engl. J. Med.* **318**:1107–1115.

37. C. Funck-Brentano, G. Thomas, E. Jacqz-Aigrain, J.M. Poirier, T. Simon, G. Bereziat, and P. Jaillon, 1992, Polymorphism of dextromethorphan metabolism: Relationships between phenotype, genotype and response to the administration of encainide in humans, *J. Pharmacol. Exp. Ther.* **263**:780–786.

38. E. Buchert, and R.L. Woosley, 1992, Clinical implications of variable antiarrhythmic drug metabolism, *Pharmacogenetics* **2**:2–11.

39. G. Mikus, A.S. Gross, J. Beckmann, R. Hertrampf, U. Gundert-Remy, and M. Eichelbaum, 1989, The influence of the sparteine/debrisoquin phenotype on the disposition of flecainide, *Clin. Pharmacol. Ther.* **45**:562–567.

40. A.S. Gross, G. Mikus, C. Fischer, R. Hertrampf, U. Gundert-Remy, and M. Eichelbaum, 1989, Stereoselective disposition of flecainide in relation to the sparteine/debrisoquine metaboliser phenotype, *Br. J. Clin. Pharmacol.* **28**:555–566.

41. A.S. Gross, G. Mikus, C. Fischer, and M. Eichelbaum, 1991, Polymorphic flecainide disposition under conditions of uncontrolled urine flow and pH, *Eur. J. Clin. Pharmacol.* **40**:155–162.

42. C. Funck-Brentano, L. Becquemont, H.K. Kroemer, K. Bühl, N.G. Knebel, M. Eichelbaum, and P. Jaillon, 1994, Variable disposition kinetics and electrocardiographic effects of flecainide during repeated dosing in humans: Contribution of genetic factors, dose-dependent clearance, and interaction with amiodarone, *Clin. Pharmacol. Ther.* **55**:256–269.

43. M. Eichelbaum and A.S. Gross, 1990, The genetic polymorphism of debrisoquine/sparteine metabolism-- clinical aspects, *Pharmacol. Ther.* **46**:377–394.

44. J. Evers, M. Eichelbaum, and H.K. Kroemer, 1994 Unpredictability of flecainide plasma concentrations in patients with renal failure: Relationship to side effects and sudden death?, *Ther. Drug Monit.* **16**:349–351.

45. G. Alvan, L.P. Balant, P.R. Bechtel, A.R. Boobis, L.F. Gram, and K. Pithan, 1990, European Consensus Conference on Pharmacogenetics, Brussels, Belgium: Commission of the European Communities 28, 167.

46. L. Bertilsson and M.L. Dahl, 1996, Polymorphic drug oxidation: Relevance to the treatment of psychiatric disorder, *CNS Drugs* **5**:200–223.

47. B.G. Pollock, B.H. Mulsant, R.A. Sweet, J. Rosen, and J.M. Perel, 1994, Pharmacogenetics of neuroleptics in the elderly, *Neuropsychopharmacology* **3S**:918S.

48. U.A. Meyer, R. Amrein, L.P. Balant, L. Bertilsson, M. Eichelbaum, T.W. Guentert, S. Henauer, P. Jackson, G. Laux, H. Mikkelsen, C. Peck, B.G. Pollock, R. Priest, F. Sjöqvist, and A. Delini-Stula, 1996, Antidepressants and drug-metabolizing enzymes - Expert group report, *Acta Psychiatr. Scand.* **93**:71–79.

49. U. Tacke, E. Leinonen, P. Lillsunde, T. Seppala, P. Arvela, O. Pelkonen, and P. Ylitalo, 1992, Debrisoquine hydroxylation phenotypes of patients with high versus low to normal serum antidepressant concentrations, *J. Clin. Psychopharmacol.* **12**:262–267.

50. S. Chen, W.H. Chou, R.A. Blouin, Z. Mao, L.L. Humphries, Q.C. Meek, J.R. Neill, W.L. Martin, L.R. Hays, and P.J. Wedlund, 1996, The cytochrome P450 2D6 (CYP2D6) enzyme plymorphism: screening costs and influence on clinical outcomes in psychiatry, *Clin. Pharm. Ther.* **60**:522–534

51. E. Spina, C. Gitto, A. Abenoso, G.M. Campo, A.P. Caputi, and E. Perucca, 1997, Relationship between plasma desipramine levels, CYP2D6 phenotype and clinical response to desipramine: a prospective study, *Eur. J. Clin. Pharmacol.* **51**:395–398

52. F. Sjöqvist, W. Hammer, C.M. Ideström, M. Lind, D. Tuck, and M. Asberg, 1967, Plasma levels of monomethylated tricyclic antidepressants and side-effects in man. In: Proceedings of the European Society for the Study of Drug Toxicity. Volume IX, Toxicity and Side-Effects of Psychotropic Drugs. Excerpta Medica International Congress Series No. 145, pp. 246–257.

53. M.L. Dahl and L. Bertilsson, 1993, Genetically variable metabolism of antidepressants and neuroleptic drugs in man, *Pharmacogenetics* **3**:61–70.

54. B. Mellström, L. Bertilsson, L. Traskman, D. Rollins, M. Asberg, and F. Sjöqvist, 1979, Intraindividual similarity in the metabolism of amitriptyline and chlorimipramine in depressed patients, *Pharmacology* **19**:282–287.

55. A.E. Balant-Gorgia, L.P. Balant, C. Genet, P. Dayer, J.M. Aeschlimann, and G. Garrone, 1986, Importance of oxidative polymorphism and levomepromazine treatment on the steady-state blood concentrations of clomipramine and its major metabolites, *Eur. J. Clin. Pharmacol.* **31**:449–455.

56. E. Spina, C. Birgersson, C. von Bahr, O. Ericsson, B. Mellström, E. Steiner, and F. Sjöqvist, 1984, Phenotypic consistency in hydroxylation of desmethylimipramine and debrisoquine in healthy subjects and in human liver microsomes, *Clin. Pharmacol. Ther.* **36**:677–682.

57. E. Spina, E. Steiner, O. Ericsson, and F. Sjöqvist, 1987, Hydroxylation of desmethylimipramine: dependence on the debrisoquin hydroxylation phenotype, *Clin. Pharmacol. Ther.* **41**:314–319.

58. M.L. Dahl, L. Iselius, C. Alm, J.O. Svensson, D. Lee, I. Johansson, M. Ingelman-Sundberg, and F. Sjöqvist, 1993, Polymorphic 2-hydroxylation of desipramine: A population and family study, *Eur. J. Clin. Pharmacol.* **44**:445–450.

59. L. Bertilsson and A. Aberg-Wistedt, 1983, The debrisoquine hydroxylation test predicts steady-state plasma levels of desipramine, *Br. J. Clin. Pharmacol.* **15**:388–390.

60. K. Brøsen and L.F. Gram, 1988, First-pass metabolism of imipramine and desipramine: impact of the sparteine oxidation phenotype, *Clin. Pharmacol. Ther.* **43**:400–406.

61. R.E. Bluhm, G.R. Wilkinson, R. Shelton, and R.A. Branch, 1993, Genetically determined drug-metabolizing activity and desipramine-associated cardiotoxicity: A case report, *Clin. Pharmacol. Ther.* **53**:89–95.

62. L. Bertilsson, A. Aberg-Wistedt, L.L. Gustafsson, and C. Nordin, 1985, Extremely rapid hydroxylation of debrisoquine: a case report with implication for treatment with nortriptyline and other tricyclic antidepressants. *Ther. Drug Monit.* **7**:478–480.

63. K.K. Nielsen, K. Brøsen, M.G.J. Hansen, and L.F. Gram, 1994, Single-dose kinetics of clomipramine: Relationship to the sparteine and S-mephenytoin oxidation polymorphisms, *Clin. Pharmacol. Ther.* **55**:518–527.

64. M. Jerling, M.L. Dahl, A. Aberg-Wistedt, B. Liljenberg, N.E. Landell, L. Bertilsson, and F. Sjöqvist, 1996, The CYP2D6 genotype predicts the oral clearance of the neuroleptic agents perphenazine and zuclopenthixol, *Clin. Pharmacol. Ther.* **59**:423–428.

65. K. Linnet and O. Wiborg, 1996, Steady-state serum concentrations of the neuroleptic perphenazine in relation to CYP2D6 genetic polymorphism. *Clin. Pharmacol. Ther.* **60**:41–47.

66. A. Llerena, C. Alm, M.L. Dahl, B. Ekqvist, and L. Bertilsson, 1992, Haloperidol disposition is dependent on debrisoquine hydroxylation phenotype, *Ther. Drug Monit.* **14**:92–97.

67. S. Nyberg, L. Farde, C. Halldin, M.L. Dahl, L. Bertilsson, (1995) D-2 dopamine receptor occupancy during low-dose treatment with haloperidol decanoate. Am. J. Psychiatry 152, 173–178.

68. Q.Y. Yue, J.O. Svensson, C. Alm, F. Sjöqvist, and J. Säwe, 1989, Codeine O-demethylation co-segregates with polymorphic debrisoquine hydroxylation, *Br. J. Clin. Pharmacol.* **28**:639–645.

69. S.H. Sindrup and K. Brøsen, 1995, The pharmacogenetics of codeine hypoalgesia, *Pharmacogenetics* **5**:335–346.

70. S.H. Sindrup, K. Brøsen, P. Bjerring, L. Arendt-Nielsen, U. Larsen, H.R. Angelo, and L.F. Gram, 1990, Codeine increases pain thresholds to copper vapor laser stimuli in extensive but not poor metabolizers of sparteine, *Clin. Pharmacol. Ther.* **48**:686–693.

71. J. Desmeules, M.P. Gascon, P. Dayer, and M. Magistris, 1991, Impact of environmental and genetic factors on codeine analgesia, *Eur. J. Clin. Pharmacol.* **41**:23–26.

72. Y. Caraco, J. Sheller, and A.J.J. Wood, 1996, Pharmacogenetic determination of the effects of codeine and prediction of drug interactions, *J. Pharmacol. Exp. Ther.* **278**:1165–1174.

73. K. Persson, S. Sjöström, I. Sigurdardottir, V. Molnar, M. Hammarlund-Udenaes, and A. Rane, 1995, Patient-controlled analgesia (PCA) with codeine for postoperative pain relief in ten extensive metabolisers and one poor metaboliser of dextromethorphan, *Br. J. Clin. Pharmacol.* **39**:182–186.

74. A.P.R. Küpfer, S. Ward, S. Schenker, R. Preisig, and R.A. Branch, 1984. Stereoselective metabolism and pharmacogenetic control of 5-phenyl-5-ethylhydantoin (Nirvanol) in humans, *J. Pharmacol. Exp. Ther.* **230**:28–33

75. P.J. Wedlund, W.S. Aslanian, E. Jacqz, C.B. McAllister, R.A. Branch, and G.R. Wilkinson, 1985, Phenotypic differences in mephenytoin pharmacokinetics in normal subjects, *J. Pharmacol. Exp. Ther.* **234**:662–669

76. D.A.P. Evans, 1993, *Genetic factors in drug therapy. Clinical and molecular pharmacogenetics.* Cambridge: Cambridge Univ. Press.

77. G.R. Wilkonson, F.P. Guengerich, and R.A. Branch, 1992, Genetic polymorphism if S-mephenytoin hydroxylation, in: *Pharmacogenetics of Drug Metabolism*, (W. Kalow, ed.) pp. 657–685, Pergamon, New York.

78. M. Romkes, M.B. Faletto, J.A. Blaisdell, J.L. Raucy, and J.A. Goldstein, 1991, Cloning and expressing of complementary DNAs for multiple members of the human cytochrome P450IIC subfamiliy, *Biochemistry* **30**:3247–3255.

79. S.A. Wrighton, J.C. Stevens, G.W. Becker, and M. VandenBranden, 1993, Isolation and characterization of human liver cytochrome P4502C19: correlation between 2C19 and S-mephenytoin 4'-hydroxylation, *Biochem. Biophys.* **306**:240–245.

80. J.A. Goldstein, M.B. Faletto, M. Romkes-Sparks, T. Sullivan, S. Kitareewan, J.L. Raucy, J.M. Lasker, and B.I. Ghanayem, 1994, Evidence for a role for 2C19 in metabolism of S-mephenytoin in humans, *Biochemistry* **33**:1743–1752.

81. C. Ged, D.R. Umbenhauer, T.M. Bellew, R.W. Bork, P. Srivastave, N. Shinriki, R.S. Lloyd, and F.P. Guengerich, 1994, Characterisation of cDNAs, mRNAs, and proteins related to humans liver microsomal cytochrome P-450 (S)-mephenytoin 4'-hydroxylase. *Biochemistry* **27**:6929–6940.

82. R.R. Meehan, J.R. Gosden, D. Rout, N.D. Hastle, and T. Friedberg, 1988, Human cytochrome P-450 PB-1: a multigene family involved in mephenytoin and steroid oxidations that maps to chromosome 10. *Am. J. Hum. Genet.* **42**:26–37.

83. S.M.F. de Moreis, H. Schweikl, J. Blaisdell, and J.A. Goldstein, 1993, Gene structure and upstream regulation regions of human CYP2C9 and CYP2C18. *Biochem. Biophys. Res. Commun.* **194**:194–201.

84. J.A. Goldstein and S.M.F. de Moreis, 1994, Biochemistry and molecular biology of the human CYP2C subfamily, *Pharmacogenetics* **4**:285–299.

85. S.M.F. de Moreis, G.R. Wilkinson, J. Blaisdell, K. Nakamura, U.A. Meyer, and J.A. Goldstein, 1994, The major defect responsible for the polymorphism of S-mephenytoin metabolism in humans, *J. Biol. Chem.* **269**:15419–15422.

86. S.M.F. de Moreis, G.R. Wilkinson, J. Blaisdell, U.A. Meyer, K. Nakamura, and J.A. Goldstein, 1994, Indentification of a new genetic defect responsible for the polymorphism of S-mephenytoin metabolism in Japanese. *Mol. Pharmacol.* **46**:594–598

87. L. Bertilsson, 1995, Geographical/interracial differences in polymorphic drug oxidation. Current state of the knowledge of cytochromes P450(CYP)2D6 and 2C19. *Clin. Pharmacokin.* **29**:192–209.

88. M.F. Fromm, H.K. Kroemer, and M. Eichelbaum, 1997, Impact of P450 genetic polymorphism on the first-pass extraction of cardiovascular and neuroactive drugs, *Adv. Drug Delivery Reviews* **27**:171–199.

89. U.A. Meyer and U.M. Zanger, 1997, Molecular Mechanisms of genetic polymorphisms of drug metabolism, *Annu. Rev. Pharmacol. Toxicol.* **37**:269–296.

9

GENETIC POLYMORPHISMS OF CYTOCHROMES P450 1A1 AND 2E1 AND OF GLUTATHIONE S-TRANSFERASE M1 AND CANCER SUSCEPTIBILITY IN THE HUMAN

Minro Watanabe[*]

Laboratory of Molecular Medicine
Faculty of Social Welfare
Iwate Prefectural University
Takizawa, Iwate 020-0173 and
Institute of Development, Aging and Cancer
Tohoku University
Sendai 980-0872, Japan

ABSTRACT

Most chemical carcinogens require metabolic activation before they interact with cellular macromolecules, nucleic acid and protein, and can cause initiation and/or promotion in chemical carcinogenesis in cells. Many cytochrome P450 (CYP) mediating oxidative enzymes and the related other drug metabolizing enzymes, cloned and characterized in human tissues,[1] show genetic and phenotypic polymorphisms. They have been suggested to contribute to individual cancer susceptibility as genetic modifiers of cancer risk against exposure to environmental chemical carcinogens.

It is well recognized that CYP1A1 and CYP2E1 mediating oxidative enzymes react with polycyclic aromatic hydrocarbons and low molecular weight nitrosamines, respectively, to form reactive intermediates similarly. Glutathione S-transferase (GST) M1 reacts with polycyclic aromatic hydrocarbon oxides to form the glutathione conjugates in the detoxification process.

The risk of lung and urinary bladder cancers was reported to be increased in individuals who carried high risk genotypes in either CYP1A1, CYP2E1 or GSTM1, and the combined genotype of both CYP1A1 and GSTM1 enzymes have an enhanced tendency of

---

[*] All correspondence should be addressed to the following: 16-36 Shihei-machi, Aoba-ku, Sendai 981-0944, Japan.

Molecular and Applied Aspects of Oxidative Drug Metabolizing Enzymes,
edited by Arinç *et al.* Kluwer Academic / Plenum Publishers, New York, 1999.

risk to lung cancer more significantly. In each category, however, there are several confused and controversial results in which no association have been found despite consideration of ethnic difference in the world.

These early prediction of genetic disposition for the incidence of such severe disease as cancer may prompt to receive their early examination and diagnosis and to expect the complete cure from the diseases to their happy life.

1. INTRODUCTION

Molecular genetic technology recently established prompt us to evoke advances in the study of oxidizing enzymes in human[1] and to allow the dissection of many of the 1molecular mechanism of cancer. Indeed, several tumors are now known to result in the establishment of heritable cancer families with high incidence of the related tumor.[2]

It is now widely recognized that most common tumors were derived from each complex sequence of mutational event, multistage carcinogenesis, each of which is consisted of initiation, promotion and progression in the consecutive process to govern normal cell behavior to ultimately form a tumor cell. In addition, many specific mutational events proceed to follow after exposure to environmental chemicals.

On the other hand, most organisms have developed complex mechanisms by which they protect themselves from environmental hazards and intoxicants. In the majority of cases, the ability to metabolize and hence either detoxify or activate xenobiotic toxins provides the first line of regulatory mechanism, especially against low molecular weight chemicals in target cells.

A number of enzymes superfamilies, including cytochromes P450 (CYPs) dependent monooxygenases, glutathione S-transferases (GSTs), N-acetyltransferases (NATs), UDP-glucuronosyltransferases (UGTs), sulfotransferases (STs), and NAD(P)H-quinone oxidoreductases (NQOs) are thought to have evolved as adaptive responses to environmental insult. These enzymes play a role in either detoxification or activation depending on the chemicals encountered. Although CYP enzyme, GST and ST generally serve as detoxifying enzymes, they also serve as activation enzymes in the metabolism of polycyclic aromatic hydrocarbons (PAHs), vicinal dihaloalkanes and methylated PAH oxides, respectively, due to formation of electrophilic metabolites.

As a result of recent advances in genome technology, we are able to separate and prepare the polymorphic gene from human materials and evaluate their segregation profile of mutational or variant alleles from wild allele. This enables us to unequivocally establish that the gene structure and expression of these enzymes are genetically polymorphic. In this manner we can therefore predict that any alteration in the activity of one of these enzymes would result in an altered susceptibility to specific chemical-induced diseases, such as cancer, and that there is an association between the presence or absence of a particular mutant or variant allele and susceptibility to cancer.

In the past, the assay of drug metabolizing enzyme activity in an individual was achieved by phenotypic assessment. The determination was performed by administration of a probe drug and measurement of the relative concentrations of unchanged drug and its metabolites in the urine and/or blood samples. Such studies, however, have inherent limitations:[2] 1) they do not allow for drug-drug interactions or 2) the effect of external factors, such as alcohol consumption or cigarette smoking or 3) disease states, such as impaired renal and liver functions or 4) malnutritional states, for example, indicating a marked effect on the rate of *in vivo* drug clearance from the body. Such difficulties accelerate the use of individual genotypic assessments,[3–6] which are carried out easily and quickly in a minimally invasive procedure by use of peripheral leukocyte. The result has allowed a much more straightforward relationship of a particular genotype to disease incidence such as cancer.

In this paper three main drug metabolizing enzymes, CYP1A1, CYP2E1 and GSTM1, are discussed briefly. Other important enzymes, such as CYP1A2, CYP2D6, NAT2, etc., should also be considered to obtain the association with cancer susceptibility.

2. POLYMORPHIC GENE EXPRESSION IN DRUG METABOLISM

Reliable genotyping approaches that can assign the genotype based on detection of germ-line mutation using either restriction fragment length polymorphism (RFLP), single stranded conformational polymorphism or polymerase chain reaction (PCR)-RFLP, are increasingly substituting for traditional phenotyping approaches to evalute a profile of genetic polymorphism. On the other hand, some problems provide in a discrepancy between allelic heterogeneity and phenotypic different expression of the gene, such as mRNA level and enzyme activity. Sequentially these genetic approaches are also hampered by multiple allelic heterogeneity in some cases, and then such a complexities can render genotype interpretation difficult.

In addition, an ethnic variation in the phenotype expression have to be considered more intensively and carefully. All expressions of genetic traits may exhibit their ethnic variation indicating of serious bias results if cases and the controls are drawn from different ethnic groups incidentally. Clear different interethnic distribution in the cases of CYP2D6[7,8] and CYP2C19[9] genes is demonstrated between Orientals and Caucasians, respectively.

Figure 1 depicts a series of polymorphic CYP genes and their representative enzyme reactions, and the related enzymes and also receptor proteins in terms of drug metabolism.

1. Cytochrome P450 enzymes
    1) CYP1A1 Gene: Benzo[a]pyrene Hydroxylation
    2) CYP1A2 Gene: Caffeine 3-Demethylation
    3) CYP2C9 Gene: Phenytoin Hydroxylation
    4) CYP2C19 Gene: Mephenytoin 4-Hydroxylation
    5) CYP2D6 Gene: Debrisoquine 4-Hydroxylation
    6) CYP2E1 Gene: Chlorzoxazone 6-Hydroxylation
    7) CYP3A4 Gene: Diazepam N-Demethylation

2. Other related enzymes
    1) Glutathion S-transferase (GST) M1 and T1
    2) N-Acetyltransferase (NAT) 1and 2
    3) Phenol Sulfotransferase (PST) TS and TL
    4) Endothelial constitutive Nitric Oxide Synthase (ec NOS)
    5) NAD(P)H Dehydrogenase (quinone), DT-Diaphorase (NQO)1
    6) Aldehyde Dehydrogenase (ALDH)2
    7) Epoxide hydrolase (microsomal) (mEH)

3. Receptor proteins
    1) Aryl Hydrocarbon Receptor (AhR)1
    2) Estrogen Receptor (ER)
    3) Androgen Receptor (AR)
    4) Vitamin D Receptor (VDR)

Figure 1. Polymorphic gene expressions in drug metabolism.

It is generally recognized in the substrate specificity of the CYP enzymes that one substrate might be concerned to react to some different CYP enzymes with their different affinity. For example, aflatoxin B_1 which is hepatocarcinogen in human and rodent is metabolized by CYPs 1A2, 2A6 and 3A4, and 4-(methylnitrosamino)-1-(3-pyridyl)-1-butanone (NNK), tobacco specific nitrosamine, is metabolized by CYPs 1A2, 2A6, 2D6 and 2E1. In benzo[a]pyrene (BP) metabolism different pathways expressed to be dependent on different CYP enzymes, that is, the reaction of BP to 3-hydroxy BP is metabolized by CYPs 1A1, 3A4, and 2C9, and the reaction of BP to BP 7,8-dihydrodiol is CYPs 1A1, 3A4, 2C9 and 2C19, and the activation metabolism of BP 7,8-dihydrodiol to BP diolepoxide is CYPs 1A1, 1A2, and 3A4, respectively.[10] Such multiple CYP dependency on the metabolism of chemical carcinogens seems to give us some trouble to interpret the association of genetic polymorphism with cancer susceptibility, even if one CYP enzyme plays a major role in a rate limiting reaction.

3. SUSPECTED PATHWAYS OF CHEMICAL CARCINOGEN METABOLISM IN TERMS OF PULMONARY CARCINOGENESIS

It was recently notified that the incidence of lung cancer and its death number have been increased markedly and that lung cancer is the leading case of cancer deaths in Japan, and parhaps world wide. While lung cancer is the malignancy associated with external chemical exposures such as environmental air pollution and tobacco smoking, and then becomes a model for human chemical carcinogenesis in general, evidence on hereditary factor in high risk group will be also concerned more intensively. Therefore, molecular events that accompany the malignant process seem to be derived from the gene-environment interactions which is a key factor for difference in susceptibility to cancer.

The suspected metabolic pathway of chemical carcinogen in human pulmonary carcinogenesis is shown in Figure 2. BP and the other PAHs, NNK and dinitropyrenes are considered as pulmonary carcinogens, and may metabolized by CYPs 1A1, 1A2, 2A6, 2D6, 2E1 and 3A4, and also NQO1 to form diolepoxides, hydroquinones and reactive oxygen species as each of the active metabolites, and then finally to proceed the formation of carcinogen-DNA adducts or 8-hydroxydeoxyguanosine. On the other hand, the reactive metabolites formed are mostly converted by conjugation enzymes, such as GST, NAT, ST and UGT, to form the respective non-toxic and water-soluble products.

4. GENETIC POLYMORPHISMS IN CYP1A1

PAH such as BP is metabolized mainly by members of the CYP1 subfamily enzymes, CYPs 1A1, 1A2 and 1B1. Variability in metabolic activity with respect to BP concordant with cancer susceptibility in the skin and lung has been well described and documented in the experiments using different strains of mice. These different susceptibility to cancer is mainly derived from PAH inducibility of CYP1A1 dependent oxidative enzyme, aryl hydrocarbon hydroxylase (AHH), which is mediated in the affinity of aryl hydrocarbon receptor(AHR) to PAHs.[11,12]

It was reported previously that the ability to induce the AHH enzyme varies widely in humans, and is divided into three phenotypes of the inducibility consistent with autosomal dominant inheritance,[13] and then a person with high inducibility is demonstrated to be linked

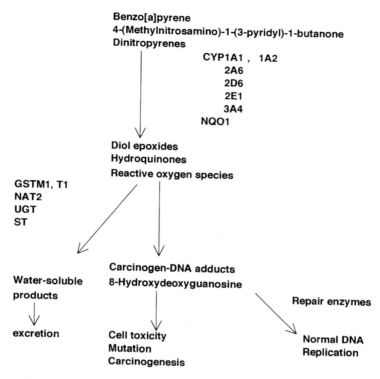

Figure 2. Role of drug metabolizing enzymes in pulmonary carcinogenesis.

with lung cancer susceptibility,[14,15] although not all studies in different laboratories in the world were supportive.[16] Methodological difficulties have been blamed for heterogeneity in some studies, e.g., seasonal variation and poor reproducibility in the assay, effect-cause bias, and poor cell survival *in vitro* in lung cancer patients. In the light of these situations there has been a great interest in developing a genotype marker that accurately reflects the human variation in this traits. An RFLP based on a *Msp1* site at 6235 nucleotide position of the CYP1A1 gene has been studied in lung cancer patients and the controls in 1990.[17] Up to this time there are many reports concerning genetic polymorphisms of human CYP1A1 gene in healthy populations, as shown in Table 1. It is of interest for us to evaluate a clear interethnic difference in CYP1A1 genetic polymorphism, especially between Japanese and Caucasians, although their mutation frequency is determined each at *Msp1* site (m1) and heme binding site (m2) of the gene, and not to be correlated well between the m1 and m2 sites.

The association of an *Msp1* polymorphism with Kreyberg type 1, especially squamous cell type of lung cancer was demonstrated in Japan,[32] and then such association is strongest in the case with low to moderate tobacco exposure. This result suggests that the genetic cofactor acts predominantly within a certain exposure range, and below that range, exposure is insufficient to cause cancer in sufficient number, and also above that range, the genetic factor is less important because the enzyme is saturated and carcinogenic effect is more efficient to act into the cells.

Table 2 depicts the summary of genetic polymorphisms of CYP1A1 gene in lung cancer patients and the controls. To date three studies in Western populations and two studies in Japanese except Saitama area where the first positive report appeared have,

Table 1. Frequency of CYP1A1 mutations in different ethnic populations

Individuals		Mutation frequency (% of alleles)				
Location assayed	Population	m1(6235)[a]	m2(4889)	m3(5639)	m4(4887)	References
Saitama, Japan	375	33.5	n.d.	n.d.[b]	n.d.	Kawajiri[17] 1990
Oslo, Norway	212	11.5	n.d.	n.d.	n.d.	Tafre[18] 1991
Saitama, Japan	358	n.d.	19.8	n.d.	n.d.	Hayashi[19] 1992
Helsinki, Finland	121	11.6	5.4	n.d.	n.d.	Hirvonen[20] 1992
Washington DC, USA	95	10.0	n.d.	n.d.	n.d.	Shields[21] 1993
Berlin, Germany	171	7.3	3.2	n.d.	n.d.	Drakoulis[22] 1994
Honolulu, USA	47	27.0	15.0	n.d.	n.d.	Sivaraman[23] 1994
Los Angeles, USA	120	15.2	n.d.	n.d.	n.d.	London[24] 1995
Sendai, Japan	43	46.5	n.d.	n.d.	n.d.	Ikawa[25] 1995
Yokohama, Japan	185	38.0	25.3	n.d.	n.d.	Kihara[26] 1995
Rio de Janeiro, Brazil	188	24.5	13.0	n.d.	n.d.	Hamada[27] 1995
Fukuoka, Japan	84	31.0	15.4	n.d.	n.d.	Kiyohara[28] 1996
New York, USA	171	12.6	9.7	0.0	n.d.	Garte[29] 1996
Berlin, Germany	880	7.7	2.8	0.0	3.0	Cascorbi[30] 1996
Posman, Poland	324	6.6	2.2	0.0	2.0	Mrozikiewicz[31] 1997

[a] Nucleotide at position of CYP1A1 gene is shown in parentheses, indicating of different mutation sites.
[b] n.d. means to be not determined at that time.

Table 2. Genetic polymorphsims of CYP1A1 in lung cancer

Authors	(Method)		Population	Mutated homozygote (%)	Mutated allele (%)
Nakachi[32]	(B)	Controls	375	10.7	55.8
		Cases			
		Kreyberg I type	91	26.4	65.8
		Kreyberg II type	60	13.3	50.0
		Total	151	21.2	59.6
Ikawa[25]	(A)	Controls	43	18.6	74.4
		Cases	71	14.1	50.7
Kihara[26]	(A)	Controls	185	17.8	57.2
		Cases	97	16.5	62.9
	(B)	Controls	182	6.0	44.5
		Cases	95	5.3	37.9
Hirvonen[20]	(A)	Controls	122	1.6	21.3
		Cases	106	0	21.7
Tefre[18]	(A)	Controls	212	0.9	21.2
		Cases	221	0.9	22.2
Shields[21]	(A)	Controls	56	3.6	23.2
		Cases	48	6.3	31.3
Drakoulis[22]	(A)	Controls	171	0	7.3
		Cases	142	0.7	8.5
	(B)	Controls	171	0	3.2
		Cases	142	1.4	6.7

Method A; RFLP analysis using Msp1 as a restriction enzyme (m1 mutation).
Method B; PCR analysis in Exon 7 indicated a change from isoleucin to valin. (m2 mutation).

Table 3. Genetic polymorphisms of CYP1A1 in mammary and colorectal cancers

	Authors	Race	(Method)		Population	Mutated homozygote (%)	Mutated allele (%)
Mammary cancer	Taioli[36]	African-American	(A)	Controls	85	3.5	40.0
				Cases	21	19.0	66.6
			(B)	Controls	20	0	0
				Cases	83	0	6.0
		Caucasian	(A)	Controls	183	2.7	20.2
				Cases	30	0	26.7
			(B)	Controls	175	1.1	17.1
				Cases	29	0	17.2
	Ambrosone[37]	Caucasian	(B)	Controls	228	0.9	14.5
				Cases	176	2.3	20.5
Colorectal cancer	Sivaraman[23]	Japanese	(A)	Controls	47	3.0	30.0
				Cases	43	22.0	46.6
			(B)	Controls	47	1.9	19.0
				Cases	43	9.0	25.6

Method A; RFLP analysis using Msp1 as a restriction enzyme (m1 mutation).
Method B; PCR analysis in Exon 7 indicating a change from isolencine to valin (m2 mutation).

however, revealed no association between lung cancer risk and the m1 genetic allele frequency of CYP1A1 enzyme. On the other hand, a genetic polymorphism in m2 mutation (A^{4889} to G) of the human CYP1A1 gene creates an amino acid substitution (Ile^{462} to Val) and suggests to cause altered enzymic properties of CYP1A1. To date two papers was evaluated the presence of an association between the m2 variant allele and *in vitro* kinetic properties,[33,34] but one paper could not express any correlation,[35] indicating of appearance of controversial results in the genetic expression.

In the case of mammary and colorectal cancers it is of interest to know whether the association of genetic polymorphisms to the cases is demonstrated or not. The results showed in Table 3 express to be obtained a correlation against mammary cancer in African-American population with only *Msp1* mutation, and also against colorectal cancer in Japanese, but not in Caucasian population. From these studies CYP1A1 gene suggests to play a potential role in PAH etiology of either mammary or colorectal cancer in the individual with a high gene frequency.

Pleiotropic biological effects of some xenobiotic chemicals including PAH and chlorinated dioxins are considered to be mediated by AHR to induce CYPs 1A1, 1A2, GSTA, UGT1 and NQO genes, as mentioned above. Then the gene encoding the AHR is characterized to molecular basis of genetic polymorphisms in the human. Although the polymorphism of AHR gene is evaluated, the relationship to CYP1A1 gene and also to lung cancer susceptibility is uncertain at present.[25,38]

5. GENETIC POLYMORPHISMS IN CYP2E1

CYP2E1 is an ethanol-inducible CYP that metabolizes mainly low molecular weight chemicals such as ethanol, acetone, nitrosamines, urethane and halogenated hydrocarbons, and therefore is of most importance to be evaluated a role in terms of chemical carcinogenesis and environmental hazards.

Table 4. Ethnic difference of allele frequency in CYP2E1 gene

Authors	Method[a]	Wild allele	Mutated allele	Population
Uematsu[40] (Japanese)	A	0.793	0.207	76
Yu[41] (Taiwanese)	A	0.774	0.226	146
Hirvonen[42] (Finn)	A	0.893	0.107	242
Uematsu[43] (Japanese)	B	0.595	0.405	42
Yu[25] (Taiwanese)	B,D[b]	0.854	0.146	150
Persson[44] (Swede)	C	0.946	0.054	130
Uematsu[43] (Japanese)	D	0.838	0.162	40
Hirvonen[42] (Finn)	D	0.988	0.012	242
Watanabe[45] (Japanese)	E	0.809	0.191	503
Persson[44] (Swede)	E	0.946	0.054	130
Kato[46] (Caucasian)	E	0.981	0.019	107
Kato[46] (African-American)	E	0.983	0.017	87
Kato[46] (Japanese)	E	0.735	0.265	49
Kato[47] (Japanese)	E	0.764	0.236	203

[a]Method A, B, C and D are PCR-RFLP analyses using Dra1, Pst1, Taq 1 and Rsa1 as restriction enzymes, respectively. Method E is PCR-RFLP analysis in 5'-flanking region using Rsa-1.
[b]Polymorphism is completely coincided between Pst1 and Rsa1.

Phenotyping studies using the muscle relaxant chlorzoxazone as a probe have been reported,[39] and these have demonstrated two to three-fold variations in the percentage of the drug excreted as the 6-hydroxy metabolites, but no evidence for a bimodal distribution with metabolic polymorphism has been obtained. It has been suggested that chlorzoxazone is not a completely specific probe for CYP2E1 because it is also metabolized by CYPs 1A1 and 1A2 with different affinity to the substrate.

Genetic polymorphism in CYP2E1 was initially reported[40] using Dra1 restriction enzyme for RFLP analysis in leucocyte DNA. Allele frequency in CYP2E1 gene in healthy population depicts in Table 4. CYP2E1 polymorphism occurs in intron 6 of the gene in the case of Dra1 RFLP, and also appears in the 5'-flanking regions of the gene with Rsa1 RFLP which is of considerable interest for the transcription activity of CYP2E1 induction. There is, however, no evidence for positive relationship in phenotype-genotype interaction of CYP2E1 enzyme. There is also a marked interethnic difference in the percentage of mutation or variant alleles between Orientals and Caucasians, and the incidence of the mutated allele is higher in Orientals than in Caucasians, relatively.

A variety of polymorphisms of CYP2E1 gene has been examined in terms of lung cancer risk in human population, as shown in Table 5. High incidence of lung cancer cases was demonstrated in heterozygotes, compared to the controls, with using Dra1, not Rsa1 and Taq1, as restriction enzymes.[40] Furthermore, it is of interest to know that in human autopsy liver, the higher level of mRNA of CYP2E1 in heterozygotes than in wild homozygotes is observed. These results of lung cancer risk could not, however, be confirmed by the other laboratories even if Rsa1 polymorphism in the 5'-flanking region of the gene was assayed in Japanese population.

It was reported precisely that some low molecular weight nitrosamines are confirmable to be carcinogenic to stomach and liver in the experimental animals. Table 6 depicts the association between genetic polymorphism of CYP2E1 and susceptibility to stomach and liver cancer in human. There is good association between genetic polymorphism using Pst1 or Rsa1 but not Dra1 as a restriction enzyme and liver cancer susceptibility espe-

Table 5. Genetic polymorphisms of CYP2E1 in lung cancer

Authors	Method[a]		Mutated homozygotes (%)	Heterozygotes (%)	Wild homozygotes (%)	Population
Uematsu[40] (Sendai, Japan, 1994)	A	Controls	11(15)	22(29)	43(57)	76
Uematsu[43] (Sendai, Japan, 1994)	B	Cases	2(2)	42(46)	47(52)	91
		Controls	6(14)	22(52)	14(33)	42
	C	Cases	9(21)	19(43)	16(36)	44
		Controls	1(3)	11(28)	28(70)	40
Hirvonen[42] (Finland 1993)	A	Cases	3(9)	10(29)	22(63)	35
		Controls	1(1)	24(20)	96(79)	121
Persson[44] (Sweden 1993)	A	Cases	1(2)	14(13)	90(85)	106
		Controls	1(<1)	28(18)	123(81)	152
	C	Cases	0(0)	33(17)	160(83)	193
		Controls	1(<0.1)	28(19)	116(80)	145
Watanabe[45] (Saitama, Japan 1995)	D	Cases	0(0)	34(19)	148(81)	182
		Controls	16(3)	160(32)	327(65)	503
Kato[46] (USA 1995)	B	Cases	13(4)	96(30)	207(66)	316
		Controls[b]	0(0)	2(5)	39(95)	41
Oyama[48] (Kitakyushu, Japan 1997)	D	Cases	0(0)	3(4)	64(96)	67
		Controls	25(4)	196(32)	391(64)	612
		Cases[c]	7(6)	32(25)	87(69)	126

[a]Methods A,B,C and D are PCR-RFLP analyses using Dra1, Pst1, Taq1 and Rsa1 as restriction enzymes, respectively.
[b]"Control" is consisted of COPD (chronic obstructive pulmonary disease) patients.
[c]"Cases" means NSCLC (non-small cell lung cancer) patients.

cially in the case of cigarette smokers in Taiwanese.[41] But no association is observed in *Rsa1* genotype and liver cancer in Japanese. These results suggest that the effect of hepatocarcinogens such as aflatoxin B_1, alcohol and cigarette smoking for the etiologic factors have to be considered in addition to hepatitis B and/or C virus infection during the multistage hepatocarcinogenesis in Taiwanese.

Table 6. Genetic polymorphisms of CYP2E1 allele in gastric and liver cancers

	Method[a]		Wild allele	Mutated allele	Population
Yu[41] (Taiwanese)	A	Controls	0.854	0.146	150
		Liver Cancer	0.828	0.172	29
	B	Controls	0.793	0.207	146
		Liver Cancer	0.900	0.100	30
Kato[47] (Japanese)	C	Controls	0.759	0.241	203
		Gastric Cancer	0.780	0.220	150
		Liver Cancer	0.800	0.200	15

[a]Method A, B and C are PCR-RFLP analyses using Pst1 (Rsa1), Dra1 and Rsa1 as restriction enzymes, respectively.

6. GENETIC POLYMORPHISM IN GSTM1

Glutathione conjugation is an important metabolic pathway for a variety of hydrophobic and electrophilic compounds and is generally detoxicating with further metabolism prior to excretion. There are four different classes of human soluble enzymes termed GSTA, M, P and T.[49] Genetic polymorphisms in GSTM1 and T1 have been demonstrated in human although rare polymorphisms may also occur in the other classes.[50] In general the main substrates for GST are cytotoxic drugs and chemical carcinogens, with possible role for GSTM1 in the metabolism of BP, nitrosourea and nitrogen mustard, and for GSTT1 in the metabolism of small organic toxicants such as dichloromethane and ethylene oxide, and lipid peroxides, which is therefore unlikely to have a significant role in pharmaceutical metabolism.

This reaction has prompted to perform a large number of case-control studies in the world. Associations between deficient GSTM1 activity or gene and lung cancer susceptibility is shown in Table 7. The GSTM1 polymorphism results in an absence of GSTM1 activity in 40 to 55% of individuals from variety of ethnic groups in the world. Most, but not all, association studies show significant roles for the polymorphism in determining the susceptibility to lung cancer. Furthermore, the studies on urinary bladder cancer are more suggestive of a possible role for the GSTM1 polymorphism in the susceptibility, as shown in Table 8. Similar profiles on susceptibility to larynx cancer and pituitary adenoma are also observed, but in the case of breast cancer the women who are less than 58 years old expressed a moderate tendency in the polymorphism to the susceptibility. Furthermore, the association in the susceptibility of GSTM1 null type to colon cancer, especially the proximal cases of colon cancer, is so significant.

Genetic polymorphisms of human GSTT1 are also demonstrated.[71] Approximately 20 percent of the population is shown as GSTT1 null type without any ethnic difference of the distribution. Contrasted to the variable significance in GSTM1 null type, GSTT1 null type does affect a susceptibility to astrocytoma and meningioma,[70] but not lung, urinary bladder and cervical cancers.[57,72,73]

Table 7. Frequency of deficient GSTM1 gene in lung cancer

Method	Authors	Cases No.	Cases %	Controls No.	Controls %
Enzyme activity	Seidegard[51]	191	63.4	192	41.7
	Roots[52]	71	64.8	101	53.5
	Heckert[53]	66	63.0	120	57.0
Gene structure	Zhong[54]	228	43.0	225	42.0
	Kihara[55]				
	Male	108	61.1	140	45.0
	Female	13	84.6	61	45.9
	Total	121	63.6	210	45.3
	Hirvonen[56]	73	52.9	142	43.7
	Brockmoeller[57]	117	53.0	155	52.9
	Hayashi[33]	212	55.7	358	46.6
	Nazar-Stewart[58]	25	64.0	29	48.3
	Heckbert[53]	99	64.6	100	58.0
	Seidegard[59]	66	65.2	78	41.0
	Alexandrie[60]	292	55.1	408	52.2

Table 8. Frequency of deficient GSTM1 gene in the patients bearing extrapulmonary tumors

Histology	Authors	Cases		Controls	
		No.	%	No.	%
Urinary Bladder Cancer	Bell[61]	130	61.0	96	48.2
	Daly[62]	45	84.9	62	56.4
	Lin[63]	52	58.4	236	49.3
	Lafuente[64]	50	66.7	57	45.4
	Brockmoeller[65]	216	58.4	187	50.7
	Zhong[66]	39	40.2	94	41.8
Urothelial Cancer	Katoh[67]	51	61.4	43	42.6
Breast Cancer	Zhong[66]	94	47.7	94	41.8
	Ambrosone[68]	93	52.5	117	50.2
(less them 58 years old		25	64.1	24	41.4)
Larynx Cancer	Lafuente[64]	52	66.7	35	44.9
Colon Cancer	Zhong[66]	110	56.1	94	41.8
(proximal site		51	70.8	94	41.8)
(Distal site		56	54.4	94	41.8)
Pituitary adenoma	Fryer[69]	113	57.9	39	43.8
Astrocytoma	Elexpuru-Camiruaga[70]	65	59.6	315	54.6
Meningioma		27	55.1	315	54.6

7. COMBINATION EFFECT OF CYP1A1 AND GSTM1 GENES ON GENETIC SUSCEPTIBILITY TO LUNG CANCER

Cancer susceptibility following chemical exposure is likely to be determined by an individual phenotype or genotype for a number of enzymes, both activating and inactivating, relevant to the chemicals or mixtures of the chemicals, and the amount of variability in the expression of carcinogen-metabolizing enzymes and the complexity of chemical carcinogen mixtures to which individual may be exposed, have to be concerned more extensively and precisely. The assessment of a single phenotype or genotype may not be sufficient to detect their reliability to susceptibility to chemical carcinogens as an individual predisposition to cancer, because multiple exposure of chemical carcinogens in the significance of multistage chemical carcinogenesis is widely proposed in human.

Recently a clear combination effect of the risk genotypes on cancer susceptibility is evaluated. Marked increases of inducibility of CYP1A1 gene transcription by 2, 3, 7,8-tetrachlorodibenzo-p-dioxin and also of PAH-DNA adduct levels were associated with GSTM1 null genotype in human.[77,78] Hayashi et al.[19] and Kihara et al.[26] describe a marked increase of the relative risk for squamons cell carcinoma, but not so significant for adenocarcinoma in the lung from Japanese individuals who carried simultaneously both the CYP1A1 Valine and GSTM1 null alleles. These findings are consistent with the hypothesis that some environmental procarcinogens are activated by CYP1A1 and inactivated by GSTM1 enzymes probably in the target lung tissue. In the case of Caucasian population

Table 9. Effect of smoking exposure in genetic susceptibility of drug metabolizing enzymes to lung cancer

Gene/enzyme	Degree of exposure	Relative risk	Reference
CYP1A1 mutated allele	Lower	Increase (squamous cell carcinoma)	Nakachi[32]
CYP1A1 mutated allele	Lower	Increase (adenocarcinoma)	Nakachi[74]
CYP2E1 heterozygote	Lower	Increase	Uematsu[40]
CYP2E1 mutated allele with CYP2D6 extensive metabolizer	Lower	(Increase in formation of 7-methyl-deoxy guanosine adduct)	Kato[75]
GST M1 null type	Lower	Increase	London[76]
CYP1A1 mutated allele with GST M1 null type	Higher	Increase (squamous and small cell carcinoma)	Kihara[26]

such a combination effect is not reported in the lung cancer risk, but is observed in higher risk to oral[79] and postmenopausal breast cancer[68] when CYP1A1 (isoleucine: valine) heterozygotes with GSTM1 null genotype are assayed in the patients.

When the combination assay of CYP1A1 induclibility in resected lung tissue and GSTM1 gene polymorphism in peripheral leucocyte are performed in the patient with lung cancer,[80] the patients detected of non-inducible CYP1A1 are associated solely with bronchial, mainly squamous, cell carcinoma. The patients carried of inducible CYP1A1 with functional GSTM1 gene have peripheral lung tumor mainly adenocarcinoma, and on the contrary, with GSTM1 null gene, equal numbers of peripheral and bronchial tumors, suspecting of modification by functional GSTM1 enzyme in the histological type of tumor and in the location of tumor development.

8. INTERACTION BETWEEN GENETIC POLYMORPHISM AND SMOKING EXPOSURE

In the viewpoint of molecular epidemiology gene-environment interaction is of the most important to evaluate their effectiveness to susceptibility to intoxication and disease states under the respective environmental conditions in the world. These association, however, are among the most difficult to demonstrate because of limitations in statistical power.

An example of an intriqueing gene-environment intraction, such as genetic polymorphism and cigarette smoking exposure, are summarized in Table 9. Interestingly, gene-environment interactions are more pronounced at lower levels of cigarette exposure in which the susceptibility to lung cancer increased in the case of patients with either CYP1A1 mutated alleles,[32,74] CYP2E1 heterozygotes[40] or GSTM1 null type.[76] Persons with both CYP2E1 mutated allele and CYP2D6 extensive metabolizer *in vivo* showed an increased level of 7-methyl-deoxyguanosine adducts in the lung with lower exposure of cigarette. These results indicate that in persons who smoke less, inherited susceptibility might be important, but in havier smoker such genetic responsibility does not lies. On the contrary, it was reported that higher exposure of cigarette smoke in the patients who are carried with both CYP1A1 mutated allele and GSTM1 null gene, proceeds to express a significant increase in the lung cancer, especially squamons cell carcinoma, but not any difference in the oral[79] and postmenopausal breast[68] cancers, indicating of controversial statistical results of smoking habits.

9. CONCLUSIONS

Human cancerization seems to be derived mostly from the environmental factors, such as the exposure via either respiratory, digestive or percutaneous routes with chemical carcinogens. On the other hand, the profile of genetic polymorphism in carcinogen metabolizing enzymes is recently estimated and seems to be essential factors for determination as a predisposition against external chemical carcinogens.

In this paper only three enzymes, such as CYP1A1, CYP2E1 and GSTM1 were concerned precisely to evaluate the association between genetic polymorphisms and susceptibility to cancer in lung, urinary bladder and the other tissues, because under the epidemiological findings lung and urinary bladder cancers may speculated to be affected significantly by environmental factor, such as smoking habit. We, however, have to consider more precisely and extensively to estimate the roles of the other carcinogen metabolizing enzymes, such as CYP1A2, CYP2C9 and 19, CYP2D6, CYP3A4, NAT, ST and UGT, for genetic susceptibility to cancer, because these enzymes are shown clearly to consist of genetic polymorphisms, as shown in Figure 1.

As a consequence we need to try to disclose a new profile of genetic polymorphism of carcinogen metabolizing enzymes and also DNA repair enzymes, and to estimate the association profile between genetic variety and the biological effects, such as the production of reactive oxygen species[83] and of carcinogen-DNA adducts,[84] and then to collaborate each other to evaluate each precise data from different laboratory in the world.

ACKNOWLEDGMENT

The author thanks Drs. Fumiyuki Uematsu, Shuntaro Ikawa, Hideaki Kikuchi, Ryunosuke Kanamaru, Masakichi Motomiya, Shigefumi Fujimura and Toshihiro Nukiwa at Tohoku University for collaborative works, and Mrs. Ikuko Imamura and Miss Noriko Yamagata for secretarial assistance. We also acknowledge support for this work from the Smoking Research Foundation, Japan.

REFERENCES

1. F.J. Gonzalez, 1993, Cytochrome P450 in humans, in : *Hundbook of Experimental Pharmacology* vol. **105** (J.B. Schenkman, and H. Greim, eds.) pp. 239–257, Springer-Verlag, Berlin.
2. N.E. Caporaso, 1995, The genetics of lung cancer, in : *The Genetics of Cancer* (B.A.J. Ponder, and M.J. Waring, eds.), pp. 21–43, Kluwer Academic Pub., Dordrecht.
3. H. Raunio, K. Husgafvel-Pursianen, S. Anttila, E. Hietanen, A. Hirvonen, and D. Pelkonen, 1995, Diagnosis of polymorphisms in carcinogen activating and inactivating enzymes and cancer susceptibility - a review, *Gene,* **159** : 113–121.
4. G. Smith, L.A. Stanley, E. Sim, R. C. Strange, and C.R. Wolf, 1995, Metabolic polymorphisms and cancer susceptibility, in : *Cancer Surveys*, vol.**25**, Genetics and Cancer, A Second Look (B.A.J. Ponder, W.K. Cavenee, and E. Solomon, eds.) pp. 27–65, Cold Spring Harbor Lab. Press, New York.)
5. P.G. Shields, 1996, Molecular epidemiology, in : *Genetics and Cancer Susceptibility : Implications for Risk Assessment*, (C. Walker, J. Groopman, T. J. Slaga, and A. Klein-Szanto, eds.) pp. 141–157, Wiley-Liss, Inc., New York.
6. D.W.Nebert, 1997, Polymorphisms in drug-metabolizing enzymes: What is their clinical relevance and why do they exist ?, *Am. J. Hum. Genet.* **60**: 265–271.
7. U.A. Meyer, R.C. Skoda, U.M.Zanger, M. Heim, and F. Broly, 1992, The genetic polymorphism of debrisoquine sparteine metabolism-Molecular mechanism, in : *Pharmacogenetics of Drug Metabolism*, (W. Kalow, ed.) pp. 609–623. Pergamon Press, New York.

8. Y. C. Lou, L. Ying, L. bertilsson, and F. Sjoqvist, 1987, Low frequency of slow debrisoquine hydroxylation in a native Chinese population, *Lancet* ii 852–853.

9. G.R. Wilkinson, F. P. Guengerich, and R. A. Branch, 1987, Genetic polymorphism of S-mephenytoin hydroxylation, *Pharmacol. Therapeut.* **43**: 53–76.

10. J.-C. Gautier, S. Lecoeur, J. Cosme, A. Perret, P. Urban, P. Beaune, and D. Pompon, 1996, Contribution of human cytochrome P450 to benzo[a]pyrene and benzo[a]pyrene-7,8-dihydrodiol metabolism, as predicted from heterologous expression in yeast, *Pharmacogenetics* **6**: 489–499.

11. D. W. Nebert, 1989, The Ah locus: Genetic differences in toxicity, cancer mutation and birth defects. *Crit. Res. Toxicol.* **20**: 153–174.

12. M. Ema, N. Obe, M. Suzuki, J. Mimura, K. Sogawa, S. Ikawa, and Y. Fujii-Kuriyama, 1994, Dioxin-binding activities of polymorphic forms of mouse and human Ah receptors, *J. Biol. Chem.* **269**: 27337–27343.

13. G. Kellerman, M. Luyte-Kellerman, and C.R. Shaw, 1973, Genetic variation of aryl hydrocarbon hydroxylase in human lymphocytes, *Am. J. Hum. Genet.* **25**: 327–331.

14. G. Kellerman, C.R. Shaw, and M. Luyten-Kellerman, 1973, Aryl hydrocarbon hydroxylase inducibility and bronchogenic carcinoma, *New Engl. J. Med.* **289**: 934–937.

15. R. E. Kouri, C.E. McKinney, D.J. Slomiany, D. R. Snodgrass, N.P. Wray, and T. L. McLemore, 1982, Positive correlation between high aryl hydrocarbon hydroxylase activity and primary lung cancer as analyzed in cryopreserved lymphocytes, *Cancer Res.* **42**: 5030–5037.

16. B. Paigen, H.L. Gurtoo, J. Minowada, 1997, Questionable relation of aryl hydrocarbon hydroxylase to lung cancer risk, *New Engl. J. Med.* **297**: 346–350.

17. K.Kawajiri, K. Nakachi, K. Iwai, A. Yoshii, N. Shinoda, and J. Watanabe, 1990, Identification of genetically high risk individuals to lung cancer by DNA polymorphisms of the cytochrome P-4501A1 gene, *FEBS Lett.* **263**: 131–133.

18. T.Tefre, D. Ryberg, A. Haugen, D. W. Nebert, V. Skaug, A. Brogger, and A. L. Borresen, 1991, Human CYP1A1 (cytochrome P(1) 450) gene: lack of association between the *Msp1* restriction fragment length polymorphism and incidence of lung cancer in a Norwegian population, *Pharmacogenetics* **1**: 20–25.

19. S.-I. Hayashi, J. Watanabe, and K. Kawajiri, 1992, High susceptibility to lung cancer analyzed in terms of combined genotypes of P450 1A1 and Mu-class glutathione S-transferase genes, *Jpn. J. Cancer Res.* **83**: 866–870.

20. A. Hirvonen, K. Husgafrel-Pursiainen, A. Karjalainen, S. Anttila, and H. Vainio, 1992, Point-mutational *Msp1* and Ile-Val polymorphisms closely linked in the CYP1A1 gene: lack of association with susceptibility to lung cancer in a Finnish study population, *Cancer Epidemiol. Biomarkers, Prev.* **1**: 485–489.

21. P.G. Shields, N.E. Caporaso, R.T. Falk, H. Sugimura, G.E. Trivers, B.F. Trump, R. N. Hoover, A. Weston, and C. C. Harris, 1993, Lung cancer, race, and a CYP1A1 genetic polymorphism, *Cancer Epidemiol, Biomarkers*, Prev. **2**: 481–485.

22. N. Drakoulis, I. Cascorbi, J. Brockmoeller, C.R. Gross, and I. Roots, 1994, Polymorphisms in the human *CYP1A1* gene as susceptibility factors for lung cancer : exon-7 mutation (4889 A to G), and a T to C mutation in the 3'-flanking region. *Clin. Investig.* **72**: 240–248.

23. L. Sivaraman, M.P. Leatham, J. Lee, L.R. Wilkens, A. F. Lau, and L. L. Marchand, 1994, *CYP1A1* genetic polymorphisms and *in situ* colorectal cancer, *Cancer Res.* **54**: 3692–3695.

24. S. J. London, A.K. Waly, K. S. Fairbrother, C. Holmes, C. L. Carpenter, W. C. Navidi, and J. R. Idle, 1995, Lung cancer risk in African-Americans in relation to race-specific *CYP1A1* polymorphism, *Cancer Res.* **55**: 6035–6037.

25. S. Ikawa, F. Uematsu, K. Watanabe, T. Kimpara, M. Osada, A. Hossain, I. Sagami, H. Kikuchi, and M. Watanabe, 1995, Assessment of cancer susceptibility in humans by use of genetic polymorphisms in carcinogen metabolism, *Pharmacogenetics* **5**: 5154–5160.

26. M. Kihara, M.Kihara, and K. Noda, 1995, Risk of smoking for squamous and small cell carcinomas of the lung modulated by combination of CYP1A1 and GSTM1 gene polymorphisms in a Japanese population, *Carcinogenesis* **16**: 2331–2336.

27. G. S. Hamada, H. Sugimura, I. Suzuki, K. Nagura, E. Kiyokawa, T. Iwase, M. Tanaka, T. Takahashi, S. Watanabe, I. Kino, and S. Tsugane, 1995, The heme-binding region polymorphism of cytochrome P450 1A1 (CYP1A1), rather than the *Rsa1* polymorphism of 11E1 (CYP11E1), is associated with lung cancer in Rio de Janeiro, *Cancer Epidemiol, Biomarkers Prev.* **4**: 63–67.

28. C. Kiyohara, T. Hirohata, and S. Inutsuka, 1996. The relationship between aryl hydrocarbon hydroxylase and polymorphisms of the CYP1A1 gene, *Jpn. J. Cancer Res.* **87**: 18–24.

29. S. J. Garte, J. Trachman, F. Crofts, P. Toniolo, J. Buxbaum, S. Bayo, and E. Taioli, 1996, Distribution of composite CYP1A1 genotypes in Africans, African-Americans and Caucasians, *Hum. Hered*, **146**: 121–127.

30. I. Cascorbi, J. Brockmoeller, and J. Roots, 1996, A C4887A polymorphism in exon 7 of human CYP1A1: population frequency, mutation linkages and impact on lung cancer susceptibility, 1996, *Cancer Res.* **56**: 4965–4969.

31. P.M. Mrozikiewicz, O. Landt, I. Cascorbi, and I. Roots, 1997, Peptide nucleic acid-mediated polymerase chain reaction clamping allows allelic allocation of CYP1A1 mutations, *Anal. Biochem.* **250**: 256–257.

32. K. Nakachi, K. Iwai, S. Hayashi, J. Watanabe, and K. Kawajiri, 1991, Genetic susceptibility to squamous cell carcinoma of the lung in relation to cigarette smoking dose, *Cancer Res.* **51**: 5177–5180.

33. S. Hayashi, J. Watanabe, and K. Kawajiri, 1992, High susceptibility to lung cancer analyzed in terms of combined genotypes of P450 1A1 and mu-class glutathione S-transferase gene, *Jpn. J. Cancer Res.* **83**: 866–870.

34. D. D. Petersen, C. E. McKinney, K. Ikeya, H. H. Smith, A. E. Bale, O. E. McBride, and D. W. Nebert, 1991, Cosegregation of the enzyme inducibility phenotype and an RFLP, *Am. J. Hum. Genet.* **48**: 720–725.

35. I. Persson, I. Johansson, and M. Ingeleman-Sundberg, 1997, *In vitro* kinetics of two human CYP1A1 variant enzymes suggested to be associated with interindividual differences in cancer susceptibility, *Biochem. Biophys. Res. Comm.* **231**: 227–230.

36. E. Taioli, J. Trachman, X. Chan, P. Toniolo, and S. J. Garte, 1995, A CYP1A1 restriction fragment length polymorphism is associated with breast cancer in African-American women, *Cancer Res.* **55**: 3757–3758.

37. C. R. Ambrosone, J. L. Freudenheim, S. Graham, J. R. Marshall, J. E. Vena, J. R. Brasure, R. Laughlin, T. Nemoto, A. M. Michalek, and A. Harrington, 1995, Cytochrome P450 1A1 and glutathione S-transferase (M1) genetic polymorphisms and postmenopausal breast cancer risk, *Cancer Res.* **55**: 3483–3485.

38. K. Kawajiri, J. Watanabe, H. Eguchi, C. Kiyohara, and S. Hayashi, 1995, Polymorphisms of human Ah receptor gene are not involved in lung cancer, *Pharmacogenetics* **5**; 151–158.

39. E. S. Vesell, T. Deangelo Seaton, and Y. I. A-Rahim, 1995, Studies on inter-individual variations of CYP2E1 using chlorzoxazone as an in vivo probe, *Pharmacogenetics* **5**: 53–57.

40. F. Uematsu, S. Ikawa, H. Kikuchi, R. Kanamaru, T. Abe, K. Satoh, M. Motomiya, and M. Watanabe, 1994. Restriction fragment length polymorphism of the human CYP2E1 (cytochrome P450 IIE1) gene and susceptibility to lung cancer: Possible relevance to low smoking exposure. *Pharmacogenetics* **4**: 58–63.

41. M.-W. Yu, A. Gladek-Yarborough, S. Chiamprasert, R. M. Santella, Y-F. Liau, and C-J. Chen, 1995, Cytochrome P450 2E1 and glutathione S-transferase M1 polymorphisms and susceptibility to hepatocellular carcinoma, *Gastroenterol.* **109**: 1266–1273.

42. A. Hirvonen, K. Husgafvel-Pursiainen, S. Anttila, A. Karjalainen, and H. Vainio, 1993, The human CYP2E1 gene and lung cancer: *Dra*I and *Rsa*I restriction fragment length polymorphisms in a Finnish study population. *Carcinogenesis* **14**: 85–88.

43. F. Uematsu, H. Kikuchi, M. Motomiya, T. Abe, I. Sagami, T. Ohmachi, A. Wakui, R. Kanamaru, and M. Watanabe, 1991, Association between restriction fragment length polymorphisms of the human cytochrome P450 IIE1 gene and susceptibility of lung Cancer, *Jpn. J. Cancer Res.* **82**: 254–256.

44. I. Persson, I. Johansson, H. Bergling, M. L. Dahl, J. Seidegard, R. Rylander, A. Rannug, J. Hoegberg, and M. I. Sundberg, 1993. Genetic polymorphisms of cytochrome P450 2E1 in a Swedish population: Relationship to incidence of lung cancer, *FEBS Lett.* **319**: 207–211.

45. J. Watanabe, J. P. Yang, H. Eguchi, S. Hayashi, K. Imai, K. Nakachi, and K. Kawajiri, 1995. An Rsa polymorphism in the CYP2E1 genes does not affect lung cancer risk in a Japanese population, *Jpn. J. Cancer Res.* **86**: 245–248.

46. S. Kato, P.G. Shields, N.E. Caporaso, R. N. Hoover, B. F. Trump, H. Sugimura, A. Weston, and C. C. Harris, 1992, Cytochrome P450 IIE1 genetic polymorphisms, racial variation and lung cancer risk. *Cancer Res.* **52**: 6712–6715.

47. S. Kato, M. Onda, N. Matsukura, A.Tokunaga, T. Tajiri, D. Y. Kim, H. Tsuruta, N. Matsuda, K. Yamashita, and P.G. Shields, 1995, Cytochrome P450 2E1 (CYP2E1) genetic polymorphism in a cancer-control study of gastric cancer and liver disease, *Pharmacogenetics* **5**: S141-S144.

48. T. Oyama, T. Kawamoto, T. Mizoue, K. Sugio, Y. Kodama, T. Mitsudomi, and K. Yasumoto, 1997. Cytochrome P4502E1 polymorphism as a risk factor for lung cancer: In relation to p53 gene mutation, *Anticancer Res.* **17**: 583–588.

49. B. Ketterer, J. M. Harris, G. Talaska, D. J. Meyer, S. E. Pemble, J. B. Jaylor, N. P. Lang, and F. F. Kadlubar, 1992. The glutathone S- transferase supergene family, its polymorphism and its effects on susceptibility to lung cancer, *Environ. Health Perspect.* **98**: 87–94.

50. P. Board, M. Coggan, P. Johnston, V. Ross, T. Suzuki, and G. Webb, 1990, Genetic heterogeneity of the human glutathione transferases: A complex of gene families, *Pharmacol. Ther.* **48**: 357–369.

51. J. Seidegard, R. W. Pero, D. G. Miller, and E. J. Beattie, 1986, Glutathione transferase in human leukocytes as a marker for the susceptibility to lung cancer, *Carcinogenesis* **7**: 751–753.

52. I. Roots, N. Drakoulis, and J. Brockmoeller, 1992, Polymorphic enzymes and cancer risk: Concepts, methodology and data review, in: *Pharmacogenetics of Drug Metabolism* (W. Kalow, ed.), pp. 815–841, Pergamon Press, New York.

53. S. R. Heckbert, N.S. Wiess, S. K. Hornung, D. L. Eaton, A.G. Motulsky, 1992, Glutathione S-transferase and epoxide hydrolase activity in human leukocytes in relation to risk of lung cancer and other smoking-related cancers, *J. Natl. Cancer Inst.* **84**: 414–422.

54. S. Zhong, A.F. Howie, B. Ketterer, J.Taylor, J.O. Hayes, G. J. Beckett, C. G. Wathen, C. R. Wolf, and N.K. Spurr, 1991, Glutathione S-transferase *mu* locus: Use of genotyping and phenotyping assays to assess association with lung cancer susceptibility, *Carcinogenesis* **12**: 1533–1537.

55. M. Kihara, M. Kihara, and K. Noda, 1994, Lung cancer risk of GSTM1 null genotype is dependent on the extent of tobacco smoke exposure, *Carcinogenesis* **15**: 415–418.

56. A. Hirvonen, K. Husgafvel-Pursiainen, S. Anttila, and H. Vainio, 1993, The GSTM1 null genotype as a potential risk modifier for squamous cell carcinoma of the lung, *Carcinogenesis* **14**: 1479–1481.

57. J. Brockmoeller, R. Kerb, N. Drakoulis, M. Nitz, and I. Roots, 1993, Genotype and phenotype of glutathione S-transferase class μ isozyme and φ in lung cancer patients and controls, *Cancer Res.* **53**: 1004–1011.

58. V. Nazar-Stewart, A. G. Motulsky, D. L. Eaton, 1993, The glutathione S-transferase μ polymorphism as a marker for susceptibility to lung cancer, *Cancer Res.* **53**: 2313–2318.

59. J. Seidegard, R. W. Pero, M.M. Markowitz, G. Roush, D. G. Miller, and E.J. Beattie, 1990, Isozyme(s) of glutathione transferase (class mμ) as a marker for the susceptibility to lung cancer: A follow up study, *Carcinogenesis* **11**: 33–36.

60. A.-K. Alexandrie, M.I. Sundberg, J. Seidegard, G. Tornling, and A. Rannug, 1994, Genetic susceptibility to lung cancer with special emphasis on CYP1A1 and GSTM1: A study on host factors in relation to age at onset, gender and histological cancer types, *Carcinogenesis* **15**: 1785–1790.

61. D.A. Bell, J.A. Taylor, D. F. Paulson, C. N. Robertson, J. L. Mohler, and G. W. Lucier, 1993, Genetic risk and carcinogen exposure: A common inherited defect of the carcinogen-metabolism gene glutathione S-transferase M1 (GSTM1) that increases susceptibility to bladder cancer, *J. Natl. Cancer Inst.***85**: 1159–1164.

62. A.K. Daly, D. J. Thomas, J. Cooper, W. R. Pearson, D. E. Neal, and J. R. Idle, 1993, Homozygous deletion of gene for glutathione S-transferase M1 in bladder cancer, *Brit. Med. J.* **307**: 481–482.

63. H.J. Lin, N.N. Probst-Hensch, S. A. Ingles, C-Y. Han, B.K.Lin, D. B. Lee, H.D. Frankl, E. R. Lee, M.P. Longnecker, and R. W. Haile, 1995, Glutathion S-transferase (GSTM1) null genotype, smoking and prevalence of colorectal adenomas, *Cancer Res.* **55**: 1224–1226.

64. A. Lafuente, F. Pujol, P. Carretero, J. P. Villa, and A. Cuchi, 1993, Human glutathione S- fransferase μ (GSTμ) deficiency as a marker for the susceptibility to bladder and laryn cancers among smoker, *Cancer Lett.* **68**: 49–54.

65. J. Brockmoeller, R. Kerb, N. Drakoulis, B. Staffeldt, and I. Roots, 1994, Glutathione S-transferase M1 and its variants A and B as host factors of bladder cancer susceptibility: A case- control study, *Cancer Res.* **54**: 4103–4111.

66. S. Zhong, A. H. Wyllie, D. Barnes, C. R. Wolf, and N.K. Spurr, 1993, Relationship between the GSTM1 genetic polymorphism and susceptibility to bladder, breast and colon cancer, *Carcinogenesis* **14**: 1821–1824.

67. T. Katoh, H. Inatomi, A. Nagaoka, and A. Sugita, 1995, Cytochrome P4501A1 gene polymorphism and homozygous deletion of the glutathione S-transferase M1 gene in urothelial cancer patients, *Carcinogenesis* **16**: 655–657.

68. C.B. Ambrosone, J. L. Freudenheim, S. Graham, J. R. Marshall, J.E. Vena, J. R. Brasure, R. Laughlin, T. Nemoto, A. M. Michalek, A. Harrington, T. D. Ford, and P. G. Shields, 1995, Cytochrome P4501A1 and glutathione S-transferase (M1) genetic polymorphisms and postmenopausal breast cancer risk, *Cancer Res.* **55**: 3483–3485.

69. A. A. Fryer, L. Zhao, J. Alldersea, M. D. Boggild, C. W. Perrett, R. N. Clayton, P.W. Jones, and R. C. Strange, 1993, The glutathione S-transferases: Polymerase chain reaction studies on the frequency of the GSTM10 genotype in patients with pituitary adenomas, *Carcinogenesis* **14**: 563–566.

70. J. Elexpuru-Camiruaga, N. Buxton, V. Kandula, P. S. Dias, D. Campbell, J. McIntosh, J. Broome, P. Jones, A. Inskip, J. Alldersea, A. A. Fryer, and R. C. Strange, 1995, Susceptibility to astrocytoma and meningioma: Influence of allelism at glutathione S-transferase (GSTT1 and GSTM1) and cytochrome P450 (CYP2D6) Loci, *Cancer Res.* **55**: 4237–4239.

71. S. Pemble, K.R. Schroeder, S. R. Spencer, D. J. Meyer, E. Hallier, H. M. Bolt, B. Ketterer, and J. B. Taylor, 1994, Human glutathione S-transferase Theta (GSTT1): cDNA cloning and the characterization of a genetic polymorphism, *Biochem. J.* **300**: 271–276.

72. M. Warholm, A. K. Alexandrie, J. Hogberg, K. Sigvaedsson, A. Rannug, 1995, Genotypic and phenotypic determination of polymorphic glutathione transferase T1 in a Swedish population, *Pharmacogenetics* **5**: 252–254.

73. A. Warwick, P. Sarhanis, C. Redman, S. Pemble, J. B. Taylor, B. Ketterer, P. Jones, J. Alldersea, J. Gilford, L. Yengi, A. Fryer, and R. C. Strange, 1994, Theta class glutathione S-transferase GSTT1 genotypes and susceptibility to cervical neoplasia: Interactions with GSTM1, CYP2D6 and smoking, *Carcinogenesis* **15**: 2841–2845.

74. V. Nakachi, S.-I. Hayashi, K. Kawajiri, and K. Iwai, 1995, Association of cigarette smoking and CYP1A1 polymorphism with adenocarcinoma of the lung by grades of differentiation, *Carcinogenesis* **16**: 2209–2213.

75. S. Kato, E. D. Bowman, A. M. Harrington, B. Blomeke, and P.G. Shields, 1995, Human lung carcinogen-DNA adduct levels mediated by genetic polymorphisms *in vivo*, *J. Natl. Cancer Inst.* **87**: 902–907.

76. S. J. London, A.K. Daly, J. Cooper, W. C. Navidi, C. L. Carpenter, and J. R. Idle, 1995, Polymorphism of glutathione S-transferase M1 and lung cancer risk among African-Americans and Caucasians in Los Angeles County, California, *J. Natl. Cancer Inst.* **87**: 1246–1252.

77. C. Vaury, R. Laine, P. Noguiez, P. de Coppet, C. Jaulin, F. Praz, D. Pompon, and M. Amor-Gueret, 1995, Human glutathione S-transferase M1 null genotype is associated with a high inducibility of cytochrome P450 1A1 gene transcription, *Cancer Res.* **55**: 5520–5523.

78. N. Rothman, P. G. Shields, M. C. Poirier, A. M. Harrington, D. P. Ford, and P. T. Strickland, 1995, The impact of glutathione S- transferase M1 and cytochrome P450 1A1 genotypes on white-blood-cell polycyclic aromatic hydrocarbon-DNA adduct levels in humans, *Mol. Carcinogenesis* **14**: 63–68.

79. J. Y. Park, J. E. Muscat, Q. Ren, S. P. Schantz, R. D. Harwick, J.C. Ste, V. Pike, J. P. Richie, Jr., and P. Lazarus, 1997, CYP1A1 and GSTM1 polymorphism and oral cancer risk, *Cancer Epidemiol. Biomark. Prev.* **6**: 791–797.

80. S. Anttila, A. Hirvonen, K. Husgafvel-Pursiainen, A. Karjakainen, T. Nurminen, and M. Vainio, 1994, Combined effect of CYP1A1 inducibility and GSTM1 polymorphism on histological type of lung cancer, *Carcinogenesis* **15**: 1133–1135.

81. K. Kawajiri, H. Eguchi, K. Nakachi, T. Sekiya, and M. Yamamoto, 1996, Association of CYP1A1 germ line polymorphisms with mutation of the p53 gene in lung cancer, *Cancer Res.* **56**: 72–76.

82. T. Oyama, T. Kawamoto, T. Mizoue, K. Sugio, Y. Kodama, T. Mitsudomi, and K. Yasumoto, 1997, Cytochrome P450 2E1 polymorphism as a risk factor for lung cancer: In relation to p53 gene mutation, *Anticancer Res.* **17**: 583–588.

83. T. Castillo, D. R. Koop, S. Kamimura, G. Triadafilopoulos, and H. Tsukamoto, 1992, Role of cytochrome P450 2E1 in ethanol-, carbon tetrachloride and iron-dependent microsomal lipid peroxidation. *Hepatol.* **16**: 992–996.

84. A. F. Budawi, S. J. Stern, N.P. Lang, and F. K. Kadlubar, 1996, Cytochrome P-450 and acetyltransferase expression as biomarkers of carcinogen-DNA adduct levels and human cancer susceptibility, in: *Genetics and Cancer Susceptibility: Implications for Risk as Assessment* (C. Walker, J. Groopman, T. J. Slaga, and A. Klein-Szanto), pp. 109–140, Wiley-Liss, New York.

10

CYTOCHROME P450 ISOFORMS

Insecticide Metabolism in Insects and Mammals and Role in Insecticide Resistance

Ernest Hodgson,[1] R. Michael Roe,[1] Joyce E. Goldstein,[2] Siming Liu,[1] Scott C. Coleman,[1] and Randy L. Rose[1]

[1]Department of Toxicology
North Carolina State University
Raleigh, North Carolina 27695-7633
[2]National Institute of Environmental Health Sciences
Box 12233
Research Triangle Park, North Carolina 27709

1. INTRODUCTION

1.1. Historical Aspects

Pesticides, like other xenobiotics, are metabolized in both target and non-target species by a number of different enzymes, including several isoforms of cytochrome P450 (P450) and the flavin-containing monooxygenase (FMO).[1–3] While P450 and FMO can both as activation as well as detoxication enzymes for xenobiotics, P450 is the most important activating enzyme, producing electrophilic metabolites that react with endogenous cellular nucleophiles, including substituents on both proteins and nucleic acids.

P450 isoforms carry out a wide variety of of monooxygenations including epoxidation, hydroxylation, N-, O-, and S-dealkylation, N-, S-, and P-oxidation, oxidative desulfuration and dehalogenation (Table 1). Pesticide substrates have been identified for all of these reaction types.[1,2,4] Insect P450s, in addition to their role in the activation and detoxication of pesticides, have also proven to be one of the most important factors in the adaptation of insects to secondary plant compounds in their diets and in the remarkable ability of insects to develop resistance to synthetic organic insecticides.

The techniques of molecular biology have been applied extensively to P450, particularly in mammals but also, increasingly, in insects. Over 500 P450 genes have been characterized and the nucleotide and derived amino acid sequences compared. In a number of

Molecular and Applied Aspects of Oxidative Drug Metabolizing Enzymes,
edited by Arinç *et al.* Kluwer Academic / Plenum Publishers, New York, 1999.

Table 1. Some examples of reactions catalyzed by P450 isoforms

Reaction	Substrate(s)
Epoxidation	Aldrin, benzo(a)pyrene, aflatoxin
N-Dealkylation	Atrazine, dimethylaniline, ethylmorphine
O-Dealkylation	Chlorfenvinphos, p-nitroanisole
S-Dealkylation	Methyl mercaptan
N-Oxidation	2-Acetylaminofluorene
S-Oxidation	Phorate, methiocarb, chlorpromazine
P-Oxidation	Diethylphenylphosphine
Desulfuration	Parathion, carbon disulfide, fonofos
Dehalogenation	Carbon tetrachloride, chloroform
Benzodioxole ring opening	Piperonyl butoxide, isosafrole

cases the chromosome location of the gene has been determined and in others the mechanism of gene expression has been investigated. A system of nomenclature based on derived amino acid sequences was proposed in 1987 and has been revised at intervals since, the most recent revision in 1996.[5]

2. INSECT CYTOCHROME P450

2.1. Introduction

The insect P450-dependent monooxygenase system is involved in many different physiological processes, including growth and development, interaction of plant feeding insects with their host plants and resistance to insecticides. The development of novel metabolic mechanisms of resistance may be viewed in the context of the adaptation, by insects, to competition, as illustrated in Figure 1. Earlier studies on insect P450 were summarized at a previous NATO Advanced Study Institute.[6] P450 was first demonstrated in insects in 1967 and is now known to be both microsomal and mitochondrial and to have wide species and organ distribution. As in mammals, insect P450 exists as multiple isoforms and requires the presence of a flavoprotein reductase. Cytochrome b_5 is either required or stimulatory for some substrates and some isoforms. Spectral interactions between insect P450 and substrates and inhibitors are similar, if not identical, to those seen with mammalian P450s. Insect P450s oxidize a similar array of synthetic and naturally occuring pesticides to mammalian P450s (Table 1).

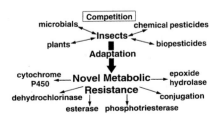

Figure 1. Competition, adaptation and resistance in the insecta.

2.2. Insect Cytochrome P450 and Naturally-Occurring Compounds

Krieger et al.[7] first demonstrated that the nature of the insect diet was reflected in levels of monooxygenase. They demonstrated that, in lepidopterous larvae of species capable of feeding on many host plants (polyphagous), the level of aldrin epoxidase activity in mid-gut preparations was more than 10-fold higher than that in similar preparations from insects adapted to feeding on a single plant species (monophagous), with insects restricted to a fairly narrow range of host plants (oligophagous) showing intermediate values.

Insect interactions with plant allelochemicals have been widely studied and have been the subject of several articles and reviews.[8–12] In most aspects interactions with synthetic organic insecticides parallel these interactions with allelochemicals and cross interactions in which allelochemicals affect the toxicity of synthetic insecticides are also seen. A typical cross interaction can be seen in the variation in LD50 values for insecticides in insects fed different diets. In our earlier studies on the larvae of *Heliothis virescens* it was shown that the P450 content varied 2–3 fold depending upon the host plant, with those insects raised on wild tomato or peppermint having the highest levels. At the same time the toxicity of diazinon varied through an almost 5-fold range, with toxicity being lowest to larvae fed wild tomato or peppermint. More detailed studies[13,14] on 2-tridecanone, an allelochemical found in wild tomato, demonstrated that larvae fed 2-tridecanone or wild tomato leaves had 2.2–3.1-fold increases in midgut P450 accompanied by the same qualitative changes in binding spectra associated with insecticide resistance. These larvae were also more tolerant of diazinon and more able to degrade diazinon and its oxon than larvae not fed 2-tridecanone.

Genetic resistance to plant insecticidal allelochemicals can also be induced, such resistance being associated in some cases with increases in P450 but not in others. To date, resistance to allelochemicals has not been observed to be extensive, usually 5-fold or less, in contrast to the high levels of resistance frequently observed following selection with synthetic organic pesticides. Details of findings involving *H. virescens* are provided by Rose et al.[8] but in summary, resistance to 2-tridecanone and nicotine involves an increase in P450 while resistance to quercetin does not.

2.3. Insecticide Resistance

2.3.1. General Aspects. Insecticide resistance is defined as an increase in the ability of an population of insects to survive an insecticide due to genetic selection exerted on previous generations by the insecticide or, in the case of cross-resistance, by another insecticide. Tolerance, on the other hand, reflects the ability of one unselected insect population to survive exposure to an insecticide more effectively than another unselected population. Many behavioral, physiological and biochemical mechanisms have been implicated in the resistance and/or tolerance of insects to insecticides.[15–17] These resistance mechanisms include avoidance behavior, reduced cuticular penetration, modification of the target site resulting in reduced sensitivity and changes in the ability of the insect to metabolize the insecticide. The most extensive early studies of resistance in insects were carried out on the housefly and have been summarized elsewhere.[6] It was apparent, however, that in many cases, multiple mechanisms contributed to the overall level of resistance.

Since insecticides are still essential for the production of food and fiber for an expanding world population, the loss of chemicals for use, in many cases due to resistance

Table 2. CYP subfamilies in which partial and complete insect cytochrome P450 (CYP) cDNA sequences have been recorded[1]

Species	Isoform
Anopheles albimanus	4C, 4D, 4H, 4J, 4K,
Blaberus discoidalis	4C, 6A,
Ceratitis capitata	4D, 4E, 6A
Culex quinquefasciatus	6E, 6F
Diploptera punctata	4C, 9A
Drosophila melanogaster	4c, 4d, 4e, 4g, 6a, 4p, 9b, 9c, 12b
Drosophila mettleri	4d, 4e
Helicoverpa armigera	4G, 4L, 4M, 4S, 6B, 9A
Heliothis virescens	4G, 9A
Manduca sexta	4G, 4L, 4M, 9A
Mastotermes darwiniensis	4C
Musca domestica	4D, 4G, 4N, 6A, 6C, 6D
Papilio glaucus	6B
Papilio polyxenes	6B,
Tribolium castaneum	4G, 4H, 4Q, 4R

[1]Information obtained from the web site of the Department of Biochemistry, University of Tennessee, Memphis, TN at http://drnelson.utmem.edu/biblioB.html#9A. Further information, references, etc., can be obtained from this site.

and in others due to regulatory action, has created problems because of the lack of safe and effective chemicals. For this reason, *H. virescens*, (a member of the genus *Heliothis*, probably the most important genus, world-wide, for pest insects), has become one of the insects of choice for studies of resistance. P450-dependent-monooxygenases, esterases and glutathione transferases have all been implicated in *H. virescens* resistance.[18-25] Monooxygenation is also important in pyrethroid resistance in this species although esterases, reduced penetration and nerve insensitivity may also be involved.[23-26]

2.3.2. Role of Cytochrome P450. The groundwork for the molecular characterization of insect P450s, particularly those of the housefly, was established by studies of the biochemical genetics of P450 carried out in the 1970s and 1980s, reviewed elsewhere.[6,15,16] In insects, recent progress has been made in the molecular characterization of P450s. The first insect P450 gene sequence was identified in an insecticide-resistant strain of housefly which had been induced with phenobarbital.[27] Two other P450 genes were isolated from Drosophila, one belonging to family 6,[28] the other to family 4.[29] Other insect P450 sequences identified include those from cockroach (family 4),[30] Papilio polyxenes (family 6)[31] as well as additional P450s from the housefly.[32] Further examples are given in Table 2.

Recently, our laboratory obtained a population of H. virescens (the Hebert strain) resistant to thiodicarb and cypermethrin. Resistance was associated with high P450 levels and at least one P450 cross reacted with a P450 antibody raised in the mouse to a P450 isolated from an insecticide-resistant strain of *Drosophila* melanogaster.[33,34] The elevated P450 content in the Hebert strain and high specific activity, compared to a susceptible strain, towards several monooxygenase substrates suggested that, at least in this strain, P450 was an important component of the insecticide resistance mechanism. Synergism studies on a similarly resistant population from the same location demonstrated that toxicity to both thiodicarb and cypermethrin were increased by the P450 inhibitor (and insecticide synergist), piperonyl butoxide[35] (Figures 2 and 3).

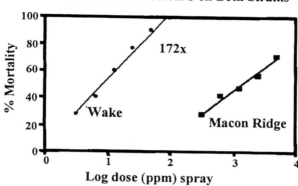

Figure 2. Toxicity of thiodicarb and cypermethrin to an insecticide-susceptible (Wake) and an insecticide-resistant (Macon Ridge) strain of *Heliothis virescens*.

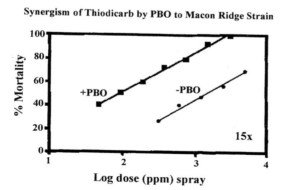

Figure 3. Synergism of thiodicarb toxicity to an insecticide-resistant (Macon Ridge) strain of *Heliothis virescens* by piperonyl butoxide.

Subsequently, a novel cytochrome P450 cDNA with its complete coding sequence and part or all of the 3' (77 nucleotides) and 5' (87 nucleotides) noncoding sequence was isolated from *H. virescens* (F).[36] A cDNA library isolated from third instar larvae of the Hebert strain was screened with a P450 antibody from the insecticide resistant Drosophila strain mentioned previously. The sequence of the longest cDNA clone was merged with a sequence obtained using 5'- RACE to obtain the final nucleotide sequence consisting of 1763 bases. The 1763 nucleotide sequence encodes a protein of 532 amino acids which includes a hydrophobic N-terminal region and the highly conserved heme binding regions typical of P450s. Low sequence similarity to other P450 sequences and the presence of a thromboxane synthase-like insertion upstream from the I helix resulted in its assignment as the first member of family 9, i.e. *CYP9A1. CYP9A1* is most similar to *CYP3A1* from the rat (34.7% identity),[37] but is also similar to the insect P450s from family 6, including *CYP6B1v1* from *Papilio polyxenes* (33.3%),[31] *CYP6A2A* from *D. melanogaster* (32.4%),[28] *CYP6A3* from *Musca domestica* (31.7%)[38] and *CYP6B2* from *Helicoverpa* armigera.[39] Comparative Western and Northern blot studies indicate that expression of *CYP9A1* in thiodicarb selected populations of tobacco budworm is associated with insecticide resistance. The pattern of RFLP (restriction fragment length polymorphism) variation in offspring of single-pair matings demonstrated autosomal inheritance of *CYP9A1* and enabled its assignment to Linkage Group 7. The coding region of *CYP9A1* occupies no more than 10 kb in the tobacco budworm genome.

3. METABOLISM OF PESTICIDES IN MAMMALS

3.1. Organophosphorus Insecticides

Although pesticides have long been known as substrates for cytochrome P450[4] there is still little known about the substrate specificity of specific isoforms, and this is particularly true in the case of human isoforms. In the case of the mouse, we demonstrated some time ago that, while several isoforms metabolized fenitrothion, one of these, induced by phenobarbital and almost certainly CYP 2B10, was most active. Probably the most interesting finding was that the ratios of the oxon to phenolic detoxication products varied from isoform to isoform, a finding that would not be predicted from the purported mechanism of activation via a phosphooxithirane ring.[40]

Our recent studies on the specificity of human P450 isoforms for sulfoxidation of phorate are shown in Figure 4, the isoforms used being expressed in *E coli* in the case of the 2C subfamily and in a human lymphoblastoid cell line in the case of CYP 1A1, 2E1 and 3A4. It is apparent that the human liver can carry out this reaction and that the activity of the 2C subfamily isoforms accounts for most of the activity of the liver.[3] It is of interest that 2C19 is polymorphic, not being expressed in 15–23% of Japanese and other oriental populations, while the polymorpism responsible for this failure to express is relatively rare in other populations.[41,42] The importance of this observation for the human metabolism of phorate and perhaps, other pesticides, is currently under investigation.

3.2. Herbicides

In animal studies the herbicide, alachlor, has been shown to be metabolized to a number of metabolites (Figure 5). Since this herbicide has been implicated as a possible carcinogen in humans[43] it appeared important to investigate its metabolism by human

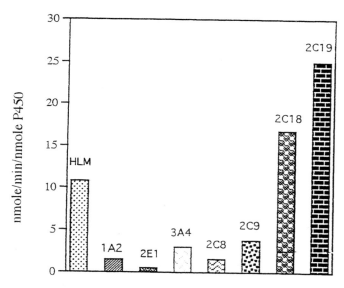

Figure 4. Metabolism of the insecticide, phorate, by human cytochrome P450 isoforms. From Hodgson et al.[3]

xenobiotic metabolizing enzymes. Our studies in progress[3] have demonstrated that three metabolites (E, G and H, Figure 5) are readily produced by human liver microsomes. Assuming the initial production of an O-demethylated intermediate, metabolite H can be produced by two different routes: the first route involves alkyl hydroxylation to yield metabolite E, followed by the elimination of formaldehyde from the -NCH$_2$OH group; the second route involves the initial elimination of formaldehyde from the -NCH$_2$OH group (metabolite G) followed by the aliphatic hydroxylation. It is of interest that human recombinant CYP 3A4 yields exactly the same metabolite profile as intact human liver microsomes and that all other P450 isoforms tested had either no or very little activity. Thus CYP 3A4 appears to be primarily responsible for the oxidative metabolism of alachlor in human liver and for the production of metabolite G, a metabolite believed, from in vivo animal studies, to be further metabolized to 2,6-diethylaniline and eventually to a reactive carcinogenic metabolite.

Another interesting finding based on studies in the mouse is that the isoform specificity for inhibition and for induction may be quite different for the same pesticide. For example, tridiphane is an inhibitor of P450 with considerable specificity for CYP 2B10. It is also an inducer, specifically inducing CYP 4A1, as is typical of peroxisome proliferators.[44,45]

4. SUMMARY AND CONCLUSIONS

While P450 is, without doubt, the most studied enzyme in the history of science there are still large gaps in our knowledge of its distribution and functions. Despite its important role in the primary metabolic processing of xenobiotics, little is known of the way individual isoforms metabolize pesticides. This omission is difficult to understand since pesticides are arguably the most important class of environment chemicals, presenting enormous benefits in the production of food and fiber and potential risks to human and environmental health.

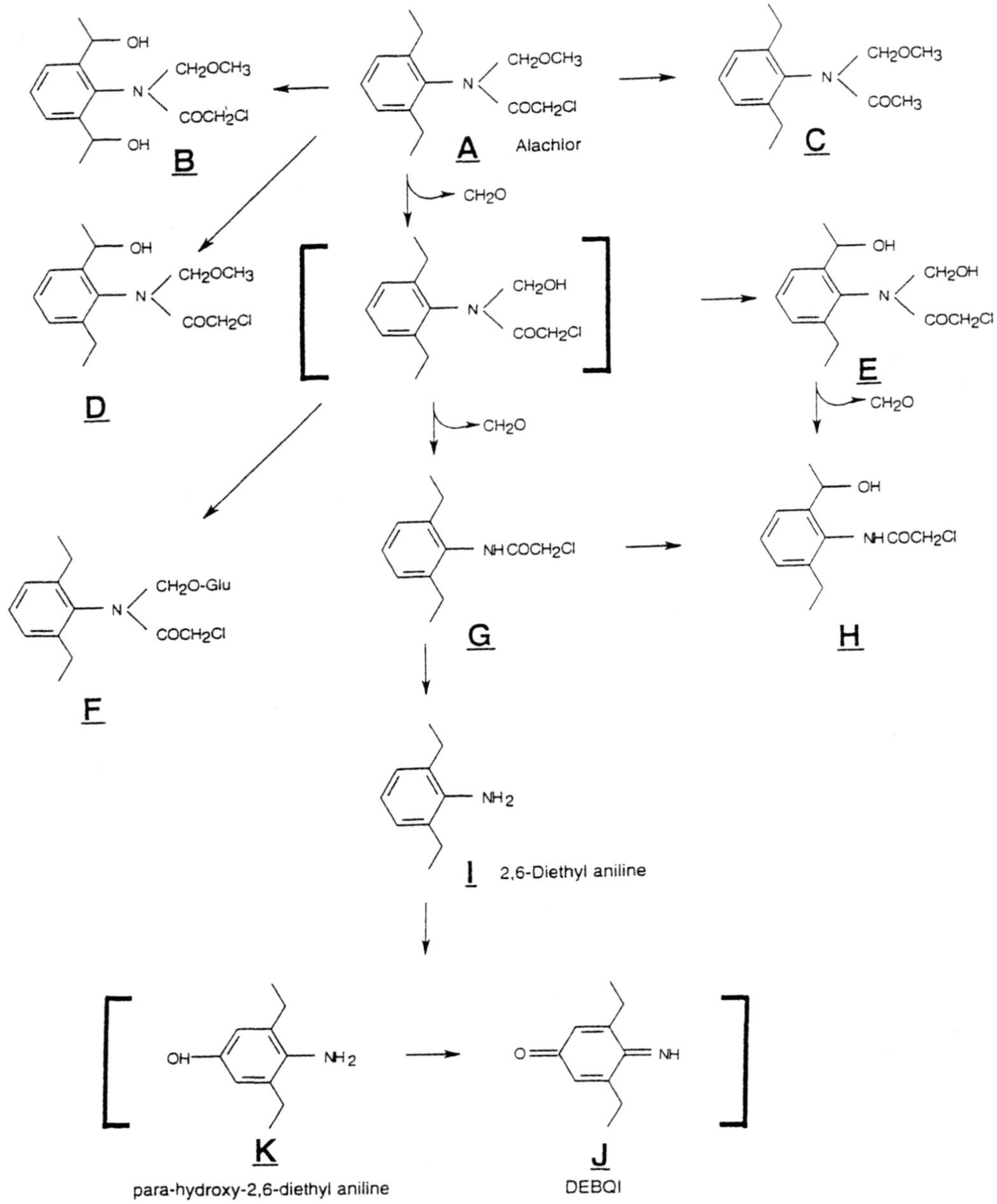

Figure 5. Metabolism of alachlor in mammals. From Hodgson et al.[3]

Resistance to insecticides is also a problem of considerable practical importance and, in recent years, progress is being made in understanding the role, in resistance, of P450 as well as other xenobiotic-metabolizing enzymes and neuroreceptors. Much further progress in both of these areas is needed if we are to obtain the benefits of pesticides and avoid the risks attendant upon their use.

ACKNOWLEDGMENTS

The studies carried out at NCSU were all supported, in part, by NIEHS grants ES-00044 and 2T32ESO7046 and the North Carolina Agricultural Research Service. We gratefully acknowledge the assistance of Mr. Re Bai, Mr. Michael Wyde and Dr. Deborah Thompson, all of NCSU, and the collaboration of Dr. Dave Heckel, of Clemson University.

REFERENCES

1. E. Hodgson, R. L. Rose, D-Y, Ryu, G. Falls and P. E. Levi, 1995, Pesticide-metabolizing enzymes, *Toxicol. Lett.* **82/83:**73–81.
2. E. Hodgson and P. E. Levi, 1996, Pesticides: an important, but underused model for the environmental health sciences, *Environ. Hlth. Perspect.* **104(Suppl. 1):**97–106.
3. E. Hodgson, N. Cherrington, S. C. Coleman, S. Liu, J. G. Falls, Y. Cao, J. G. Goldstein and R. L. Rose, 1998, Flavin-containing monooxygenase and cytochrome P450 mediated metabolism of pesticides: from mouse to human, *Revs. Toxicol.* **2:**in press.
4. A. P. Kulkarni and E. Hodgson, 1984, Metabolism of insecticides by mixed-function oxidase systems, in:Differential Toxicities of Insecticides and Halogenated Aromatics, (F. Matsumura, ed), *Int. Encycl. Pharmacol. Ther.* **113:**27–128.
5. D. R. Nelson, L. Koymans, T. Kamataki, J. J. Stegeman, R. Feyereisen, D. J. Waxman, M. R. Waterman, O. Gotoh, M. J. Coon, R. W. Estabrook, I. C. Gunsalus and D. W. Nebert, 1996, P450 superfamily: update on new sequences, gene mapping, accession numbers and nomenclature, *Pharmacogenetics,* **6:**1–42.
6. E. Hodgson and R. L. Rose, 1991, Insect cytochrome P450, in Molecular Aspects of Monooxygenases and Bioactivation of Toxic Compounds (E. Arinc, J. B. Schenkman and E. Hodgson, eds) Plenum Press. New York, pp. 75–91.
7. R. I. Krieger, P. P. Feeny and C. F. Wilkinson, 1971, Detoxication enzymes in the guts of caterpillers: an evolutionary answer to plant defenses, *Science* **172:**579–581.
8. R. L. Rose, F. Gould, P. Levi, T. Konno and E. Hodgson, 1992, Resistance to plant allelochemicals in *Heliothis virescens (Fabricius),* in Molecular Mechanisms of Insecticide Resistance, (C. A. Mullin and J. G. Scott, eds), ACS Symposium Series, number 505, American Chemical Society, Washington DC, pp. 137–148.
9. R.L. Lindroth, 1991, Differential toxicity of plant allelochemicals to insects: Roles of enzymatic detoxication systems. In Insect Plant Interactions, Volume III. (E. Bernays, editor) CRC Press, Boca Raton, FL
10. M.R. Berenbaum, 1991, Comparative processessing of allelochemicals in the Papilionidae (Lepidoptera). Archives Insect Biochem. Physiol. 17:213–221.
11. Berenbaum, M.R. 1995. Turnabout is fair play: secondary roles for primary compounds. J. Chem. Ecol. 21:925–940.
12. Schuler, M.A., 1996, The role of cytochrome P450 monooxygenases in plant insect interactions. Plant Physiol. 112: 1411–1419.
13. M. R. Riskallah, W. C. Dauterman and E. Hodgson, 1986, Nutritional effects on the induction of cytochrome P-450 and glutathione transferase in larvae of the tobacco budworm, *Heliothis virescens (F.), Insect Biochem.* **16:**491–499.
14. M. R. Riskallah, W. C. Dauterman and E. Hodgson, 1986, Host plant induction of microsomal monooxygenase activity in relation to diazinon metabolism and toxicity in larvae of the tobacco budworm, *Heliothis virescens (F.),* 25:233–247.
15. E. Hodgson, 1983, The significance of cytochrome P450 in insects, *Insect. Biochem.* **13:**237–246.
16. E. Hodgson, 1985, Microsomal monooxygenases, in: Complete Insect Physiology, Biochemistry and Pharmacology, volume 11, (G. A. Kerkut and L. I. Gilbert, eds), Pergamon Press, Oxford.
17. L. B. Brattsten, C. W. Holyoke, Jr., J. R. Leeper and K. F. Rappa, 1986, Insecticide resistance: challenge to pest management and basic research, *Science,* **231:**1225–1260.
18. D. L. Bull, 1981, Factors that influence tobacco budworm resistance to organophosphorus insecticides. *Entomol. Soc. Amer. Bull.* **27:**193–197.
19. D. L. Bull and C. J. Whitten, 1972, Factors influencing organophosphorus insecticide resistance in the tobacco budworm, *J. Agric. Food Chem.* **20:**561–564.

20. C. J. Whitten and D. L. Bull, 1974, Comparative toxicity, absorption, and metabolism of chlorpyrifos and its dimethyl homologue in methyl parathion-resistant and -susceptible tobacco budworms. *Pestic. Biochem. Physiol.* **4**:266–274.

21. R. L. Williamson and M. S. Schector, 1970, Microsomal epoxidation of aldrin in lepidopterous larvae, *Biochem. Physiol.* **19**:1719–1727.

22. W. T. Reed, 1974, *Heliothis* larvae: variations in mixed-function oxidase activity as related to insecticide tolerance, *J. Econ. Entomol.* **67**:150–152.

23. A. R. McCaffery, E. J. Little, R. T. Gladwell, G. J. Holloway, and C. H. Walker, 1989, Detection and mechanisms of resistance of resistance in *Heliothis virescens, Proc. Beltwide Cotton Prod. Res. Conf., National Cotton Council of America,* Nashville TN, pp. 207–211.

24. E. J. Little, A. R. McCaffery, and C. H. Walker, 1989, Biochemistry of insecticide resistance in *Heliothis virescens, Proc. Beltwide Cotton Prod. Res. Conf., National Cotton Council of America,* Nashville TN, pp. 335–337.

25. P. F. Dowd, C. C. Gagne and T. C. Sparks, 1987, Enhanced pyrethroid hydrolysis in pyrethroid resistant larvae of the tobacco, Heliothis virescens (F.), *Pestic. Biochem. Physiol.* **28**:9–16.

26. R. T. Gladwell, A. R. McCaffery and C. H. Walker, 1990, Nerve insensitivity to cypermethrin in field and laboratory strains of *Heliothis virescens, Proc. Beltwide Cotton Prod. Res. Conf., National Cotton Council of America,* Las Vegas NV, pp. 173–176.

27. R. Feyereisen, J. F. Koener, D. E. Farnsworth, and D. W. Nebert, 1989, Isolation and sequence of cDNA encoding a cytochrome P450 from an insecticide-resistant strain of the housefly, *Musca domestica, Proc. Natl. Acad. Sci. USA,* **86**:1465–1469.

28. L. C. Waters, A. C. Zelhof, B. J. Shaw and L. Y. Ch'ang, 1992, Possible involvement of the long terminal repeat of transposable element 17.6 in regulating expression of an insecticide resistance-associated P450 gene in *Drosophila. Proc. Natl. Acad. Sci. USA,* **89**:4855–4859.

29. R. Ghandi, E. Varak, and M. L. Goldberg, 1992, Molecular analysis of a cytochrome P450 gene of family 4 on the *Drosophila* X chromosome. *DNA Cell Biology,* **11**:397–404.

30. J. Y. Bradfield, Y. H. Lee and L. L. Keeley, 1991, Cytochrome P450 family 4 in a cockroach: molecular cloning and regulation by hypertrehalosemic hormone. *Proc. Natl. Acad. Sci. USA,* **88**:4558–4562.

31. M. B. Cohen, M. A. Shuler and M. R. Berenbaum, 1992, A host-inducible cytochrome P-450 from a host-specific caterpillar: molecular cloning and evolution, *Proc. Natl. Acad. Sci. USA,* **89**:10920–10924.

32. T. Tomita and J. G. Scott, 1955, cDNA and deduced protein sequence of CYP6D1: the putative gene for a cytochrome P450 responsible for pyrethroid resistance in the housefly. *Insect Biochem. Molec. Biol.* **25**:275–283.

33. S. Sundseth, S. Kennel and L. C. Waters, 1989, Monoclonol antibodies to resistance-related forms of cytochrome P450 in *Drosophila melanogaster, Pestic. Biochem. Physiol.* **33**:176–188.

34. S. Sundseth, C. E. Nix and L. C. Waters, 1990, Isolation of resistance related forms of cytochrome P-450 from *Drosophila melanogaster, Biochem. J.* **265**:213–217.

35. G. Zhao, R. L. Rose, Ernest Hodgson and R. M. Roe, 1996, Biochemical mechanisms and diagnostic microassays for pyrethroid, carbamate and organophosphate insecticide resistance/cross resistance in the tobacco budworm, *Heliothis virescens, Pestic. Biochem. Physiol.,* **56**:183–195.

36. R. L. Rose, D. Goh, D. M. Thompson, K. D. Verma, D. G. Heckel, L. J. Gahan, R. M. Roe and E. Hodgson, 1997, Cytochrome P450 (CYP) 9A1 in *Heliothis virescens:* the first member of a new CYP family, *Insect Biochem. Molec. Biol.,* **27**:605–615.

37. D. Strotkamp, P. H. Roos and W. G. Hanstein, 1995, A novel CYP3 gene from female rats. *Biochem. Biophys. Acta,* **1260**:341–344.

38. M. B. Cohen and R. Feyerreisen, 1995, A cluster of cytochrome P450 genes of the CYP6 family in the house fly. *DNA Cell Biol.,* **14**:73–82.

39. X. P. Wang and Hobbs, 1995, Isolation and sequence analysis of a clone for a pyrethroid inducible cytochrome P450 from *Helicoverpa armigera, Insect Biochem. Molec. Biol.,* **25**:1001–1009.

40. P. E. Levi R. M. Hollingworth and E. Hodgson, 1988, Differences in oxidative dearylation and desulfuration of fenitrothion by cytochrome P450 isozymes and in the subsequent inhibition of monooxygenase activity, *Pestic. Biochem. Physiol.,* **32**:224–231.

41. J. A. Goldstein and S. M. F. de Morais, 1994, Biochemistry and molecular biology of the human CYP2C subfamily, *Pharmacogenetics,* **4**:285–299.

42. J. A. Goldstein, T. Ishizaki, K. Chiba, S. M. F. de Morais, D. Bell, P. M. Krahn and D. A. Proce Evans, 1997, Frequencies of the defective CYP2C19 alleles responsible for the mephenytoin metabolizer phenotype in various oriental, causasian, saidi arabian and american black populations, *Pharmacogenetics,* **7**:59–64.

43. A. Blair, M. Dosemeci and E. F. Heineman, 1993, Cancer and other causes of death among male and female farmers from twenty-three states. *Am. J. Indust. Med.,* **23:**729–741.

44. D. E. Moreland, W. P. Novitsky and P. E. Levi, 1989, Selective inhibition of cytochrome P450 isozymes by the herbicide synergis, tridiphane, *Pestic. Biochem. Physiol.,* **35:**42–49.

45. P. E. Levi, R. L. Rose, N. H. Adams and E. Hodgson, Induction of cytochrome P450 4A1 in mouse liver by the herbicide synergist, tridiphane, *Pestic. Biochem. Physiol.,* **44:**1–19.

11

INHIBITORS OF CYP51 AS ANTIFUNGAL AGENTS AND RESISTANCE TO AZOLE ANTIFUNGALS

Steven L. Kelly, David C. Lamb, and Diane E. Kelly

Institute of Biological Sciences
University of Wales Aberystwyth
Aberystwyth SY23 3DA, Wales

1. INTRODUCTION

Sterol biosynthesis is an essential metabolic pathway in animals (cholesterol), fungi (ergosterol) and plants (sitosterol) (Figure 1) and requires the removal of the C32-methyl group at position 14α- from precursor sterols. This reaction is catalysed by a microsomal cytochrome P450, the sterol 14α-demethylase (cytochrome P45051, CYP51) and is the only cytochrome P450 family found so far in plants, fungi and animals. Fungal CYP51 is the target of the commercially important azole drugs and fungicides, which are central to therapy (>$2bn pa) and represent about one-third of agrochemical fungicides used ($1bn pa). The importance of the compounds in the clinic stems from the emergency of fungal infections, but many have also developed resistance and cross-resistance to the agents available.

2. THE CYP51s

2.1. Roles for CYP51

The rat[1,2] and human CYP51 proteins[3] are 93% identical, and show 35%–42% identity to fungal[4,5,6,7] and plant[8] CYP51 enzymes. The high level of homology across phyla reflects the housekeeping nature of the CYP51 gene. A single functional CYP51 gene[9] and two processed pseudogenes[10] are present in the human genome. The CYP51 mRNA is ubiquitously expressed with particularly high levels in testis[11]. CYP51 is also responsible for production of the meiosis activating sterols FF-MAS and T-MAS in oocytes and testis.[12,13] This fundamental role for CYP51 may be reflected in other Kingdoms, is the subject of current study and appears seperate from the bulk production of membrane sterols. It will be interest-

Molecular and Applied Aspects of Oxidative Drug Metabolizing Enzymes,
edited by Arinç *et al.* Kluwer Academic / Plenum Publishers, New York, 1999.

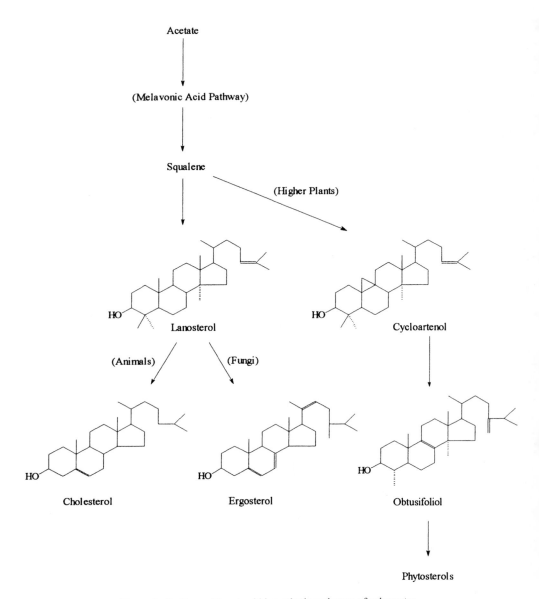

Figure 1. Outlines of the sterol biosynthetic pathways of eukaryotes.

ing to learn if CYP51 is regulated differently across tissues to reflect the involvement in meiotic progression, with applications in treating fertility possibly emerging.

2.2. Blocking CYP51 Activity

Fungal CYP51 is located in the microsomal fraction as an endoplasmic reticulum associated enzyme. Three sequential reactions by this monooxygenase result in the removal of sterol 14α-methyl group during ergosterol biosynthesis. This activity also requires the

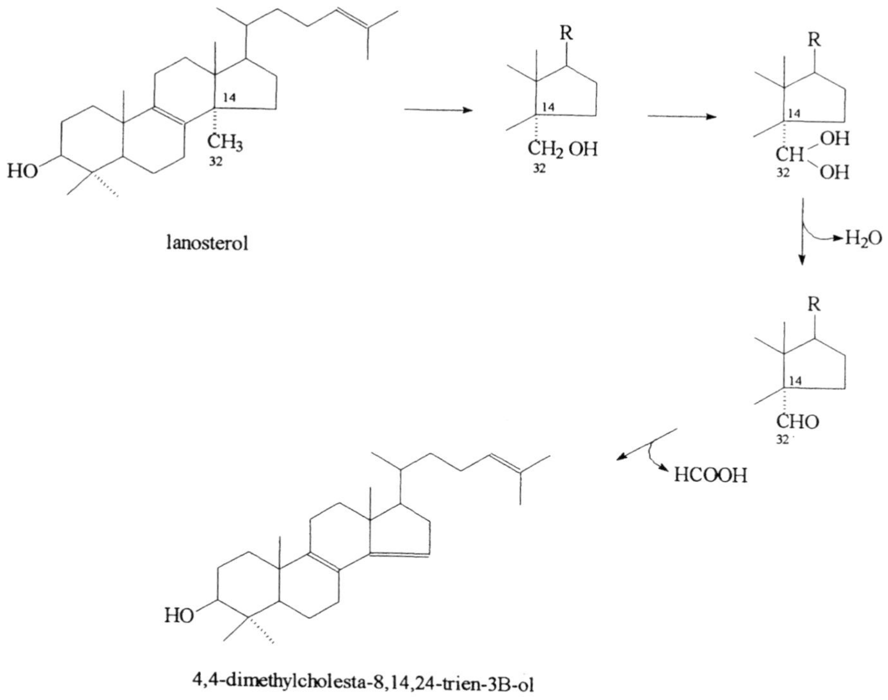

4,4-dimethylcholesta-8,14,24-trien-3B-ol

Figure 2. Chemical process of lanosterol 14α-demethylation catalysed by CYP51. The reaction process consists of three successive mono-oxygenations and they are catalysed sequentially by a single molecule of CYP51.

electron donor NADPH-cytochrome P450 reductase (Figure 2), although in a gene disrupted host for the reductase ergosterol can still be synthesised at a slightly reduced level, probably via the cytochrome b_5 system.[14]

Blocking CYP51 from fungi with antifungal agents inhibits ergosterol biosynthesis and is used in treatment of patients suffering from various fungal infections.[15] In agriculture a similar differential effect is produced i.e. inhibition of fungal sterol biosynthesis, but not plant growth. Other azole compounds have been developed as plant growth regulators which inhibit kaurene oxidation by CYP during giberellin biosynthesis.[16] The therapeutic azole antifungal compounds were introduced during the 1980's in orally administered forms, firstly with ketoconazole and then later with fluconazole and itraconazole (Figure 3). These drugs are used extensively due to the dramatic increased incidence of fungal infection associated with AIDS, but also associated with organ transplantation, cancer chemotherapy and in the intensive care unit. Infections include not only the previously known pathogens, but also fungi previously associated with plant disease e.g. *Fusarium* and even bakers/brewers yeast *Saccharomyces cerevisiae*.

Blocking mammalian CYP51 activity with specific inhibitors results in inhibition of cholesterol biosynthesis[17] and inhibition of steroidogenic CYPs have been observed after antifungal (ketoconazole) therapy.[18] Thus, more detailed information on the activity of CYP51 inhibitors is important toxicologically so that further applications may emerge. For example, ketoconazole has been considered for effect on testosterone biosynthesis as a

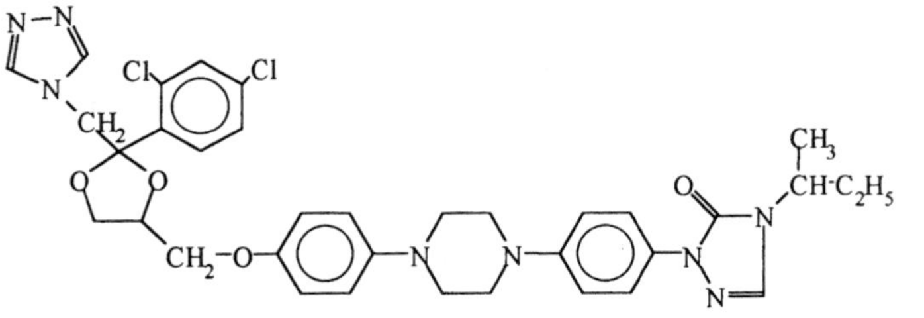

Figure 3. The molecular structure of the major orally active azole antifungal drugs. For AIDS patients fungal infection often represents the presenting symptom and prophylactic treatment and lifetime maintenance therapy is common. The use of ketoconazole has declined due to the more favourable characteristics associated with fluconazole and itraconazole and other CYP51 inhibitors are in clinical trial and can be anticipated to enter use in the future. However, resistance to fluconazole is observed for candidiasis in up to 25% of late-stage AIDS patients and in the field resistance is widespread.

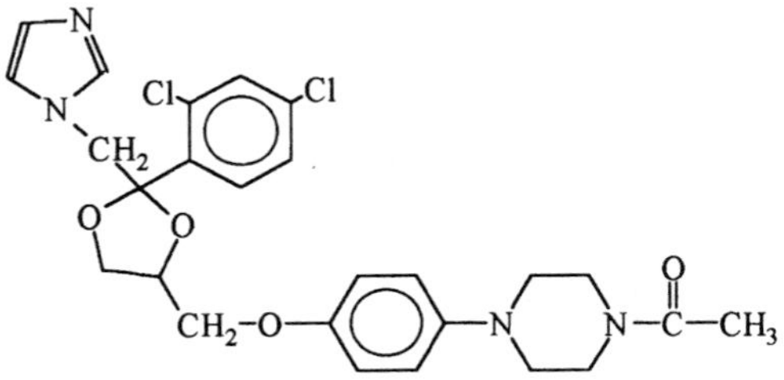

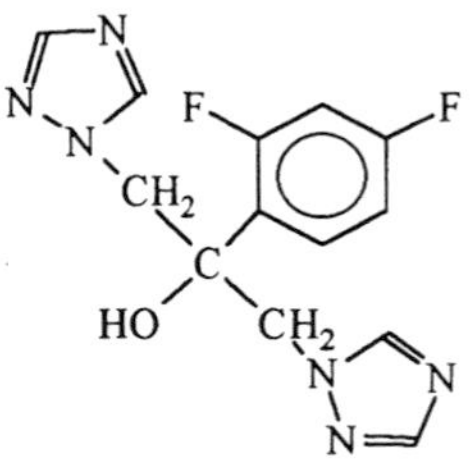

possible therapy for prostate cancer, inhibition of aromatase could be valuable in treating breast cancer and anti-cholesterol agents of this type may emerge.

In agriculture a very wide variety of similar compounds are economically important in treating various infections e.g. cereal infections with powdery mildew. Other compounds, e.g. the pyrimidine fenarimol, are not azole compounds, but have the same mode of action through an unhindered nitrogen which can bind to the haem of CYP51. As agrochemicals these compounds are often called demethylase inhibitors (DMI's) or sterol demethylase inhibitors (SDI's).

3. AZOLE ANTIFUNGAL MODE OF ACTION

3.1. Azole Binding to CYP51

The mode of action of azole antifungals has been subject to genetic and biochemical investigation.[19] They bind to sterol 14α-demethylase (CYP51) as a sixth ligand to the haem iron via the N-4 of the triazole ring (fluconazole and itraconazole) or the N-3 of the imidazole ring (ketoconazole). This interaction is commonly monitored through the associated Type II binding spectrum produced on the shift to low spin in the haem of CYP.[20] The N-1 substituent group is presumed to bind to the protein in an as yet not understood

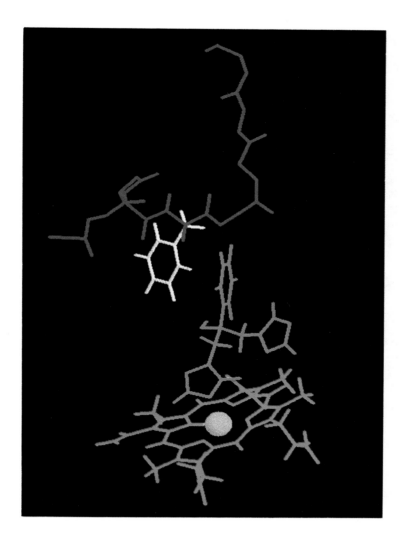

Figure 4. A molecular model of fluconazole (green) bound to CYP51 of *C. albicans*. The azole group is bound to haem (red) as a sixth ligand with the aromatic group predicted to interact with the aromatic group (white) of Phe235 (blue) or Phe233

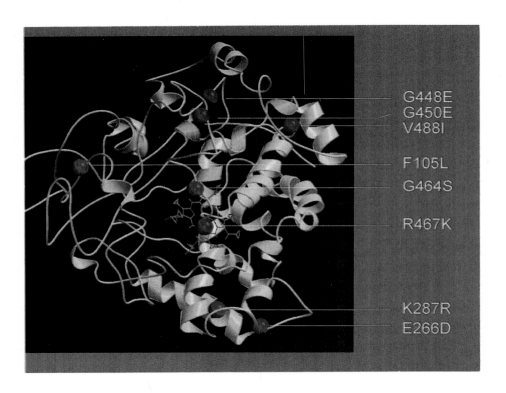

Figure 9. The alterations observed in CYP51 of *C. albicans* from fluconazole resistant strains isolated in the clinic. The ribbon diagram of the model structure shows the positions of altered amino acids as spheres of 6Å radius. The figure shows a view from beneath the haem (red).

```
Human   CYP51    GGFSHAAWLLPGWLPLP
Rat     CYP51    GGFSHAAWLLPGWLPLP
S.c.    CYP51    KGFTPINFVFPN-LPLE
C.t.    CYP51    KGFTPINFVFPN-LPLP
C.a.    CYP51    KGFTPINFVFPN-LPLP
C.g.    CYP51    KGFTPLNFVFPN-LPLE
C.k.    CYP51    KGFTPLNFVFSY-LPLP
                     **        **
```

Figure 5. Sequence alignment of amino acids predicted to be involved in binding sterol substrate 8-double bond. Phe 233/235 are predicted to bind the double bond in *C. albicans* and are shown to be conserved in *S. cerevisiae*, *C. tropicalis*, *C. glabrata* and *C. krusei*. However, hydrophobic aromatic amino acids are also conserved in the human and rat enzymes.

manner, although molecular modelling has suggested interaction of aromatic substituents of the antifungals with the aromatic group of Phe 233 or 235 of CYP51 from *Candida albicans*[21] (Figure 4) This prediction has been refractory to testing due to inactive enzyme being produced when probed by *in vitro* mutagenesis followed by high level heterologous expression. Interestingly this region is not conserved in mammalian or plant CYP51, which may contribute to the selective inhibition of fungal CYP51 (Figure 5).

3.2. Fungal Growth Arrest after CYP51 Inhibition

The major sterol accumulating in *C. albicans* after azole treatment is 14α-methyl-ergosta-8,24(28)-dien-3β,6α-diol (14-methyl-3,6-diol). It is unable to support growth[22] and genetic studies on *S. cerevisiae* have pinpointed this sterol accumulation as being central to the mode of action of these compounds[23]. Strains not synthesising this sterol due to mutation in sterol C5,6-desaturation are resistant and this has been observed clinically in fluconazole-resistant isolates of *C. albicans* from AIDS patients[22,24] (Figure 6). This latter observation shows that the accumulation of the 14-methyl-3,6-diol is responsible for growth arrest of *C. albicans* under clinical conditions of treatment confirming our original hypothesis.[23] The structures of these sterols are shown in Figure 6 and the role of sterol C5,6-desaturase in resistance is discussed later. Other pathogens including *Cryptococcus neoformans* and *Histoplasma capsulatum* accumulate 3-ketosteroid products (Figure 7) under treatment which could arise through interference in sterol C4-demethylation, a step which preceeds sterol C5,6-desaturation. Limited, or no further metabolism occurs which is required for the formation of 14-methyl-3,6-diol.[25,26] Mechanisms of resistance in *C. neoformans* include alteration of the target, CYP51, and reduced accumulation of antifungal within the cell.[27] Presumably in these species retention of the 14-methyl group inhibits the activity of enzymes normally undertaking subsequent metabolic steps in the pathway. Ketosteroid reduction is the penultimate step in sterol C4-demethylation,[28] although the observation of eburicone under treatment in some fungi[29] i.e., a ketosteroid product preceeding initiation of C4-demethylation, indicates other steroid dehydrogenase enzymes may be involved in ketosteroid accumulation under azole treatment.

3.3. Selective Inhibition of CYP51 in the Fungal Pathogen

Antifungals which are used in humans and are targeted to inhibit the fungal CYP51, have been evaluated for possible interactions with endogenous (human) CYP enzymes, especially the human lanosterol 14α-demethylase. The therapeutic effect of these compounds was thought to lie in the selective inhibition of their fungal target over the human counterpart, but these have not proved to be very different in direct comparison of inhibition using the purified enzymes (Table 1). The experimental approach used for this included heterologous expression of both the *C. albicans* and human CYP51, their purification and studies using the azole antifungals from the clinic to inhibit reconstituted enzyme assays. It seems likely that the selective inhibitory action on the fungal enzyme is due to other factors such as the presence of other cytochrome P450s in the human microsomes and metabolism acting as a sink for azole binding and/or CYP-mediated metabolic detoxification. With the isolation of plant CYP51, a similar comparison to fungal CYP51 azole affinity is underway. The small differences in affinity for azole antifungals

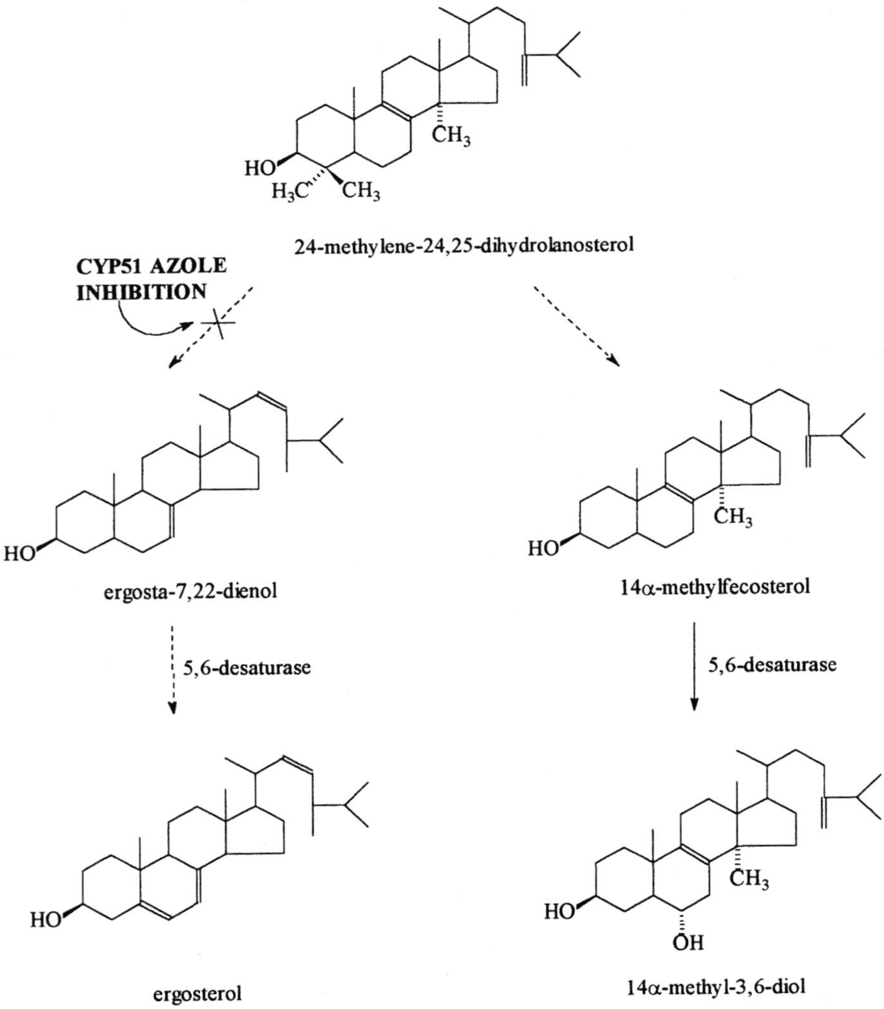

Figure 6. The normal biosynthetic route to ergosterol and the consequence of CYP51 inhibition in *Candida albicans*, including the centrally important step for antifungal effect of 14-methylergosta-8,24(28)-dien-3β,6α-diol formation after metabolism of 14-methylfecosterol by C5,6-desaturase.

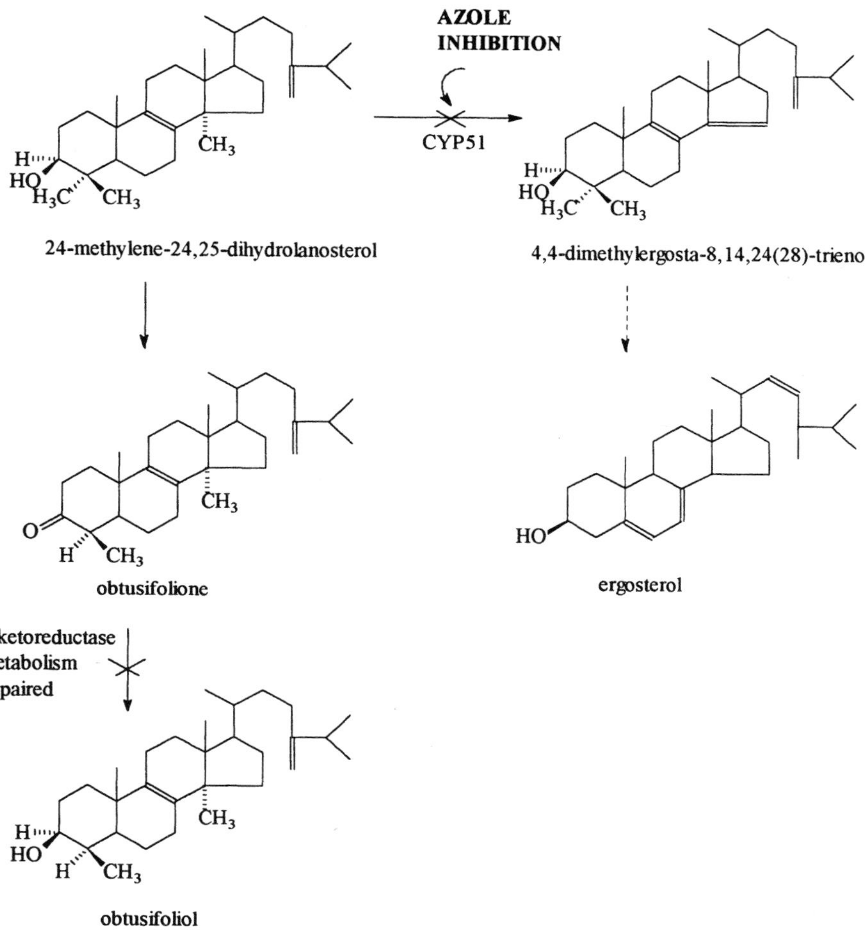

Figure 7. Alternative sterol conversion pathway that occurs upon sterol 14α-demethylase inhibition in the pathogen *Cryptococcus neoformans.*

Table 1. Summary of comparative inhibition by azole antifungals.

Azole antifungal	K_i (nM)	
	C. albicans	Human
Ketoconazole	0.02	0.12
Itraconazole	0.02	0.13
Fluconazole	0.03	0.16

K_i values for ketoconazole, itraconazole, and fluconazole inhibition of sterol 14α-demethylase activity of purified *C. albicans* and human CYP51, respectively, were determined from the plots of activity against azole concentration.

betwen the fungal and human CYP51 is an interesting precedent for drug discovery where targets are normally restricted to novel biochemistry of the pathogen or highly selective inhibitor binding to an enzyme of the pathogen.

4. RESISTANCE TO AZOLE ANTIFUNGALS

Studies on azole resistance are very important currently due to the high incidence in the clinic and in the field. In the clinic up to 33% of late-stage AIDS patients become affected by resistant candidiasis. In field populations of *Rhynchosporium secalis* infecting barley it has become difficult to isolate sensitive isolates indicating a resistant population has evolved (Hollomon, personal communication). Several mechanisms of resistance are known and these are discussed below. No evidence of metabolic detoxification has been observed as a cause of resistance, but many of the novel species of higher fungi have not been assessed for evidence of metabolism of the drug. Increased cellular levels of the target, CYP51, has been proposed as a mechanism of resistance,[30] but supporting evidence for this as a cause of resistance remains weak. Vast overexpression of CYP51 by heterologous expression only modifies sensitivity of yeast by less than ten-fold and small changes have been observed to cause negligible shifts in sensitivity.[31]

4.1. Sterol C5,6-Desaturase Defective Mutants

One mechanism of resistance involves avoidance (suppression) of the effect of CYP51 inhibition and was described in part above. The isolation of firstly ketoconazole[32] and subsequently fluconazole resistant *S. cerevisiae*[23] was in all cases examined (>20) found to be genetically linked with a defect in sterol C5,6-desaturase (*erg*3 mutants). Analysis of the sterols of treated and resistant wild-type cells indicated a clear reason for this. Wild-type parental cultures accumulated 14-methyl-3,6-diol to a high level, but this 6-hydroxylated sterol was not produced by the C5,6-desaturase defective strains. Although it is yet to be proved biochemically, it is assumed that the 6-OH is introduced during a defective attempt to introduce a C5,6-double bond by this non-CYP monooxygenase. In the treated sterol C5,6-desaturase mutants the precursor, 14-methylfecosterol, was produced which allows functional membrane biosynthesis and growth (Figure 8).

The previously reported CYP51 defective mutant of yeast, SG1,[33] had been observed to be resistant to azole antifungals. We now know this resistance can be attributed to the associated second defect in sterol C5,6-desaturase which suppresses the lethal effect of such a *CYP51* mutation or gene disruption.[34,35] This has also been observed for *CYP51* defective strains of *C. albicans*[36] and *C. glabrata*.[37]

Relevance of defective sterol C5,6-desaturase as a cause of resistance in another organism came from *Ustilago maydis*,[38] but no clear information supporting relevance in a practical setting arose until analysis of two *C. albicans* from AIDS patients undertaking long-term fluconazole therapy over several years.[22,24] These patients had been on increasing fluconazole therapy, in one case rising from 100mg/day to 400mg/day and in another up to 800mg/day. Analysis of the sterol profile of the strains causing infection showed accumulation of ergosta-7,22-dienol and ergosta-7-enol i.e., a defect in sterol C5,6-desaturation. Table 2a shows sensitivity of the isolates to antifungal agents and Figure 2b the sterol compositions with and without fluconazole treatment. The latter show that under treatment the resistant strains contain 14-methylfecosterol, and not 14-methyl-3,6-diol. This explained

Table 2a. The minimum inhibitory concentrations of fluconazole and amphotericin B observed on treating the various clinical isolates of *Candida albicans* from AIDS patients

C. albicans isolate	Minimum inhibitory concentration (μg/ml)	
	Fluconazole	Amphotericin B
S1	4.0	0.5
S2	5.0	1.0
R1	62.0	4.0
R2	78.0	4.5

their continued growth, unlike the wild-type, and pin-pointed the role of 14-methyl-3,6-diol in a clinical setting as well as in previous experimental studies with *S. cerevisiae*.

One observation of our original work on C5,6-desaturase defective *S. cerevisiae*, was that cross-resistance to other azole antifungals occurs, as might be expected, but more alarmingly, resistance also occurred for the other major antifungal amphotericin B. This is related to the mode of action of amphotericin B, which involves binding to ergosterol in the membrane resulting in pores being produced. The interaction between of amphotericin B and membranes containing ergosta-7,22-dienol and ergosta-7-enol is presumably reduced as observed for the clinical *C. albicans* sterol C5,6-desaturase defective strains we have analysed.[22] One of these isolates (R1) also failed to respond to amphotericin B therapy at 30mg/day prior to death of the patient. Two molecular events are needed for the emergence of such strains in *C. albicans* to account for defects in both alleles of this diploid yeast. However, we had observed that in a heterozygous state *S. cerevisiae* showed partial resistance to fluconazole.[23] Development of the phenotype observed in the clinical isolates may have involved firstly a mutation of an allele to a defective state and then either forward mutation at the second allele, or gene conversion to give a more resistant strain. The frequency with which defective alleles can arise is quite high due to the multiplicity of events which are likely to produce a defective gene via forward mutation.

Table 2b. The percentage composition of various sterols of the *C. albicans* clinical isolates with and without treatment with 16μg/ml fluconazole. Sensitive strains have undergone growth arrest with this treatment unlike the resistant strains

Sterol	C. albicans isolate							
	Without treatment				Following fluconazole treatment			
	S1	S2	R1	R2	S1	S2	R1	R2
Ergosterol	98.0	97.0			2.0	1.6		
Ergosta-7,24(28)-dienol				11.0				
Ergosta-7,22-dienol			77.0	74.0				
Ergosta-7-enol			22.0	12.0				
Euburicol					16.1	12.3	7.0	12.0
Obtusifoliol					34.5	13.1	10.0	14.0
14α-methylfecosterol							82.0	71.0
14α-methyl-3,6-diol					45.2	72.0		
Unknown	2.0	3.0	1.0	3.0	2.2	1.0	1.0	3.0

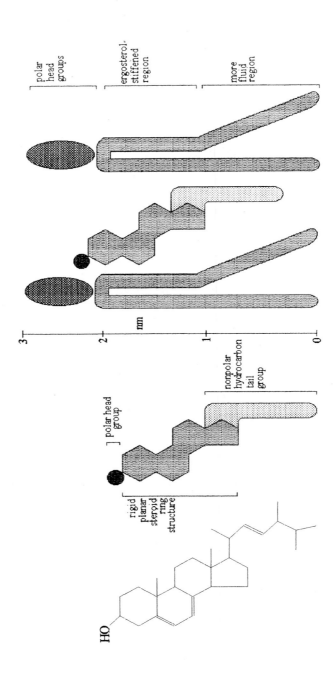

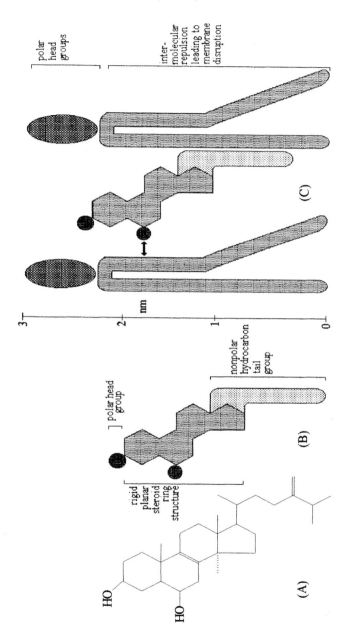

Figure 8. Schematic drawing representing the interaction of ergosterol (**A**) and 14α-methyl-3,6-diol (**B**) with the membrane phospholipids. The hydroxyl group at position 3 interacts with the phospholipid polar head whilst the ergosterol tail interacts with flexible phospholipid tails allowing membrane fluidity. However, the presence of the hydroxyl group at the 6 position in 14α-methyl-3,6-diol on the sterol ring results in inter-molecular repulsion and membrane disruption would result from its incorporation.

Other clinical isolates exhibiting defective sterol C5,6-desaturase and cross-resistance to fluconazole and amphotericin B have been recovered from blood samples of two leukaemic patients undergoing treatment for fungemia. In this case the therapy was adjusted to include a high dose of amphotericin B and flucytosine, which proved successful.[39] Previously, a patient suffering chronic mucocutaneous candidiasis was also observed to have a ketoconazole resistant strain which contained fecosterol. This sterol is indicative of a different block in sterol biosynthesis at C8-isomerisation,[40] but in the light of the results described this analysis should be repeated. It is clear that in a medical setting sterol C5,6-desaturase defective strains are a common mechanism of resistance, but in the field the relevance is unclear. Leaky mutations in sterol C5,6-desaturation need to be screened for carefully. Initial analysis showing the presence of significant levels of ergosterol can be superficially designated sterol wild-type, but these are still resistant.[38,23]

4.2. Altered Cellular Accumulation of Azole Antifungal

Considerable interest has appeared in resistance mechanisms changing cellular antifungal efflux since the discovery of drug transporters in yeast. Alterations in these are correlated with resistance in some clinical isolates of *C. albicans*[41,42] and other species such as *Candida krusei* appear to pump out fluconazole at a constitutively high rate.[43] In analysis of azole resistance in a field isolate of *Septoria tritici* reduced accumulation of fungicide was also observed.[44] The first association of changed efflux with fungicide resistance was in fenarimol resistant filamentous fungi[45] which implicated an energy dependent process. Following the discovery that the ATP-dependent ABC transporter PDR5 of *S.cerevisiae*[46] was involved in azole transport, related genes of *Candida albicans* were found to be expressed at higher levels in certain strains of fluconazole resistant clinical isolates. The genes CDR1–5 (Candida Drug Resistance) have been identified as part of the PDR5 family. Another gene Ben[R] has also been studied and belongs the the major facilitator family of transporters.[47] Altered expression of this gene appears to cause fluconazole resistance, but not azole cross-resistance. Recently, another member of the major facilitator family FLP1 has also been identified as being involved in fluconazole transport and hence potentially in resistance.[48]

Where all the parameters which cause resistance have been assessed[42] it appears that altered drug accumulation alone can account for high level resistance, as can sterol C5,6-desaturase defects. Further transporters should be identified in parallel approaches or involving PCR methods to identify for instance ABC transporters.[49] It should be noted that the *S. cerevisiae* genome project is undertaking the systematic disruption of all (approximately 200) transport genes and from these strains comprehensive information on fungal drug transport will emerge.

4.3. Alterations to the Target CYP51 which Causes Resistance

Alterations in the target site also appear to be important as a mechanism of resistance and for which molecular details are only recently emerging. Certain strains e.g. SG1 containing defective CYP51 were known to be azole resistant, but contain additional blocks in sterol C5,6-desaturase which allow their viability. For many such fungal strains like SG1 this may also occur and has needs close examination as in some of these strains these additional mutations could be leaky. Of course, the requirement of CYP51 activity for growth is at the centre of it's value as an antifungal target and single CYP51 defective strains would not be expected to arise.

In one strain of *C. albicans* rapid displacement of azole antifungal from microsomal CYP was observed, but resistance of the isolate was lost from subcultures.[50] Subsequent analysis of this strain, including assessment of CYP51 inhibition, sterol analysis and drug accumulation pinpointed drug accumulation as the cause of resistance.[51] Evaluation of a series of clinical isolates of *Cryptococcus neoformans* for such differences indicated increased tolerance to azole through reduced inhibition of CYP51.[27] The first clear information that altered CYP51 could give rise to significant resistance through reduced affinity for antifungal was based on a congenic comparison in laboratory isolates of *U. maydis*[53] which developed resistance to triadimenol. Recently we have demonstrated the first point mutation responsible for altering the affinity of CYP51 for azole and subsequently causing resistance, T315A in CYP51 from *C. albicans*.[53] This mutation altered enzyme activity as expected for this residue being involved in hydrogen bonding to the 3-OH of sterol substrate and also gave rise to an altered active site which led to resistance (Figure 8).

From an investigation of the sequence of *CYP51* in nineteen fluconazole resistant strains of *C. albicans* other changes were observed, but not consistently.[54] The alteration F105L occured in five isolates, E266D in five isolates, K287R in one, G448E in one, G450E in one, G464S in three and V488I in one (Figure 9 , see color insert facing page 161). Any change in azole affinity for CYP51 in these strains has not yet been determined, but many show no changes in drug efflux pumps (Loeffler, personal communication) and also exhibit normal ergosterol biosynthesis (our own data). Another alteration in CYP51 has recently been reported within a series of isolates taken during the emergence of resistance in *C. albicans* from a patient treated with fluconazole.[55] The alteration R467K, three amino acids towards the N-terminus of the cysteine responsible for haem binding, correlated with an increase in resistance. As for the mutation G464S observed elsewhere,[54] no obvious dramatic effect might have been anticipated for this change, but resistance may be caused by a slight alteration in positioning of the haem changing the assocation between residues in the active site and the N-1 substituent group of the azole. However, it could be argued that such a change might also be anticipated to alter substrate binding and activity and this has not been observed for these strains as far as ergosterol biosynthesis was concerned (our observations).

Elsewhere, the CYP51 mutation F136Y has been found in sixteen out of nineteen resistant isolates of the obligate biotrophic grape powdery mildew,[56] which is a compelling correlation. This mutation may alter antifungal access to the active site as for the F105L mutation above. As with the clinical *C. albicans* isolates, biochemical and/or genetic proof of any role for the mutations in resistance is still required. It is apparent from other studies that altered target site mutation may cause azole resistance on it's own, but the series of strains in which the R467K mutation arose was also accompanied by changes in the expression of efflux pumps and by altered transcription of CYP51 which might also have contributed to resistance. Recently, we also examined resistance in *C. tropicalis* isolates from an intensive care unit as the incidence had risen dramatically as a proportion of infections during the 1990's. In this setting all the resistant strains showed CYP51 activity which was azole resistant when assessed in cell-free extracts, with no change in drug accumulation or in ergosterol production (unpublished). Ultimately, better appreciation of the structure of CYP51 is needed to understand how such protein changes reduce azole affinity for the target.

5. CONCLUSIONS

The presence of CYP51 in animals, plants and fungi indicate an ancestral function for these enzymes within the CYP superfamily. Further investigations on *CYP51* from other organisms will assist in understanding the molecular phylogeny of the superfamily.

The identification of a role for the enzyme activity in stimulating meiotic progression is also intriguing, but again may reflect an ancient role for CYP51.

As described above the major azole antifungal resistance mechanisms for fungi have been identified, but a biochemical screen to identify the proportions in the clinic is overdue. In agriculture this is more problematic for pathogens which only grow on plants. Each of the individual mechanisms described can cause high-level resistance on their own, but more than one molecular event may be needed in each case. With better understanding of the molecular basis of resistance it can be envisaged that new diagnostic tools will be developed to allow rapid and appropriate changes in treatment.

The use of combinations of fungicides is common in agriculture and increasing in medicine and could be a route of choice for overcoming resistance. Alternatively inhibitors of the fungicide pumps could produce increased cellular levels of azole and overcome some resistance problems. Although further introduction of azole compounds from discovery programmes is unlikely new compounds are under consideration and in clinical trial and will be central to therapy in the medium term. The emergence of resistance to these agents can be predicted and needs advance assessment to provide assistance for their proper integration into the antifungal drugs regimes available.

REFERENCES

1. Aoyama, Y., Funae, Y., Noshiro, M., Horiuchi, T. & Yoshida, Y. (1994). Occurrence of a P450 showing high homology to yeast lanosterol 14-demethylase (P45014DM) in the rat liver. Biochem Biophys Res Commun **201**, 1320–1326.
2. Sloane, D. L., So, O.-Y., Leung, R., Scarafia, L. E., Saldou, N., Jarnagin, K. & Swinney, D. C. (1995). Cloning and functional expression of the cDNA encoding rat lanosterol 14 demethylase. Gene **161**, 243–248.
3. Strömstedt, M., Rozman, D. & Waterman, M. R. (1996). The ubiquitously expressed human CYP51 encodes lanosterol 14-demethylase, a cytochrome P450 whose expression is regulated by oxysterols. Arch Biochem Biophys **329**, 73–81.
4. Kalb, V. F., Woods, C. W., Turi, T. G., Dey, C. R., Sutter, T. R. & Loper, J. C. (1987). The ubiquitously expressed human CYP51 encodes lanosterol 14-demethylase, a cytochrome P450 whose expression is regulated by oxysterols. DNA **6**, 529–537.
5. Chen, C., Kalb, V. F., Turi, T. G. & Loper, J. C. (1988). Primary structure of the cytochrome P450 lanosterol 14α-demethylase gene from *Candida tropicalis*. DNA **7**, 617–626.
6. Lai, M. H. & Kirsch, D. R. (1989). Nucleotide sequence of cytochrome P450LIA1 (lanosterol 14-demethylase) from *Candida albicans*. Nucleic Acids Res **17**, 804.
7. Burgener-Kairuz, P., Zuber, J. P., Jaunin, P., Buchman, T. G., Bille, J. & Rossier, M. (1994). Rapid detection and identification of *Candida albicans* and *Torulopsis (Candida) glabrata* in clinical specimens by species-specific nested PCR amplification of a cytochrome P-450 lanosterol-a-demethylase (LIA1) gene fragment J Clin Microbiol **32**, 1902–1907.
8. Bak, S., Kahn, R. A., Olsen, C.-E. & Halkier, B. A. (1997) Cloning and expression in E. coli of the obtusifoliol 14a-demethylase of *Sorghum bicolor* (L.) Moench, a cytochrome P450 orthologous to the sterol 14á-demethylases (CYP51) from fungi and mammals. Plant. J . **11**, 191–201.
9. Rozman, D., Stromstedt, M., Tsui, L.-C., Scherer., S.W. & Waterman, M.R. (1996) Structure and mapping of the human lanosterol 14α-demethylase gene (CYP51) encoding the cytochrome P450 involved in cholesterol biosynthesis; comparison of exon/intron organization with other mammalian and fungal CYP genes. Genomics **38**, 371–381.
10. Rozman, D., Strömstedt, M. & Waterman, M. R. (1996) The three human cytochrome P450 lanosterol 14-demethylase (CYP51) genes reside on chromosomes 3,7 and 13: Structure of the two retrotransposed pseudogenes, association with a LINE-1 element and evolution of the human CYP51 family. Arch Biochem Biophys **333**, 466–474.
11. Strömstedt, M., Waterman, M. R., Haugen, T. B., Taskén, K., Parvinen, M. & Rozman, D. (personal communication)

12. Byskov, A. G., Andersen, C. Y., Nordholm, L., Thogersen, H., Guoliang, X., Wassman, O., Guddal, J. V. A. E. & Roed, T. (1995) Chemical structure of sterols that activte oocyte meiosis. Nature **374**, 559–562.

13. Swinney, D. C., So, O.-Y., Watson, D. M., Berry, P. W., Webb, A. S., Kertesz, D. J., Shelton, E. J., Burton, P. M. & Walker, K. A. M. (1994) Selective inhibition of mammalian lanosterol 14α-demethylase by RS-2106 in vitro and in vivo. Biochemistry **33**, 4702–4713.

14. Sutter, T. R., and Loper, J. C. (1989) Disruption of the *Saccharomyces cerevisiae* gene for NADPH-cytochrome P-450 reductase causes increased sensitivity to ketoconazole. Biochem. Biophys. Res. Commun. **160**, 1257–1266.

15. Kelly, S. L., Arnoldi, A. & Kelly, D. E. (1993). Molecular genetic analysis of azole antifungal mode of action. Biochem.Soc.Trans. **21**, 1034–1038.

16. Buchenauer, H. (1990) Physiological reactions in the inhibition of plant pathogenic fungi. In: Haug, G. and Hoffman, H. (Eds.) Chemistry of plant protection Vol. 6: controlled release, biochemical effects of peasticides and inhibition of plant pathogenic fugi. Springer-Verlag, Berlin, pp. 217–292.

17. Pont, A., Graybill, J. R., Craven, P. C., Galgiani, J. N., Dismukes, W. E. and Reitz, R. E. (1984) High dose ketoconazole therapy and adrenal and testicular function in humans. *Arch. Int. Med.* **144**, 2150–2153.

18. Warnock, D. W. and Campbell, C. K. (1996) Medical mycology; Centenary review. *Mycol. Res.* **100**, 1153–1162.

19. Baldwin B.C. (1983) Fungicidal inhibitors of ergosterol biosynthesis. Biochem.Soc.Trans. **11**, 659–663.

20. Jefcoate, C. R., Gaylor, J. L. and Calabrese, R. L. (1969). Ligand interactions with cytochrome P450. I. Binding of primary amines. Biochemistry. **8**, 455–3463.

21. Boscott, P. E. and Grant, G. H. (1994). Modelling cytochrome P450 14α-demethylase (*Candida albicans*) from P450$_{cam}$. *J. Mol. Graph.* **12**: 185–193.

22. Kelly, S.L., Lamb, D.C., Kelly, D.E., Manning, N.J., Loeffler, J., Hebart, H., Schumacher, U. & Einsele, H. (1997) Resistance to fluconazole and cross-resistance to amphotericin B in *Candida albicans* from AIDS patients caused by defective sterol Δ^{5,6}-desaturase. FEBS Lett. **400**, 80–82.

23. Watson, P. F., Rose, M. E., King, D. J., Ellis, S. W., England, H. and Kelly, S. L. (1989). Defective sterol C5–6 desaturation and azole resistance: A new hypothesis on the mode of action of azole antifungals. Biochem. Biophys. Res. Commun. **164**, 1170–1175.

24. Kelly, S. L., Lamb, D. C., Kelly, D. E., Loeffler, J. and Einsele, H. (1996) Resistance to fluconazole in *Candida albicans* from AIDS patients involving cross-resistance to amphotericin. Lancet **348**, 1523–1524.

25. Vanden Bossche, H., Marichal, P., Le Jeune, L., Coene, M. C., Gorrens, J. and Cools, W. (1993). Effects of itraconazole on cytochrome P450-dependent sterol 14α-demethylation and reduction of 3-ketosteroids in *Cryptococcus neoformans*. Antimicrob. Ag. Chemother. **37**, 2101–2105.

26. Wheat, J., Marichal, P., Vanden Bossche, H., Le Monte, A. and Connolly, P. (1997) Hypothesis on the mechanism of resistance to fluconazole in *Histoplasma capsulatum*. Antimicrob. Ag. Chemother. **41**, 410–414.

27. Lamb, D.C., Corran, A., Baldwin, B.C., Kwon-Chung, K.J. & Kelly, S.L. (1995) Resistant sterol 14α-demethylase associated with clinical failure during fluconazole chemotherapy of *Cryptococcus neoformans* in AIDS patients. FEBS Lett., **368**, 326–330.

28. Nes, D.W., Janssen, G.G., Cromley, F.G. Kalinowska, M. and Akihisa, T. (1993) The structural requirements of sterols for membrane function in *Saccharomyces cerevisiae*. Arch. Biochem. Biophys. **300**, 724–733.

29. Joseph-Horne, T., Loeffler, R.S.T., Hollomon, D. and Kelly, S.L. (1995) Cross resistance to polyene and azole drugs in *Cryptococcus neoformans*. Antimicrob. Ag. Chemother. **39**, 1526–1529.

30. White, T.C. (1997) Increased mRNA levels of *ERG16, CDR* and *MDR1* correlate with increases in azole resistance in *Candida albicans* isolates from an HIV-infected patient. Antimicrob. Ag. Chemother. **41**, 1482–1487.

31. Kenna, S., Bligh, H.F.J., Watson, P.F. and Kelly, S.L. (1989) Genetic and physiological analysis of azole sensitivity in *Saccharomyces cerevisiae*. J. Med. Vet. Mycol. **27**, 397–406.

32. Watson, P.F., Rose, M.E. and Kelly, S.L. (1988) Isolation and analysis of ketoconazole resistant mutants of *Saccharomyces cerevisiae*. J. Med. Vet. Mycol. **26**, 153–162.

33. Aoyama, Y., Yoshida, Y., Hata, S., Nishino, T., Katsuki, H., Maitra, S., Mohan, S. P. and Sprinson, D. B. (1983). Altered cytochrome P450 in a yeast mutant blocked in demethylating C32 of lanosterol. J. Biol. Chem. **258**, 9040–9042.

34. Kelly, S.L., Lamb, D.C., Corran, A., Baldwin, B.C. and Kelly, D.E. (1995) Mode of action and resistance to azole antifungals is associated with the formation of 14α-methylergosta - 8,24(28)-dien-3β, 6α-diol. Biochem. Biophys. Res. Commun. **207**, 910–915.

35. Lamb, D.C., Kelly, D.E., Corran, A.J., Baldwin, B.C. and Kelly, S.L. (1996) Role of sterol Δ^{5,6}-desaturase in azole antifungal mode of action and resistance. Pestic. Sci. **46**, 294–298.

36. Shimokawa, O., Kato, Y., Kawano, K. and Nakayama, H. (1989) Accumulation of 14α-methylergosta - 8,24(28)-dien-3β, 6α-diol. in 14α-demethylation mutants of *Candida albicans*: genetic evidence for the involvement of 5-desaturase. Biochem. Biophys. Acta. **1003**, 15–19.

37. Geber A., Hitchcock C.A., Swartz J.E., Pullen F.S., Marsden K.E., Kwon-Chung K.J. and Bennett J.E. (1995) Deletion of the *Candida glabrata ERG3* and *ERG11* Genes: Effect on cell viability, cell growth, sterol composition and antifungal susceptibility. Antimicrob.Ag.Chemother. 39, 2708–2717.

38. Joseph-Horne, T., Manning, N., Hollomon, D. and Kelly, S.L. (1995) Defective $\Delta^{5,6}$-desaturase as a cause of azole resistance in *Ustilago maydis*. FEMS Lett. **127**, 29–34.

39. Nolte F.S., Parkinson T., Falconer D.J., Dix S., Williams J., Gilmore C., Geller R. and Wingard J.R. (1997) Isolation and characterisation of fluconazole- and amphotericin B-resistant *Candida albicans* from blood of two patients with leukaemia. Antimicrob.Ag.Chemother. **44**, 196–199.

40. Howell, S.A., Mallet, A.I. and Noble, W.C. (1990) A comparison of the sterol content of multiple isolates of the *Candida albicans* Darlington strain with other clinically azole-sensitive and -resistant strains. J. Appl. Bacteriol. **69**, 692–696.

41. Sanglard, D., Kuchler, K., Ischer, F., Pagani, J.-L., Monod, M. and Bille, J. (1995) Mechanisms of resistance to azole antifungal agents in *Candida albicans* isolates from AIDS patients involve specific multidrug transporters. Antimicrob. Ag. Chemother. **39**, 2378–2386.

42. Venkateswarlu, K., Denning, D.W., Manning, N.J. and Kelly, S.L. (1995) Resistance of fluconazole in *Candida albicans* from AIDS patients correlated with reduced intracellular accumulation of drug. FEMS Lett. **131**, 337–341.

43. Venkateswarlu, K., Denning, D.W., Manning, N.J. and Kelly, S.L. (1996) Reduced accumulation of drug in *Candida krusei* accounting for itraconazole resistance. Antimicrob. Ag. Chemother.

44. Joseph-Horne, T., Manning, N., Hollomon, D. and Kelly, S.L. (1995) Investigation of the sterol composition and azole resistance in a field isolate of *Septoria tritici*. App. Enviro. Microbiol.

45. De Waard, M.A. and Gieskes, S.A. (1977) Characterisation of fenarimol resistant mutants of *Aspergillus nidulans*. Netherl. J. Plant. Path. **83**, 177–188.

46. Balzi, E., Wang, M., Leterme, S., Van Dyck, L. and Goffeau, A. (1994) PDR5, a novel yeast multidrug resistance conferrinng transporter controlled by the transcription regulator PDR1. J. Biol. Chem. **269**, 2206–2214.

47. Fling, M.E., Kopf, J., Tamarkin, A., Gorman, J.A., Smith, H.A. and Koltin, Y. (1991) Analysis of a *Candida albicans* gene that encodes a novel mechanism for resistance to benomyl and methotrexate. Mol. Gen. Genet. **227**, 318–329.

48. Alarco A-M., Balan I., Talibi D., Mainville N. and Raymond M. (1997) AP1-mediated multidrug resistance in *Saccharomyces cerevisiae* requires *FLR1* encoding a transporter of the major facilitator superfamily. J.Biol.Chem. **272**, 19304–19313.

49. Marger, M.D. and Saier, M.H. (1993) A major superfamily of transmembrane facilitators that catalyse uniport, symport and antiport. Trends in Biomedical Sciences. **18**, 13–20.

50. Vanden Bossche, H., Marichal, P., Gorrens, J., Bellens, D., Moereels, H. and Jannssen, P. A. J. (1990) Mutation in cytochrome P450-dependent 14α-demethylase results in decreased affinity for azole antifungals. Biochem. Soc. Trans. **18**, 56–59.

51. Lamb D.C., Manning N.J., Kelly D.E. and Kelly S.L. (1997) Reduced intracellular accumulation of azole antifungal results in resistance in *Candida albicans* isolate NCPF 3363. FEMS Lett. **147**, 189–193.

52. Joseph-Horne, T., Hollomon, D., Loeffler, R.S.T. & Kelly, S.L. (1995) Altered P450 activity associated with direct selection for fungal azole resistance. FEBS Lett., **374**, 174–178.

53. Lamb D.C., Kelly D.E., Schunck W.H., Sheyadehi A., Akhtar M., Baldwin B.C. and Kelly S.L. (1997) The mutation T315A in *Candida albicans* sterol 14α-demethylase causes reduced enzyme activity and fluconazole resistance through reduced affinity. J.Biol.Chem. **272**, 5682–5688.

54. Loeffler, J., Kelly, S.L., Hebart, H., Schumacher, U., Lass-Florl, C. and Einsele, H. (1997) Molecular analysis of *CYP51* from fluconazole-resistant *Candida albicans* strains. **151**, 263–268.

55. White, T.C. (1997) The presence of an R367K amino acid substitution and loss of allelic variation correlate with an azole-resitant lanosterol demethylase in *Candida albicans*. Antimicrob. Ag. Chemother. **41**, 1488–1494.

56. Delye C., Laigret F. and Corio-Costet M-F. (1997) A mutation in the 14α-demethylase gene of *Uncinula necator* that correlates with resistance to a sterol biosynthesis inhibitor. App.Env.Microbiol. **63**, 2966–2970.

12

THE ROLE OF OXIDATIVE DRUG METABOLIZING ENZYMES IN LIVER AND LUNG SPECIFIC TOXICITY

Richard M. Philpot

Molecular Pharmacology Section
Laboratory of Signal Transduction
National Institute of Environmental Health Science
Research Triangle Park, North Carolina 27709

1. INTRODUCTION

Numerous tissue-specific and tissue-selective toxic responses and pathologies result-ing from both acute and chronic exposures to environmental chemicals have been de-scribed. In many of these cases, metabolic "activation" of the parent compound is known to be a required step in the process leading to the deleterious biological effect. In addition to being a requirement, it seems reasonable to hypothesize that tissue-selective charac-teristics of metabolism could actually be a determinants of tissue-selective toxic effects. In a few cases involving acute toxicity we know this to be the case. However, a role for me-tabolism as a determinant of complex tissue-specific and tissue-selective effects, like chemical carcinogenesis, has not been established with any certainty.

Differences between the responses of the lung and liver to a variety of acutely acting toxicants and toxins provide an excellent example of the role of metabolism in selectivity. Actually, the liver/lung comparison is a misnomer; the difference is between cells not tis-sues. Specifically, most examples of pulmonary-specific toxicity involve the noncilitated bronchiolar (Clara) cell. In species that do not exhibit an acute pulmonary response to these chemicals, the hepatocyte is usually the target. In this article the basis for the pulmo-nary toxicity of substituted furans will outlined, and the extrapolation of findings with laboratory animals to humans will be discussed.

Molecular and Applied Aspects of Oxidative Drug Metabolizing Enzymes,
edited by Arinç *et al.* Kluwer Academic / Plenum Publishers, New York, 1999.

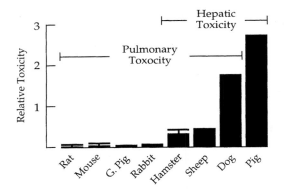

Figure 1. Relative toxicity and tissue-selectivity of furan derivatives in several species. Data were taken from the work of Dutcher and Boyd[3] and Garst et al.[5] Values for 4-ipomeanol and perilla ketone were given the same relative weights by equating their LD50s in mice.

2. TOXICITY OF FURAN DERIVATIVES IN LABORATORY ANIMALS AND LIVESTOCK

2.1. Tissue Selectivity and Species Sensitivity

A variety of furan derivatives have been found to be toxic in laboratory animals and livestock ranging from mice to cows.[1] With most species the target for the toxic insult is the Clara cell of the lung where massive covalent binding of metabolite occurs leading to pulmonary edema and death.[2] However, several exceptions to this generalization have been noted. In addition to pulmonary damage, hepatotoxicity in the hamster and renal necrosis in adult male mice are observed following treatment with 4-ipomeanol.[3] In quail and roosters (with 4-ipomeanol)[4] and in pig and dog (with perilla ketone)[5] the toxic furan effect is confined to the liver.

Species with toxic responses confined almost exclusively to the lung, which include rats, mice (female or immature), guinea pigs, rabbits and cows, exhibit the greatest sensitivity to the furan compounds (Figure 1). LD50s for 4-ipomeanol, the most studied of the toxic furans, range from 12 to 60 mg/kg in these animals.[3] In contrast, the LD50 in hamsters, which exhibit both hepatic and pulmonary damage following treatment with 4-ipomeanol, is 140 mg/kg. The same trend is seen with perilla ketone, a more potent toxic furan than 4-ipomeanol; LD50s in species exhibiting pulmonary damage exclusively (rabbits, sheep, mice) are less than 20 mg/kg, whereas the values in species showing hepatotoxicity (pig and dog) are greater than 100.[5]

2.2. Factors Contributing to Tissue-Selective Toxic Effects of Furan Derivatives

2.2.1. General Considerations. The overall process by which exposure to some exogenous chemicals can result in toxic biological responses can be divided into three categories: uptake and distribution, metabolic activation and reactivity, and the nature and consequence of the resulting biochemical lesion. For most toxicants and toxins the first of these is reasonably well understood and its contribution to species- and tissue-selective ef-

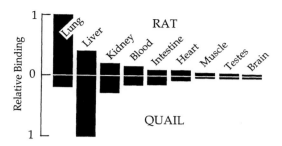

Figure 2. Tissue distribution of the binding of 4-ipomeanol in vivo in rat and quail. Data taken from Buckpitt, Statham and Boyd[4] and presented as relative values.

fects has been documented in a number of cases. For example, the pulmonary-specific effects of paraquat are known to result from uptake pathways unique to the lung.[6] In contrast, the specific events that transpire following the reaction between metabolite and biological target molecules (the third category) are unclear in most cases. In the case of toxic furan derivaties, knowledge about the second category of events, biotransformation, seems to provide reasonable answers to the questions surrounding species- and tissue-selective effects. However, because we know very little about the mechanism of furan toxicity, a role for end point targets in tissue-selectivity cannot be dismissed.

2.2.2. Morphology. In any discussion of the metabolic factors that contribute to the pulmonary selectivity of furan toxicity, it is important to understand a few basic elements of lung morphology. First of all, the lung is composed of over forty cell types and, in principle, characteristics of any of these could dictate susceptibility. However, the characteristics of a single cell type are easily masked when whole lung is examined. This is not the case for the characteristics of hepatocytes when the whole liver is considered. Second, if oxidative metabolism is part of the toxic process, the distribution of endoplasmic reticulum in the lung needs to be considered. This centers attention on the Clara cell.[7–9] As will be noted later in this article, consideration of species differences in Clara cell morphology is enlightening in extrapolation of findings with laboratory animals to humans.

2.2.3. Levels of Covalent Binding. A relationship between covalent binding and toxicity is well established for the toxic furans. Indeed, this relationship can be used in a semi-quantitative manner as a direct indicator of tissue-selective toxicity. A comparison of the in vivo patterns of covalent binding of 4-ipomeanol in the rat and quail is shown in Figure 2. Results for the rat demonstrate that the tissue with the highest level of binding is the lung. This is the case even though binding in the lung is confined primarily to a single cell type that accounts for only a small percentage of the lung tissue. In contrast, binding to the liver, which is less than half that seen in lung, is to the hepatocyte, the cell type that accounts for the majority of the tissue. This distribution is consistent with the lung being the target tissue in rats. On the other hand, the distribution of binding in the quail, a species where 4-ipomeanol is a hepatotoxin, clearly favors the liver.

Covalent binding of 4-ipomeanol to microsomal proteins following incubation with NADPH, also reflects a relationship between binding and tissue-selective toxicity. In the rat covalent binding in vitro favors the lung over the liver or kidney by at least 5-fold,[3] whereas in vitro binding in the quail or rooster clearly favors the liver.[10] Within the lung, in vitro re-

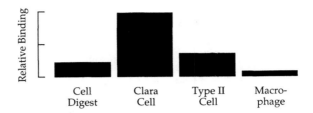

Figure 3. Covalent binding of metabolites of 4-ipomeanol to pulmonary microsomal proteins in preparations from the total cell digest, and isolated Clara cells, alveolar type II cells, and alveolar macrophages. Data taken from Devereux et al.[11]

sults are also consistent with the Clara cell being the target for the effects of 4-ipomeanol.[11] Covalent binding of 4-ipomeanol is much greater with preparations from isolated Clara cells than from type II cells, alveolar macrophages, or the whole tissue cell digest (Figure 3).

2.2.4. Metabolic Activation of Furans by Cytochrome P450. A role for cytochrome P450 in the conversion of furans to reactive products was established initially by the observation that inhibitors of this drug-metabolizing family of enzymes blocked covalent binding of 4-ipomeanol both in vivo and in vitro.[1] Subsequently, it was shown by antibody inhibition[12] and reconstitution studies[13] that 4-ipomeanol was a substrate for two forms of cytochrome P450 present in rabbit lung, form 2 and form 5 (P450 2B4 and P450 4B1). When examined in preparations from isolated cells,[14] the distribution of these enzymes was found to parallel the ability of the preparations to activate 4-ipomeanol (Figure 4). The high concentrations of both enzymes in the Clara cell were confirmed by the results of ultrastructural immunohistochemical studies of the lung.[15]

Orthologs of the same P450s present in rabbit lung are found in rat lung. However, metabolism of 4-ipomeanol in rat lung appears to be accounted for almost entirely by 4B1.[16] The lack of P450 2B involvement in the reaction in rat is due to a decreased affinity for the substrate in comparison with the situation in rabbit. P450 4B1 exhibits a unique pattern of expression that is consistent with the pulmonary effects of 4-ipomeanol. This P450 isoform has been detected immunochemically (immunoblotting and antibody inhibition of activity) in lungs of rabbits, rats, mice, guinea pigs, hamsters and monkeys.[17]

In these species the substrate specificity of 4B1 remains consistent. Except in rabbit and hamster, however, expression of P450 4B1 in liver is not detected. The expression of

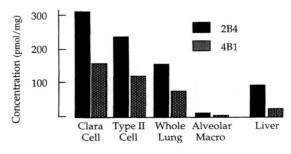

Figure 4. Concentrations of cytochrome P4350 2B4 and 4B1 in microsomal preparations from rabbit lung and liver.

P450 4B1 (and P450 2B) in rabbit liver does not contradict the conclusion that activation of furans by these enzymes is directly responsible for their pulmonary-selectivity effects. The data show clearly that the concentrations in lung are higher than in liver and that expression in lung is highly localized to the Clara cell. The case with the hamster, which is somewhat different, is discussed below.

2.2.5. Explanations for Exceptions to the Rule. Although substituted furans are generally thought of as pulmonary-selective toxins, that is not always the case. Several avian species[4] along with the pig and dog[5] suffer damage to the liver following exposure to 4-ipomeanol or perilla ketone, respectively. Two exceptions among rodents have also been reported; in addition to pulmonary damage, hamsters exhibit liver damage and male mice exhibit kidney damage.[1] Where most rodent species show significantly greater in vivo covalent binding of 4-ipomeanol to lung than liver, equal levels of binding in the two tissues is seen with hamster.[1] In vitro, the ratio of pulmonary to hepatic binding to microsomal proteins with most rodents is 1 or greater except with the hamster where the ratio is about 0.3. These results can be explained by the unique distribution of P450 4B1 in the hamster, the only species other than the rabbit where it is expressed in liver. However, unlike the case with the rabbit, the concentration of 4B1 in hamster liver is greater than in lung.[17] The renal toxicity seen with male mice can also be explained by the distribution of P450 4B1. High concentrations of P450 4B1 are present in male mouse kidney but not in the kidney of the female.[18] Avian species are characterized by lung to liver covalent binding ratios that highly favor the liver both in vivo and in vitro.[4,10] Likely this accounts for the liver-specific toxicity of 4-ipomeanol in birds. However, as pointed out by Buckpitt et al.,[4] the level of covalent binding of 4-ipomeanol metabolites in quail lung in vivo is similar to that observed with the rat and, therefore, the difference between the pulmonary responses of the two species can't be explained by a lack of oxidative metabolic capacity in quail lung. More likely the binding in quail lung is diffuse rather than highly concentrated in a single cell type. This would be consistent with morphological differences between the lungs of the two species.[7,19]

3. TISSUE-SELECTIVE TOXICITY OF FURAN DERIVATIVES IN HUMANS

3.1. Introduction

In laboratory animals pulmonary-selective toxicity of furan derivatives depends on a number of metabolic and morphological factors, many of which relate to oxidative metabolism catalyzed by cytochrome P450. Lung necrosis associated with 4-ipomeanol occurs in species that have one or more specific isoforms of P450 present at some minimum concentration in the Clara cell. Lung-selectivity is lost when similar isoforms are present in liver or other organs at higher concentrations than found in lung, or when the lung activity is diffuse. The placement of humans in this scheme is discussed below.

3.2. Pulmonary P450 in Humans

Although transcripts for a number of P450 isoforms have been identified by analysis of RNA samples from human lung or by cloning, only 1A1, 2E1 and 3A have been detected immunochemically in microsomal preparations.[19,20] The identity of the 3A isoform appears to be primarily 3A5[21] which is present chiefly in the mucous glands and the ciliated and

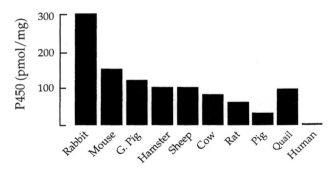

Figure 5. Microsomal concentrations of cytochrome P450 in lungs of various species.

goblet cells of the bronchial epithelium.[22] The concentration of P450 in human pulmonary microsomal preparations has been determined to be between 1 and 2 pmol/mg protein.[23]

3.3. Activation of 4-Ipomeanol by Human Pneumocytes and by Human Forms of P450

The ability of a wide variety of human lung cancer cell lines, normal lung tissue, and lung cancer tissue to metabolize 4-ipomeanol to a reactive intermediate has been demonstrated. Covalent binding in cell cultures was as high as 1 nmol per mg protein per 30 min. and in normal lung tissue was as high as 2 nmol per mg protein per 30 min.[24] The human P450 most active in the metabolism of 4-ipomeanol is 1A2. Activity with 3A3, 3A4, 2B7 and 2F1 was 20 to 40% of that with 1A2, and activity with 2A3, 2C7, 2C8, 2C9, 2F1, 3A5, and 4B1 was very low but was detectable.[25]

3.4. Comparison of Data from Laboratory Animals and Humans

3.4.1. Pulmonary P450 Systems. Although it has been established that several isoforms of cytochrome P450 are present in human lung and that human lung cell cultures and tissue samples are capable of metabolizing 4-ipomeanol to products that bind covalently, the data from humans has not always been put into the context of the parameters for the lung-selective toxicity of 4-ipomeanol that have been derived from findings with labo-

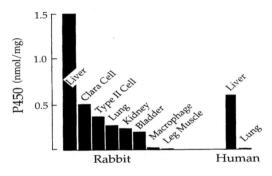

Figure 6. Comparison of cytochrome P450 concentrations in rabbit and human.

ratory animals. Several factors relating to cytochrome P450, including concentration, isoforms, specific activities and localization are important in this respect. Human lung does contain P450, but at concentrations that are remarkably low when compared with other species (Figure 5). Human lung actually contains only a fraction of the P450 found in pig, a species that exhibits hepatotoxicity nearly exclusively with perilla ketone.[5]

The exceptionally low concentration of cytochrome P450 in human lung is further emphasized when compared with the P450 content of various tissues and cell types from rabbit (Figure 6). The content of P450 in human lung is less than 1% of that present in rabbit lung; microsomal preparations from rabbit alveolar macrophages[26] and leg muscle[27] contain amounts of P450 that are similar to those reported for human lung.[23]

The potential effectiveness of the human lung to activate compounds like 4-ipomeanol is further diminished by the specific P450 isoforms present. The human P450s most active in the metabolism of 4-ipomeanol (1A2, 3A3 and 3A4) have not been detected in human lung and the isoforms that have been detected (1A1, 2B7, 2E1, 3A5 and 4B1) are either marginally active or inactive. In addition, the activities of the human isoforms suffer in comparison with that of rabbit 4B1; human 1A2 is about 10% as active and human 4B1 is less than 1% as active. The distribution of P450 in human lung also suggests a lack of potential to produce critical concentrations of reactive metabolites in a single cell type. There appears to be no equivalent in humans to the concentration of highly active P450s in the Clara cell of various laboratory animals. This difference seems easily accounted for: first, as noted above, the overall content of P450 in human lung is exceedingly low; second, the Clara cell of human lung has little or no smooth endoplasmic reticulum.[9]

As has been noted with laboratory animals, shifts in the target tissue for furan toxicity occur when the concentration of specific P450 isoforms, 4B1 in particular, in liver or kidney is greater than found in lung. In human this is clearly the case when liver is compared to lung; the hepatic concentrations of P450 isoforms active in the metabolism of 4-ipomeanol are orders of magnitude greater than the pulmonary concentrations.

3.4.2. Covalent Binding of 4-Ipomeanol Metabolites. The covalent binding of 4-ipomeanol metabolites to samples from human and rabbit lung tissue and pneumocytes is compared in Figure 7. It is evident that the covalent binding observed with human samples[24] is markedly less than with samples from rabbit.[11] This is the case even when the hu-

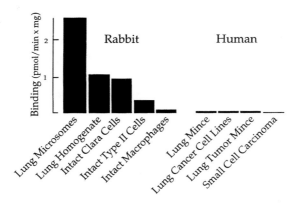

Figure 7. Covalent binding of 4-ipomeanol metabolites following incubation with samples from human[24] or rabbit lung.[11]

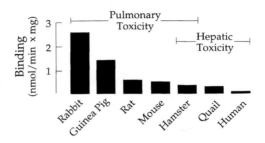

Figure 8. Covalent binding of 4-ipomeanol metabolites to pulmonary samples from various species.[3,10,24]

man data are compared to results with rabbit type II cells, which are not targets for 4-ipomeanol toxicity.

In a comparison of all species for which the in vitro binding of 4-ipomeanol has been determined, binding in human samples is by far the least and is significantly less than in species where hepatotoxicity is observed (Figure 8). (The human data have been adjusted upward to account for differences between minced tissue and microsomal preparations).

3.5. Toxicity of 4-Ipomeanol in Humans

Phase I clinical trails have been carried out to investigate the potential of 4-ipomeanol as a therapeutic agent for lung cancer.[28] The results of the study show that the dose is limited by hepatotoxicity.

4. CONCLUSIONS

Predicting the human case from results with laboratory animals is a difficult proposition at best. However, it is one of the primary justifications for doing the work in the first place and a necessary endeavor given the ethical considerations associated with human experimentation. This is a particularly perplexing problem when dealing with toxic agents that may have some medical application, and especially when the toxicity of the compound is being exploited directly for some therapeutic end. In these instances, as was pointed out in the case of 4-ipomeanol,[25] the available laboratory data should be interpreted and evaluated as rigorously as possible before any clinical application of the compound is considered.

A number of parameters that contribute to the lung-selective toxicity of furans have been identified. Of all species examined, the human scores lowest in each category—lowest pulmonary P450 concentration, lowest lung to liver P450 ratio, lowest pulmonary rate of binding in vitro, lowest turnover numbers of active P450 isoforms, and lowest concentration of smooth endoplasmic reticulum in the Clara cell. Even in comparison with the quail, a species where the toxicity of 4-ipomeanol is essentially liver-specific, the metabolic capacity of human lung to activate furans is marginal. We now know, of course, that 4-ipomeanol is a hepatotoxin in humans. It seems clear that the experimental work with laboratory animals provided a clear indication that this would be the case. In spite of the clinical data and years of experimental evidence to the contrary, however, the notion that furans may pose a human health hazard as pulmonary toxins persists. In a recent review 4-ipomeanol is listed as chemical that causes lung necrosis following activation by CYP3A (P4503A).[29]

REFERENCES

1. M.R. Boyd, 1980, Biochemical mechanisms in pulmonary toxicity of furan derivatives, *in Reviews in Biochemical Toxicology* (E. Hodgson, J. Bend, and R. Philpot, eds.) Elsevier/North Holland, NY, 1980.

2. M.R. Boyd, 1977, Evidence for the Clara cell as a site of cytochrome P-450-dependent mixed-function oxidase activity in lung, *Nature* **269:**713–715.

3. J. Dutcher and M.R. Boyd, 1979, Species differences in target organ alkylation and toxicity by 4-ipomeanol: predictive value of covalent binding in studies of target organ toxicities by reactive metabolites, *Biochem. Pharmacol.* **28:**3367.

4. A.R. Buckpitt, C.N. Statham, and M.R. Boyd, 1982, In vivo studies on the target tissue metabolism, covalent binding, glutathione depletion, and toxicity of 4-ipomeanol in birds, species deficient in pulmonary enzymes for metabolic activation, *Toxicol. Appl. Pharmacol.* **65:**38–52.

5. J.E. Garst, W.C. Wilson, N.C. Kristensen, P.C. Harrison, J.E. Corbin, J. Simon, R.M. Philpot, and R.R. Szabo, 1985, Species susceptibility to the pulmonary toxicity of 3-furyl isoamyl ketone (perilla ketone): *in vivo* support for involvement of the lung monooxygenase system, *J. Animal Sci.* **60:**248–257.

6. J.R. Bend, C.J. Serabjit-Singh, and R.M. Philpot, 1985, The pulmonary uptake, accumulation, and metabolism of xenobiotics, *Ann. Rev. Pharmacol. Toxicol.* **25:**97–125.

7. C.G. Plopper, A.T. Mariassy, and L.H. Hill, 1980, Ultrastructure of the nonciliated bronchiolar epithelial (Clara) cell of mammalian lung: 1. A comparison of rabbit, guinea pig, rat, hamster, and mouse, *Expt. Lung Res.* 1:139–154.

8. C.G. Plopper, A.T. Mariassy, and L.H. Hill, 1980, Ultrastructure of the nonciliated bronchiolar epithelial (Clara) cell of mammalian lung: 1. A comparison of horse, steer, sheep, dog, and cat, *Expt. Lung Res.* 1:155–169.

9. C.G. Plopper, L.H. Hill and A.T. Mariassy, 1980, Ultrastructure of the nonciliated bronchiolar epithelial (Clara) cell of mammalian lung: 1. A study of man with comparison of 15 mammalian species, *Expt. Lung Res.* 1:171–180.

10. A.R. Buckpitt, and M.R. Boyd, 1982, Metabolic activation of 4-ipomeanol by avian tissue homogenates, *Toxicol. Appl. Pharmacol.* **65:**53–62.

11. T.R. Devereux, K. Jones, J.R. Bend, J.R. Fouts, C.N. Statham, and M.R. Boyd, 1981, *In vitro* metabolic activation of the pulmonary toxin, 4-ipomeanol, in nonciliated bronchiolar epithelial (Clara) and alveolar type II cells isolated form rabbit lung, *J. Pharmacol. Exp. Therap.* **220:**223–227.

12. C.R. Wolf, C.N. Statham, C. McMenamin, M.G., J.R. Bend, M.R. Boyd, and R.M. Philpot, 1982, The relationship between the catalytic activities of rabbit pulmonary cytochrome P450 isozymes and the lung-specific toxicity of the furan derivative, 4-ipomeanol, *Molec. Pharmacol.*, **22:**738–744.

13. S.R. Slaughter, C.N. Statham, R.M. Philpot, and M.R. Boyd, 1983, Covalent binding of metabolites of 4-ipomeanol to rabbit pulmonary and hepatic microsomal proteins and to the enzymes of the pulmonary cytochrome P-450-dependent monooxygenase system, *J. Pharmcol. Expt. Therap.*, **224:**252–257.

14. C.J. Serabjit-Singh, S.J. Nishio, R.M. Philpot, and C.G. Plopper, 1988, The distribution of cytochrome P-450 monooxygenase in cells of the rabbit lung: An ultrastructural immunohistochemical characterization, *Molec. Pharmacol.* **33:**279–289.

15. B.A. Domin, T.R. Devereux, and R.M. Philpot, 1986, The cytochrome P-450 monooxygenase system of rabbit lung: Enzyme components, activities, and induction in the nonciliated bronchiolar (Clara) cell, alveolar type II cell, and alveolar macrophage, *Molec. Pharmacol.*, **30:**296–303.

16. R.D. Verschoyle, R.M. Philpot, C.R. Wolf, and D. Dinsdale, 1993, CYP4B1 activates 4-ipomeanol in rat lung, *Toxicol. Appl. Pharmacol*, **123:**193–198.

17. R.R. Vanderslice, B.A. Domin, G.T. Carver, and R.M. Philpot, 1987, Species-dependent expression and induction of homologues of rabbit cytochrome P450 isozyme 5 in liver and lung, *Molec. Pharmacol.* **31:**320–325.

18. R. Ryan, S.W. Grimm, K.M. Kedzie, J.R. Halpert, and R.M. Philpot, 1993, Cloning, sequencing and functional studies of phenobarbital-inducible forms of cytochrome P450 2B and 4B expressed in rabbit kidney, *Arch. Biochem. Biophys.* **304:**454–463.

19. C.W. Wheeler, S.S. Park, and T. M. Guenthner, 1990, Immunochemical analysis of a cytochrome P-4501A1 homologue in human lung microsomes, *Mol. Pharmacol.* **38:**634–643.

20. C.W. Wheeler, S.A. Wrighton, and T.M. Guenthner, 1992, Detection of human lung cytochromes P450 that are immunochemically related to cytochrome P450IIE1 and cytochrome P450IIIA, *Biochem. Pharmacol.* **44:**183–186.

21. K.T. Kivistö, E-U. Griese, P. Fritz, A. Linder, J. Hakkoola, H. Raunio, P. Beaune, H.K. Kroemer, 1996, Expression of cytochrome P 450 3A enzymes in human lung: a combined RT-PCR and immunohistochemical analysis of normal tissue and lung tumors, *Naunyn-Schmiedeberg's Arch Pharmacol.* **353**:207–212.

22. K.T. Kivistö, P. Fritz, A. Linder, G. Friedel, P. Beaune, and H.K. Kroemer, 1995, Immunohistochemical localization of cytochrome P450 3A in human pulmonary carcinomas and normal bronchial tissue, *Histochem.* **103**:25–29.

23. C.W. Wheeler and T.M. Guenthner, 1990, Spectroscopic quantitation of cytochrome P-450 in human lung microsomes, *J. Biochem. Toxicol.* **5**:1–4.

24. T.L. McLemore, C.L. Litterst, B.P. Coudert, M.C. Liu, W.C. Hubbard, S. Adelberg, M. Czerwinski, N.A. McMahon, J.C. Eggleston, and M.R. Boyd, 1990, Metabolic activation of 4-ipomeanol in human lung, primary pulmonary carcinomas, and established human pulmonary carcinoma cell lines, *J. Natl. Can. Inst.* **82**:1420–1426.

25. M. Czerwinski, T.L. McLemore, R.M. Philpot, P.T. Nhamburo, K. Korzekwa, H.V. Gelboin, and F.J. Gonzalez, 1991, Metabolic activation of 4-ipomeanol by complementary DNA-expressed human cytochromes P450: Evidence for species-specific metabolism, *Cancer Res.* **51**:4636–4638.

26. B.A. Domin, T.R. Devereux and R.M. Philpot, 1986, The cytochrome P-450 monooxygenase system of rabbit lung: Enzyme components, activities, and induction in the nonciliated bronchiolar (Clara) cell, alveolar type II cell, and alveolar macrophage, *Molec. Pharmacol.,* **30**:296–303.

27. C.J. Serabjit-Singh, J.R. Bend and R.M. Philpot, 1985, Cytochrome P-450 monooxygenase system: Localization in smooth muscle of rabbit aorta, *Mol. Pharmacol.* **28**:72–79.

28. E.K. Rowinsky, D.A. Noe, D.S. Ettinger, M.C. Christian, B.G. Lubejko, E. K. Fishman, S.E. Sarttorius, M.R. Boyd, and R.C. Donehower, 1993, Phase I and pharmacological study of the pulmonary cytotoxin 4-ipomeanol on a single dose schedule in lung cancer patients: Hepatotoxicity is dose limiting in humans, *Cancer Res.* **53**:1794–1801.

29. O. Pelkonin and H. Raunio, 1997, Metabolic Activation of Toxins: Tissue-specific expression and metabolism in target organs, *Environ. Hth. Perspect.* **105**:767–774.

13

IMPORTANCE OF DRUG METABOLISM IN DRUG DISCOVERY AND DEVELOPMENT

Rodolfo Gasser

F. Hoffmann-La Roche
PRNK 69/146
CH-4070 Basel, Switzerland

The metabolism of a drug can have important consequences on its therapeutic effect or its toxicity. For this reason, early assessments of metabolic pathways in man help to foresee interindividual variation in drug response and elimination due to metabolism. To address these issues, in vitro techniques based on human material are increasingly used as the basis for rational decisions during drug selection and development and to assess risk and potential drug interactions in man.

Many aspects of drug metabolism studies deal with variability in drug response due to differences in the content and specificity of drug metabolizing enzymes. These are modulated by exogenous and environmental factors, as well as endogenous factors that include physiological and pathological states, as well as the genetic make up of drug metabolizing enzymes in an individual. Thus, it is very important to understand the metabolic processes of a given drug, in order to predict whether they will be harmless and clinically irrelevant, or whether they could lead to severe, potentially even fatal effects.

Fortunately our knowledge of these factors has increased dramatically in the last decades, so that we can apply this large body of knowledge to eliminate those compounds exhibiting the potential for harmful interactions at an early stage of drug development. Safety and risk assessments rely heavily on animal experimentation. Although animals exhibit little variation in drug response, drug-metabolizing enzymes evolved differently among species, to produce large variability in drug disposition.[1] Considering animal experimentation alone it is often difficult to foresee the human response to a drug, whether metabolism differences are involved or not.

1. VARIATION IN DRUG METABOLIZING ENZYMES AND GENETIC POLYMORPHISM

Over the years, samples of human liver have been collected at surgery. These have been used to isolate, characterize and determine the structures of drug metabolizing en-

Molecular and Applied Aspects of Oxidative Drug Metabolizing Enzymes,
edited by Arinç *et al.* Kluwer Academic / Plenum Publishers, New York, 1999.

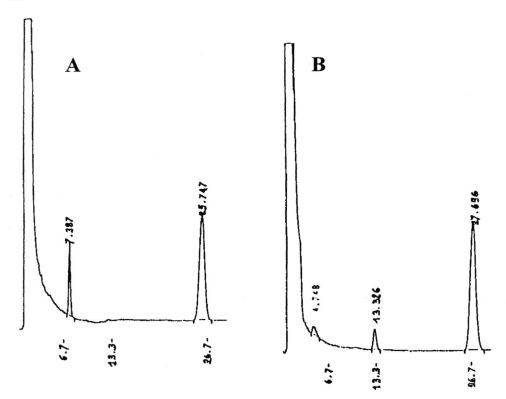

Figure 1. Comparison of HPLC metabolite patterns of S-mephenytoin incubated with human liver microsomes from extensive and poor metabolizers of mephenytoin. S-mephenytoin is metabolized to S-4-OH-mephenytoin by 95 % of the Caucasian population (Figure 1A). Some individuals, which do not express the metabolizing enzyme, CYP 2C19, are unable to produce this metabolite. Instead, the contribution of additional P450 isozymes becomes more prominent and the formation of minor metabolites becomes the major (only) elimination pathway by the patients with this genetic deficiency (Figure 1B). To test if a particular drug is metabolized by CYP 2C19 it can be incubated with liver microsomes from a known poor metabolizer of S-mephenytoin and compared to an extensive metabolizer.

zymes.[2] Based on these studies two major classes of biological transformation have been identified. They are commonly referred to as phase I (or functionalizing) reactions and phase II (conjugative) reactions. The most important phase I enzymes are the cytochrome P-450 monooxygenases. Cytochrome P-450 isoenzymes (P-450) were almost unknown thirty years ago. Now, however, they are among the most widely studied enzymes. This is because of their ubiquity (plants and animals, prokaryotes and eukaryotes all have P-450s) and the diversity of their substrates, which confer on these enzymes a central role in the biotransformation of exogenous chemicals (xenobiotics) as well as endogenous compounds, such as steroids or fatty acids.[3]

There are several reasons why it is advantageous to identify specific P-450s that interact with drug candidates prior to their entrance into clinical trials or the market.

- If a drug is metabolized by a polymorphic enzyme, like CYP2D6, then adverse reactions may occur in those patients who lack this enzyme.
- If a drug inhibits a specific enzyme then interactions might occur with other compounds that are metabolized by that enzyme.

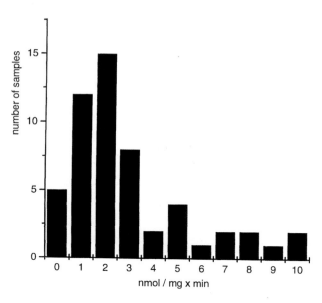

Figure 2. Variability of 6-ß-hydroxy-testosterone formation by microsomes from different human livers. Testosterone was incubated with microsomes from individual livers and the CYP 3A mediated product, 6-ß-hydroxy-testosterone, was determined by HPLC. A ten-fold variation in enzymatic activities is readily observed, which correlates with the variability in disposition observed for drugs that are dependent on this P450 for their elimination.

- If a drug with a narrow therapeutic index is metabolized by an enzyme that is known to be highly variable in human populations, differences in metabolic clearance might reduce therapeutic efficiency or increase the incidence of adverse side effects.

In this context it is extremely useful to know which particular drug metabolizing enzymes are involved in the rate limiting pathway of elimination. This is particularly important where a low capacity enzyme system is involved, as it is the case for many cytochrome P-450-mediated reactions. P-450s involved in particular reactions can be easily identified in vitro: an extensive body of knowledge on distribution, regulation and substrate specificity of individual human P-450s has accumulated over the last decades. Identification techniques involve the cumbersome microsome panel-correlations (Figure 1), inhibition by antibodies, which are often unspecific or hard to procure, or with the use of isozyme specific inhibitors. The latter are simple to use, but some limitations in their specificity and, sometimes, selectivity has to be taken into account. In all cases, the result of a particular enzyme-identification approach needs to be confirmed in another system—it is here where recombinant enzyme preparations complement other procedures ideally and several in vitro experimental elements need to be combined to anticipate human drug-metabolism correctly.

This approach is useful where a defined genetic deficiency contributes to the metabolism of a drug, but engenders large quantity of work if the metabolizing enzyme does not exhibit a clear-cut genetic polymorphism or is extremely variable in its hepatic content, as in the case of CYP3A (Figure 2).

Experiments with microsomes are limited in that not all drug-metabolizing enzymes are being tested simultaneously. This is why human hepatocytes are often employed to

Table 1. List of useful inhibitors of specific human
cytochrome P-450s

P-450 Isozyme	Inhibitor compound
CYP 1A2	furafylline
CYP 2A6	coumarin
CYP 2C9	sulphaphenazole
CYP 2C8	taxol
CYP 2C19	S-mephenytoin
CYP 2D6	quinidine
CYP 2E1	chlorzoxazone
CYP 3A4	midazolam
CYP 4A1	lauric acid

identify metabolic pathways and to estimate hepatic drug clearance. While the later extrapolation often matches the clearances seen in vivo, metabolic pathways are often not correctly identified in this system, where the drug-substrate in the closed system might be converted into metabolites not seen in vivo.

To assess the contribution of a particular P-450 enzyme to a metabolic pathway, enzymological procedures are extremely valuable. Initially, exhaustive enzyme kinetic analysis of microsomal metabolism can provide insight how P-450-mediated pathways contribute to the elimination of a drug, as illustrated by Figure 3. In this case the metabolic product is formed by two catalytic processes, which can be visualized by the Eadie-Hofstee transformation (Figure 3B) of the kinetic data presented in Figure 3A. One enzymatic process is characterized by a low capacity (Vmax) of product formation with a high affinity (Km) for the substrate and the second exhibits a high capacity (higher Vmax), but low affinity (higher Km). When the in vivo concentration of the substrate is known, it can be deduced which of the enzymatic components contributes to what extent to product formation. This can be done simply by substituting the corresponding substrate concentration (S in Figure 3A) in the corresponding Michaelis-Menten equation

A variety of chemical compounds have been described to be specific inhibitors for certain P-450 isoforms and a few commonly used are described in Table 1. The validity of the use of selective chemical inhibitors rests on the specificity of the different substrates and inhibitors. These however are often poorly characterized in published reports, and the results therefore have to be evaluated cautiously. For example, several CYP2E1 inhibitors (i.e., disulfiram, dihydrocapsaicin) and substrates (i.e. chlorzoxazone, p-nitrophenol) reported to be "rather" specific do also competitively inhibit CYP3A4 mediated reactions (i.e. midazolam hydroxylation).

When inhibitors (described in Table 1) are co-incubated with a test substrate, product is reduced in the case where the inhibited P-450 is responsible for its formation, as seen for example in Figure 4.

At the substrate concentration (25 μM in this case) used to test for inhibition (Figure 4), it appears that midazolam (a CYP 3A substrate) inhibits the reaction to a large extent and sulphaphenazole (a CYP 2C9 inhibitor) to a lesser degree. From this experiment it can be concluded that both CYP 3A and CYP 2C9 correspond to the two enzymes described by Figure 3 and that both contribute to product formation. In this case the low affinity/high capacity component corresponds to CYP 3A and the high affinity/low capacity component to CYP 2C9.

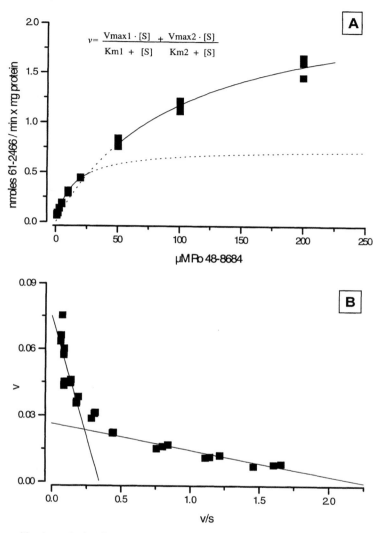

$$v = \frac{Vmax1 \cdot [S]}{Km1 + [S]} + \frac{Vmax2 \cdot [S]}{Km2 + [S]}$$

Figure 3. Enzyme kinetic analysis of a metabolic reaction with human liver microsomes in the presence of NADPH. In this case, the experimental drug Ro 48-8684 was incubated at various concentrations with human liver microsomes and the metabolic product, Ro 61-2466, was determined by HPLC.

To extrapolate from these results to the in vivo situation it has to be remembered that with human liver microsomes or subcellular fractions only a partial picture of the metabolic pathway might be derived because not all drug-metabolizing enzymes are being tested simultaneously. Relative and qualitative information can be gained despite some limitations, such as difficulties to determine optimal reaction conditions for product formation. Also, drug-substrate affinities in these systems are skewed because of futile drug-protein binding, which results in larger apparent Km values. In addition, to analyze drug-products often concentrations in excess of those attained in vivo are used in vitro. This often results in the identification of the high capacity reaction processes when multiple enzymes participate in the metabolism of the test drug. Thus, the limiting pathway is overlooked and the results might be of no rele-

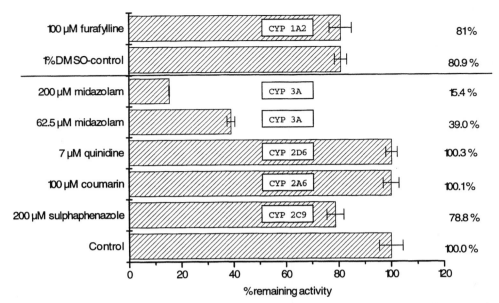

Figure 4. Inhibition of product formation by P-450 inhibitors with human liver microsomes. Ro 48-8684 was incubated with human liver microsomes in the absence (control) or presence of P-450 inhibitors at the concentrations indicated on the left of the figure. The decrease of the product (Ro 61-2466) in the presence of the inhibitor was determined.

vance to the in vivo behavior of the drug at therapeutic concentrations. In addition, often no good estimates of the actual drug concentrations at the metabolizing enzyme can be obtained, so that metabolic predictions based on in vitro drug concentrations might not bear any clinical relevance. Even then it has to be remembered that many metabolic reactions are easily saturated in vivo and that additional comedication might compete for the enzyme.

2. THE USE OF RECOMBINANT P-450 PREPARATIONS IN DRUG METABOLISM

The result of a particular enzyme-identification approach needs to be confirmed in another system: several in vitro experimental elements need to be combined to anticipate human drug-metabolism correctly.

Recombinant P-450s can be used to confirm the results of inhibition-competition studies, but the product formation rates should be high enough to detect them. Heterologous expressed drug metabolizing enzymes have gained more and more importance in biotransformation studies.[4] In contrast to classical in vitro systems, like tissue slices, hepatocytes or subcellular fractions, the use of expressed enzymes allows to study single enzymes. Enzymes, expressed only in minor amounts in human tissue can be assessed. Applications in drug development include:[5,6,7]

- Identification of metabolizing enzymes
- Identification of metabolites
- Determination of enzyme kinetics (Km, Vmax, kcat)
- Mechanistic structure activity studies
- Use as a bioreactor to produce small quantities of metabolites

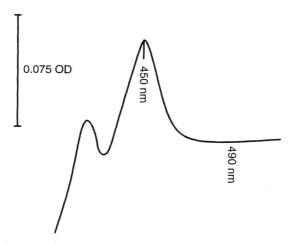

Figure 5. Reduced Fe^{+2}-CO vs. Fe^{+2} difference spectrum from membranes from E. coli expressing CYP3A4. Spectra were recorded according to Omura and Sato.[9]

Heterologous expression systems offer a constant and reproducible source of human drug metabolizing enzymes that are easily available by standard laboratory techniques. Considering the constrain on the availability and use of human tissue it is likely that these systems will be widely used in the future.

Expression of mammalian P450s in heterologous systems has been described for different hosts, like bacteria, yeast, insect cells and mammalian cell lines. Each system has advantages and disadvantages, and depending on the questions raised, one or the other system might be more advantageous. Bacterial expression systems yield large quantities of recombinant enzymes. However, functional activity is often low, due to the generation of insoluble precipitates (lack of intracellular membranes in bacteria into which P450, a membrane bound enzyme system can sequester) and lack of P450 reductase, the requisite electron transfer partner for microsomal P450. Therefore, functional activity of the expressed P450 had to be reconstituted by isolation and solubilization of the membranes, followed by the addition of cytochrome P450 reductase.

In our hands, the expression systems described by Blake[8] have provided a convenient tool for drug metabolism investigations. Functionally active P450 isoenzymes in E. coli were generated by coexpression of P450 and P450 reductase. For example, replacement of part of the CYP3A4 N-terminus with the pelB leader sequence directed enzyme expression to the periplasma, where an active and membrane bound P450 was expressed. Expression levels of 200–340 pmol CYP3A4/mg protein were obtained in the membrane preparations (Figure 5). The specific content of CYP3A4, determined by spectrophotometrical analysis was higher in membranes from E. coli than in human liver microsomes. Catalytic activity of the expressed enzymes compares favorably to the original enzyme activities present in human liver microsomes.

The cDNA expressed enzyme exhibited catalytic activity for testosterone and midazolam biotransformation. Metabolism of these two compounds is catalyzed by CYP3A4 in human liver microsomes. cDNA expressed CYP3A4 transformed testosterone in presence of NADPH to 6 β-hydroxytestosterone and testosterone hydroxylation was higher for the expressed enzyme than for human liver microsomes, with turnover numbers of 19 min^{-1}.

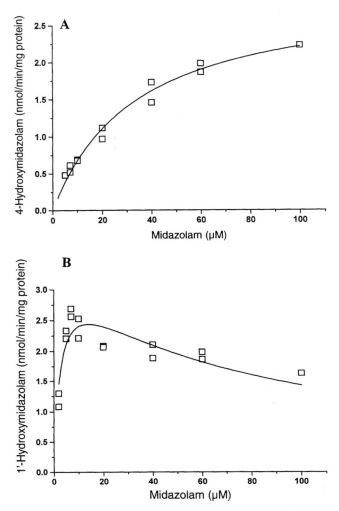

Figure 6. Kinetics of 1'-hydroxymidazolam and 4-hydroxymidazolam formation from cDNA expressed CYP3A4 from E. coli. The squares represent 1'-hydroxymidazolam, respectively 4-hydroxymidazolam, formation rates, and the curve represent the best fit calculated under the assumption of Michaelis Menten kinetics for 4-hydroxymidazolam and with noncompetitive substrate inhibition for 1'-hydroxymidazolam.

Midazolam was transformed to its two metabolites, 1'-hydroxymidazolam and 4-hydroxymidazolam by the expressed enzyme. Formation rates and kinetic parameters for 4-hydroxymidazolam were $K_m = 33 \pm 4$ μM and $V_{max} = 2950 \pm 170$ pmol/min/mg protein, compared to those determined with human liver microsomes, $K_m = 40 \pm 6$ μM and $V_{max} = 1800 \pm 117$ pmol/min/mg protein.

For 1'-hydroxymidazolam pronounced "substrate/product"-inhibition was observed for the cDNA expressed enzyme (Figure 6). Enzyme kinetic data were best fitted to Michaelis Menten kinetics with uncompetitive substrate inhibition, resulting in a K_m of 2.5 ± 1 μM, V_{max} of 3318 ± 450 pmol/min/mg protein and a K_i of 76 ± 29 μM for substrate inhibition. Apart from this, formation kinetics of 1'-hydroxymidazolam were equivalent to those, determined with human liver microsomes, $Km = 4.2 \pm 0.8$ μM and $Vmax = 2508 \pm 115$ pmol/min/mg protein. "Substrate inhibition" of 1'-hydroxymidazolam formation had also been observed[10] in human liver microsomes, but for substrate concentrations higher than 64 μM. α-Naphthoflavone activated the catalytic activity of the expressed enzyme for midazolam metabolism, similar as with human liver microsomes.[11]

3. DRUG INTERACTIONS

Increasingly patients are confronted with a variety of drugs, each one with the potential to interact with other coadministered compounds. An important factor of unexpected side effects during therapy with multiple drugs is mutual competition for or inhibition of the rate-limiting drug-metabolizing enzymes responsible for elimination of the drug. Such metabolic interactions lead to changes of plasma levels of drugs, with the potential for concomitant toxic effects in case of compounds with narrow therapeutic indices.[12] Therefore, as candidate drugs enter the clinical development phase, it needs to be considered which products are likely to be co-prescribed. The potential of these compounds to interact with the metabolism of the candidate drug can be assessed in vitro, as the basis for specific and focused trials in vivo.

Metabolic drug-drug interactions can occur when compounds are metabolized by the same drug metabolizing enzyme(s), when the differences in the affinities of the drugs for a metabolizing enzyme are large, and when the concentration of the inhibitory species at the enzyme site exceeds its affinity for the enzyme. But the results from, for example, in vitro competition studies, necessitate careful interpretation, in view that they relate only to free concentrations at the enzyme site. The concentrations of the interacting drugs should mirror the therapeutic concentrations expected in vivo and, often, the predictions only apply to the actual free (non-bound) drug fractions.

In many cases the effect of several clinically important drugs on the metabolism of cytochrome P-450 probe-substrates (for example, midazolam is often used as a probe for CYP 3A4 activity) are investigated *in vitro* with human liver microsomes. Inhibitory drug-effects can be classified into three categories, according to their mechanism. A first category comprises compounds exhibiting competitive inhibition mechanism. These compounds exhibit concentration-dependent inhibition in vivo. The second category encompasses non-competitive and uncompetitive inhibitors which often act by disrupting the microsomal electron transfer system, as in the case of many amphipathic molecules, or by tightly binding to the heme moiety present in P-450 as in the case of imidazole-containing chemicals. The third category includes drugs that elicit a preincubation time dependent decrease of product formation, like the macrolide antibiotics (for example, erythromycin), as an indication for irreversible or metabolism based inhibition mechanism.

The potential for, and the degree of metabolic inhibition of the compounds within a category can be estimated from the inhibition constant Ki and related to the estimated concentration of the compound in vivo in the liver.

4. ENZYME INDUCTION

Induction of drug metabolizing enzymes was first described in the late 50s by Conney et al.[13] The administration of structurally diverse chemicals increases the activity of one or more metabolic enzymes: nowadays, hundreds of compounds are known to elicit this effect and many drugs stimulate their own metabolism by induction of the P-450 enzyme involved. Clearance is enhanced and the duration and intensity of the pharmacological response is reduced. For example, many barbiturates induce their own P-450 mediated metabolism, resulting in increased tolerance and decreased duration of sedation.[14] As a consequence, to maintain the same pharmacological effect the dosage needs to be increased. In addition these compounds often stimulate the metabolism of unrelated compounds, both exogenous and endogenous, in particular those which are substrates of the induced P-450.

In many cases drug interactions based on P-450 induction might be of little clinical significance. However, where drugs with narrow therapeutic indices are involved, induction may have severe or fatal consequences. A well-known example is the decrease in warfarin plasma levels in patients receiving phenobarbital. The dosage of warfarin has to be increased in order to maintain adequate anticoagulant levels. When phenobarbital is discontinued, the P-450 levels return to the normal, non-induced state. If the dose of warfarin remains unchanged, higher plasma levels will result and severe bleeding ensues, with a potentially fatal outcome.[15] Because of this, the efficacy and/or plasma levels of certain compounds (for example, cyclosporine) have to be continuously monitored during therapy, resulting in increased costs and it will be always necessary to recognize if induction occurs and if so, which isoenzymes are involved.

P-450 inducers are categorized on the basis of which cytochrome(s) are increased. Thus, polycyclic aromatic hydrocarbons induce P450s 1A1 and 1A2, whereas phenobarbital, glucocorticoids, ethanol, and clofibrate increase the levels of P450s 2B, 3A, 2E, and 4A, respectively. Examples are also known where a single agent induces two or more cytochromes (phenobarbital) and where several compounds induce the same cytochrome. The mechanisms by which P-450 induction occurs have been extensively studied, but are not yet fully understood.[16] Most of our knowledge to date has been gained from experiments in rats, and far less is known about the effects in humans where ethical reasons frequently preclude such studies. Most of the human data has been derived from clinical observations and the analysis of surgical specimens.[17]

The potency of inducing agents varies between species. For example, rifampicin is a potent inducer in man but not in rats. Fortunately, induction by new drug candidates can be assessed during the first human studies. Increased levels of P-450 are mirrored by increases in certain excreted metabolites. For example, in addition to xenobiotics, CYP3A also metabolizes endogenous steroids. The inactivation of cortisol (corticosterone) to 6ß-hydroxycortisol can be used to measure this isoenzyme, and the changes in urinary excretion patterns can be used to assess CYP3A induction. The determination of cortisol and its metabolite is well documented and does not usually require the collection of any additional samples from the volunteers in a Phase 1 study.[18] Similarly, caffeine metabolism to paraxanthine in humans depends on CYP1A2, and the urinary metabolite pattern can again be used as a non-invasive test for this type of induction.[19]

REFERENCES

1. W. Kalow, 1993, Pharmacogenetics: Its biological roots and the medical challenge, *Clin Pharmacol Ther*, **54**:235–241.
2. L.M. Dislerath, F.P. Guengerich, 1994, Enzymology of human liver cytochrome P-450, in: *Mammalian Cytochromes P-450*, Volume 1 (F.P. Guengerich, ed.) pp. 133–198, CRC Press, Boca Raton, Florida.
3. F.P. Guengerich, 1987, Cytochrome P-450 and drug metabolism, *Progress in Drug Metab.* **10**:1–44.

4. A.D. Rodrigues, 1994, Use of in vitro human metabolism studies in drug development, *Biochem. Pharmacol.* **48**:2147–2156.

5. R.P. Remmel and B. Burchell, 1993, Validation and use of cloned, expressed human drug metabolizing enzymes in heterologous cells for analysis of drug metabolism and drug-drug interactions, *Biochem. Pharmacol.* **46**:559–566.

6. A. Wiseman, 1993, Genetically- engineered mammalian cytochromes P-450 from yeast-potential applications. *TIBTECH* **11**:131–136.

7. J. Doehmer, W.A. Schmalix and H. Greim, 1994, Genetically engineered in vitro systems for biotransformation studies, *Meth. Exp. Clin. Pharmacol.* **16**:513–518.

8. J.A.R. Blake, M. Pritchard, S. Ding, G.C.M Smith, B. Burchell, C.R. Wolf and T. Friedberg, 1996, Coexpression of a human P450 (CYP3A4) and P450 reductase generates a highly functional monooxygenase system in Escherichia coli, *FEBS Lett.* **397**:210–214.

9. T. Omura and R. Sato, 1964, The carbon monoxide-binding pigment of liver microsomes. *J. Biol. Chem.* **239**:2370–2378.

10. T. Kronbach, D. Mathys, M. Umeno, F.J. Gonzales and U.A. Meyer, 1989, Oxidation of midazolam and triazolam by human liver cytochrome P450IIA4. *Mol. Pharmacol.* **36**:89–96.

11. M. Shou, J. Grogan, J.A. Mancewicz, K.W. Krausz, F.J. Gonzalez, H.V. Gelboin and K.R. Korzekwa, 1994, Activation of CYP3A4: Evidence for the Simultaneously Binding of Two Substrates in a Cytochrome P450 Active Site, *Biochemistry* **33**:6450–6455.

12. C. Peck, R.T. Temple and J.M. Colllins, 1993, Understanding consequences of concurrent therapies, *JAMA*, **269**:1550–1552.

13. A.H. Conney, E.C. Miller, J.A. Miller, 1956, The metabolism of methylated amino dyes. Evidence for induction of enzyme synthesis in the rat by 3-Methylcholanthrene. *Cancer Research* **16**:450–459.

14. H. Remmer, H.J. Merker, 1963, Drug induced changes in the liver endoplasmic reticulum: association with drug metabolizing enzymes, *Science* **142**:1657–1658.

15. B.K. Park, A.M. Breckenridge, 1981, Clinical implications of enzyme induction and enzyme inhibition, *Clin. Pharmacokinet.* **6**:1–24.

16. A.B. Okey, 1990, Enzyme induction in the cytochrome P-450 system, *Pharmac. Ther.* **45**:241–298.

17. A.R. Boobis, D. Sesardic, B.P. Murray, R. J. Edwards, A.M. Singleton, K.J. Rich, S. Murray, R. de la Torre, J. Segura, O. Pelkonen, 1990, Species variation in the response of the cytochrome P-450 dependent monooxygenase system to inducers and inhibitors, *Xenobiotica* **20**:1139–1161.

18. C. Ged, J.M. Rouillon, L. Pichard, J. Lombalbert, N. Bressot, P. Bories, H. Michel, P. Beaune, P. Maurel, 1989, The increase in urinary excretion of 6ß-hydroxycortisol as a marker of human hepatic cytochrome 3A induction. *Br. J. Clin. Pharmacol.* **28**: 373–387.

19. W. Kalow, B.K. Tang, 1991, Use of caffeine metabolite ratios to explore CYP1A2 and xanthine oxidase activities, *Clin. Pharmacol. Ther.* **50**:508–519.

14

AFLATOXIN BIOTRANSFORMATION AND TOXICOLOGY[*]

David L. Eaton

Department of Environmental Health
University of Washington
4225 Roosevelt Way, NE, Suite 100
Seattle, Washington 98105-6099

1. INTRODUCTION

Aflatoxins represent a group of closely related mycotoxins produced by the common fungal molds, *Aspergillus flavus* and *Aspergillus parasiticus* (Figure 1). The major host crops for aflatoxigenic *Aspergillus* strains are corn, peanuts and cottonseed. In corn, contamination occurs most commonly in insect-damaged ears, and growth in all crops is favored by high heat and humidity.[1]

In contrast to contamination in the United States, the concentration of aflatoxins in human food in some developing countries is quite high. Groopman and colleagues[2,3] reported average daily intakes of aflatoxin in the Guangxi Autonomous Region of the Peoples Republic of China to be approximately 50 to 75 µg/day. Corn contaminated with 100–200 ppb aflatoxin was the primary dietary source.

Of the four aflatoxins that commonly occur in food and feeds (AFB$_1$, AFB$_2$, AFG$_1$, and AFG$_2$), AFB$_1$ occurs in the highest concentrations and is also the most toxic and carcinogenic. AFB$_1$ is acutely toxic in all animal species studied and death generally occurs as a result of hepatoxicity. Susceptibility to AFB$_1$ toxicity can differ by up to two orders of magnitude between species. For example, in one study comparing the toxicity of AFB$_1$ between species the LD$_{50}$s for duckling and mouse were found to be 0.5 mg/kg and 60 mg/kg, respectively. Rats are more susceptible than mice to AFB$_1$ toxicity and are extremely sensitive to the hepatocarcinogenic effects of AFB$_1$ relative to mice. Acute toxicity and carcinogenicity vary within species as a result of differences in strain, sex, age, route of administration, and nutritional status.[4]

---

[*] Adapted from: Eaton, DL and Heinonen, JT: Aflatoxin Toxicology: in Comprehensive Toxicology, Vol. 9, Chp. 26, pp. 407–422, 1997, with permission from Elsevier Science Ltd., The Boulevard, Landford Lane, Kidlington 0X5 1GB, United Kingdom.

Molecular and Applied Aspects of Oxidative Drug Metabolizing Enzymes,
edited by Arinç *et al.* Kluwer Academic / Plenum Publishers, New York, 1999.

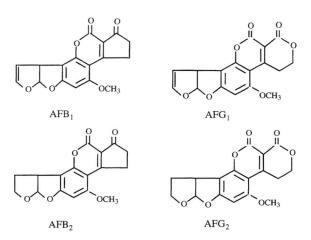

Figure 1. Structures of naturally occurring aflatoxins.

There have been over 20 different chronic studies in rats alone demonstrating the potent carcinogenic effects of aflatoxins (see[5] for a detailed review). Wogan & Newberne[6] provided the most dramatic demonstration of the potency of aflatoxin as a hepatocarcinogen. These investigators found that a dietary concentration of 15 ppb (μg/kg) fed continuously for 68–80 weeks to Fisher rats resulted in a 100% incidence of hepatic tumors. Although a later dose-response study found only a 20% incidence of tumors at 15 ppb, a 100% incidence of tumors was obtained at 100 ppb for 54–88 weeks.[7] From a comparative experimental point of view, these results place aflatoxin among the most potent carcinogens of all chemicals ever tested.

In contrast to rats and trout, mice are highly resistant to the hepatocarcinogenic effects of aflatoxin. In one study, Swiss-Webster mice fed peanut meal containing 100,000 or 150,000 ppb of a mixed aflatoxin preparation (AFB$_1$ + AFG$_1$) for 80 weeks (post-weaning) did not develop any hepatic tumors.[5] However, intraperitoneal administration of aflatoxin to newborn (pre-weaning) mice can induce liver tumors at a high rate.[8]

Epidemiological studies of liver cancer incidence in populations where dietary aflatoxin exposures are high have provided much circumstantial evidence for the development of AFB$_1$-induced liver cancer in humans.[9–11] However, in these same populations there also exists a high incidence of hepatitis B virus (HBV) infection which, along with aflatoxin exposure, has been associated with liver cancer.[12,13] Recent reports from a continuing nested case-control study in China now indicate that the unusually high incidence of liver cancer in this study region may be the result of a synergistic interaction between AFB$_1$ exposure and HBV infection.[14,15] The most recent report from this study found a relative risk of 3.4 for aflatoxin exposure alone, a relative risk of 7.3 for HBV alone, but a relative risk of 60 for combined aflatoxin exposure and HBV.[15]

2. MECHANISTIC BASIS FOR AFLATOXIN CARCINOGENESIS

2.1. Formation of DNA Adducts

AFB$_1$ is metabolized to a reactive intermediate capable of covalent modification of DNA[16–19] (Figure 2). The epoxide which serves as the ultimate genotoxic metabolite of aflatoxin exists in two stereoisomeric forms, of which the *exo* conformation of aflatoxin B$_1$-8,9-

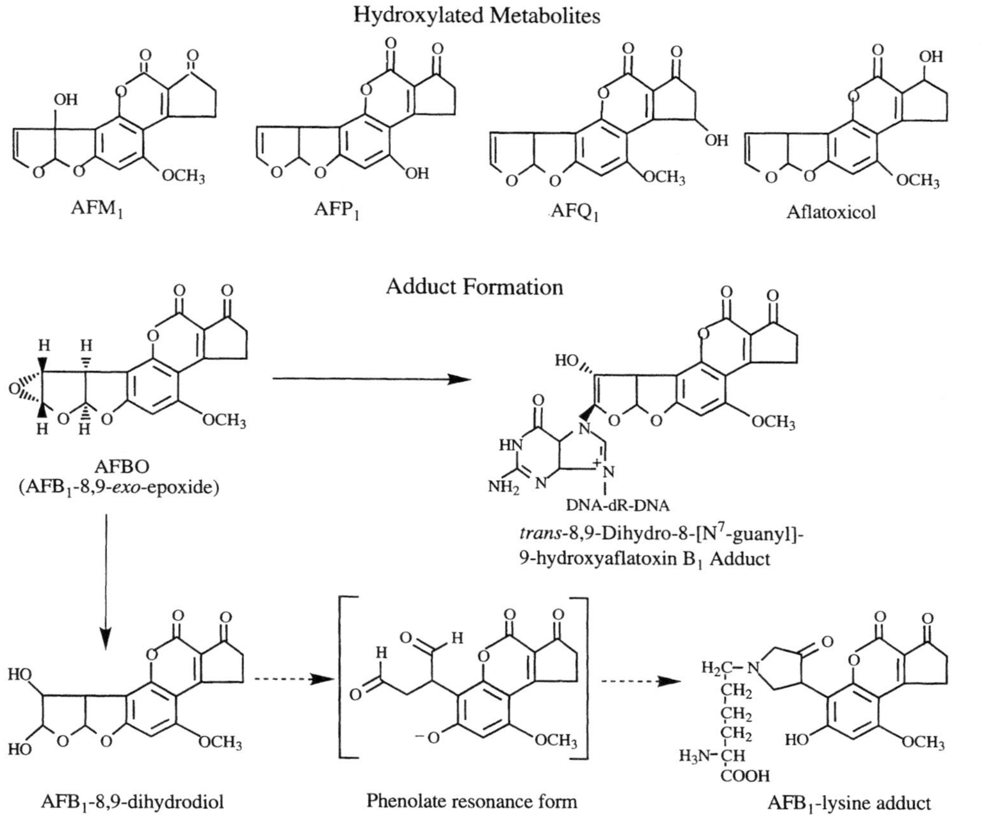

Figure 2. Structures of aflatoxin B_1 metabolites and adducts.

epoxide appears to be most important.[20] Although other aflatoxin metabolites, such as the epoxides of AFM_1, AFP_1 and AFQ_1, may contribute to DNA binding, the evidence to date strongly indicates that such secondary oxidation products are of minor importance.[21,22] Although other DNA bases such as adenosine[23] and cytosine[24] are covalently modified by aflatoxin in vitro, there is no evidence to indicate that these minor base modifications have any functional importance in aflatoxin carcinogenesis. One reason the N^7-guanine adduct may be so important is the relatively rapid rate at which it undergoes rearrangement to the ring-opened, formamidopyrimadine (FAPY) form. The FAPY-aflatoxin adduct appears to be the most stable of all AFB-DNA adducts, and is relatively resistant to DNA repair processes.[22] However, the relative importance of the FAPY adduct versus other more labile DNA adducts in the ultimate development of tumors in aflatoxin exposed animals remains uncertain.

Numerous studies have demonstrated that carcinogenic potency is highly correlated with the extent of total DNA adducts formed in vivo. When the administered dose is normalized to target dose (e.g., DNA adducts per 10^8 nucleotides), a highly linear relationship between DNA adduct formation and tumor response is obtained, even when using combined data from both rats and rainbow trout.[25] In both rats and trout there is also a highly linear relationship between administered dose and DNA adducts levels. Buss et al.[26] demonstrated a near perfect linear relationship between AFB_1 dose (ng/kg/day) and AFB-DNA adducts (adducts/10^9 bases) over a 5 order of magnitude dose range that extended from a high dose which was expected to yield a 50% tumor incidence in rats to a low dose that encompasses the range of expected human exposure. Dashwood et al.[27] also found a large range of linearity between total administered dose and AFB-DNA adduct levels in rainbow trout give aflatoxin for 2–4 weeks. These studies thus fail to provide evidence in support of a threshold hypothesis for aflatoxin genotoxicity at low doses, at least in two highly sensitive yet diverse species, rats and rainbow trout.

2.2. Activation of Oncogenes

Given the strong mutagenic effects of aflatoxin, there has been considerable interest in identifying possible oncogenes and tumor suppressor genes which may serve as critical molecular targets in aflatoxin carcinogenesis. Numerous studies have demonstrated that AFB_1 produces mutations in codon 12 of all three types of c-*ras* oncogenes (Ha-*ras*, Ki-*ras*, and N-*ras*).[28–34] The major mutation was a G-->A transition mutation at the second base of codon 12 (GGT to GAT), although G-->T transversions at both the first and second base of codon 12 were also found.[32]

Few, if any, studies have identified activated c-*ras* oncogenes in human hepatocellular carcinomas obtained from aflatoxin-endemic areas. However, both Ha-*ras* and Ki-*ras* oncogenes are frequently activated in experimental liver tumorigenesis models, and activated *ras* oncogenes have been implicated in the early stages of carcinogenesis in a variety of human cancers.[35] Whether activation of human *ras* protoocogenes is an important factor in the etiology of aflatoxin-induced liver cancer in humans remains to be demonstrated conclusively, although there is supportive evidence of such a role.

2.3. Inactivation of Tumor Suppressor Genes

A number of recent studies have demonstrated a relatively high prevalence of G-->T transversion mutations in codon 249 of the p53 tumor suppressor gene in hepatocellular carcinomas from patients residing in areas of the world with high dietary exposure to aflatoxin (see Eaton and Gallagher[36] for a review). The frequency of mutated p53 (any site) is about twice as prevalent in aflatoxin-endemic areas relative to the prevalence in developed countries with low aflatoxin exposure. However, the prevalence of a specific mutation at codon 249 in p53 (usually, but not always, a G-->T transversion in the third base) is more than 10 times that in low aflatoxin exposure regions. In fact, less than 3% of liver tumors examined from patients residing in areas of the world considered to have low aflatoxin exposure contained a mutation at codon 249. Thus, there is relatively strong circumstantial evidence to support the hypothesis that mutations at codon 249 of p53 are indicative of aflatoxin-induced hepatocellular carcinoma.

Further sub-classification of p53 codon 249 mutation data in liver tumors by hepatitis B virus status is also of interest, although caution should be used in interpreting such summary statistics, as the presence of hepatitis B virus surface antigen, while indicative of exposure to the virus, is not necessarily indicative of a current or previously active viral infection. With this qualification, it is interesting to note that, collectively, the prevalence of p53 codon 249 mutations in liver tumors from individuals residing in high aflatoxin exposure areas is about 3 times greater among those individuals that were hepatitis B-positive.[36] Taken together, these data are consistent with the hypothesis that hepatitis B virus infection acts synergistically with aflatoxin in inducing mutations at codon 249. This perhaps provides a conceptual, mechanistic basis for the apparent synergistic interaction between aflatoxin and hepatitis B virus reported epidemiologically by Groopman and co-workers.[14,15]

Although the association between p53 codon 249 mutations and liver cancer in afla-toxin-endemic regions of the world is certainly suggestive of a causal connection, thus far there is no direct evidence that aflatoxin-DNA adducts occur at codon 249 in human liver tumor samples. However, a plasmid containing a full length p53 cDNA could be mutated at codon 249 (as well as other guanine "hot spots") in vitro by incubation with by AFB_1–8,9-epoxide. No codon 249 mutations in p53 were identified in aflatoxin-induced hepatocellular carcinomas obtained experimentally from non-human primates, and only one out of nine tumors examined was found to have a p53 mutation (codon 175, G-->T).[37] These results demonstrate that, at least in non-human primates, mutation of p53 is not re-quired for aflatoxin-induced hepatocellular carcinoma. Aflatoxin has been shown to mod-ify p53 gene structure and expression in rat tumors induced with aflatoxin B_1,[38] although no rat p53 mutations in the region corresponding to human p53 codon 249 were detected in aflatoxin-induced preneoplastic foci.[39]

The high prevalence of a specific mutation in codon 249 of the human *p53* gene in human liver cancers from aflatoxin-endemic areas of the world is intriguing, and may rep-resent the first example of a "carcinogen-specific" biomarker that remains permanently fixed in the tumor tissue. Although the evidence to date does not allow the unequivocal conclusion that all tumors with p53 codon 249 mutations are aflatoxin-derived, future pro-spective studies which combine biomarker assays for aflatoxin exposure with assessment of p53 codon 249 sequence in liver tumors may ultimately allow such a remarkable con-clusion to be inferred with some confidence.

3. BIOTRANSFORMATION OF AFLATOXINS

As noted above, aflatoxins must first be activated to a reactive epoxide intermediate prior to exerting carcinogenic effects, but numerous competing pathways also function to detoxify this mycotoxin. AFB_1 is rapidly biotransformed in the liver to a variety of meta-bolites, but there are substantial species differences in the relative activities of the various biotransformation pathways (Figure 3). Because both activation and detoxification path-ways function at the same time, it is the ratio of activation : inactivation, rather than the absolute rate of overall biotransformation, that is the critical determinant of species sus-ceptibility to AFB_1 hepatocarcinogenesis and, to a lesser extent, acute toxicity. AFB_1 and/or its primary metabolites undergo oxidation, reduction, hydrolysis and conjugation via a multitude of different enzymatic pathways. To understand the mechanistic basis for aflatoxin carcinogenicity, and the reasons for such large differences in species susceptibil-ity, one must carefully examine the qualitative and quantitative aspects of each step in the overall biotransformation of AFB_1.

3.1. Oxidation of AFB_1 via Cytochromes P450

AFB_1 biotransformation occurs principally by the microsomal cytochrome P450 (CYP450) mixed function monooxygenase system and involves multiple CYP450 isozymes that show considerable variation in kinetic characteristics and product specific-ity between species (see[36] for a recent review). The predominant AFB_1 metabolites from CYP450 catalyzed reactions in mammals are aflatoxin M_1 (AFM_1), aflatoxin Q_1 (AFQ_1), aflatoxin P_1 (AFP_1), and aflatoxin B_1–8,9-epoxide (AFBO, also referred to as AFB-2,3-epoxide in older literature) (Figure 3).

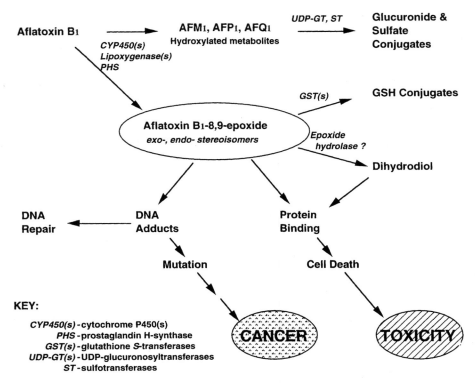

Figure 3. Overview of aflatoxin B_1 biotransformation.

3.1.1. CYP450-Mediated Detoxification of AFB$_1$. Oxidation of AFB$_1$ to AFQ$_1$, AFM$_1$, and AFP$_1$ represent detoxification reactions, as these monohydroxy metabolites are considerably less toxic and carcinogenic than the parent AFB$_1$.[40–44] AFQ$_1$ is formed via 3a hydroxylation of AFB$_1$ and AFM$_1$ is produced by 9a hydroxylation of AFB$_1$. In human liver, AFQ$_1$ is formed predominantly, if not exclusively, by CYP3A4. AFM$_1$ is formed by CYP1A2 in human liver, although cDNA-expressed human CYP1A1 is also capable for forming AFM$_1$. The O-demethylation of AFB$_1$ produces AFP$_1$ in mice, rats, and monkeys.[45] Although not a prominent AFB$_1$ metabolite in human liver microsomes,[46,47] microsomal fractions from primary human liver tumors from Thai patients produced higher levels of AFP$_1$ than observed in microsomes prepared from normal liver tissue.[48] AFP$_1$ is also produced by human liver slices incubated with AFB$_1$[49] and is also a common urinary metabolite in humans exposed to dietary AFB$_1$ in vivo.[2] In this regard, AFP$_1$ was the most highly correlated of all urinary AFB$_1$ metabolites in humans with liver cancer.[14] It remains uncertain which hepatic enzyme(s) is/are responsible for AFP$_1$ production in human or rodent liver.

A dihydroxy metabolite of aflatoxin B$_1$, AFM$_1$-P$_1$, can be formed by the 9a-hydroxylation of AFP$_1$ or the 4-O-demethylation of AFM$_1$.[50] This metabolite is formed in vivo in rats and is excreted directly in bile or as the glucuronide conjugate.[50] Overall, the secondary oxidation of AFP$_1$ is probably of little toxicological importance. Because AFM$_1$ is both mutagenic and carcinogenic, it is possible that secondary oxidative demethylation of AFM$_1$ and subsequent glucuronidation are significant routes for AFM$_1$ detoxification in vivo.

3.1.2. CYP450-Mediated Activation of AFB1. Microsomal cytochrome P450 (CYP450)-dependent epoxidation of the terminal furan ring of AFB_1 generates the highly reactive epoxide which is responsible for nucleic acid alkylation.[19] Chemical and enzymatic epoxidations of AFB_1 yields exo- and endo-AFB_1 epoxide stereoisomers.[51] The epoxide ring is positioned above the plane and trans to the 5a and 9a protons in the endo-stereoisomer, whereas the epoxide ring points below the plane and cis to the 5a and 9a protons in the exo-epoxide.[51] The metabolic activation of AFB_1 by human or rat microsomes produces a mixture of endo-and exo-epoxides, which can be trapped with GSH and identified by comparison with standards prepared by the reaction of the endo- or *exo*-stereoisomers with GSH.[51] Both isome rs are produced by human liver microsomes, although the *exo*-epoxide predominates.[51] Rat microsomes also efficiently form the exo-epoxide, but are far less efficient at forming endo-epoxide when compared to human microsomes. Although the endo-epoxide is less susceptible to hydrolysis compared to the *exo*-conformation,[51] the exo-epoxide is much more efficient at forming DNA adducts, and is much more mutagenic than the endo-epoxide.[52]

The extremely short half life of the AFBO in aqueous solution has precluded its isolation from biological systems. Until relatively recently, AFBO formation could only be inferred by interception with trapping agents such as DNA and glutathione *S*-transferase (GST) catalyzed GSH conjugation. It has now been successfully synthesized and chemically characterized.[53,54] When added to aqueous solutions, synthetic AFBO undergoes rapid non-enzymatic hydrolysis to AFB_1–8,9-dihydrodiol.[51] AFB_1,-8,9-dihydrodiol may exist in a phenolate resonance form that is capable of forming Schiff bases which react with amino acids (Fig. 3). Sabbioni et al.[55] demonstrated that AFB_1-lysine adduct is the predominant adduct found in rats after in vivo AFB_1 exposure. The rapid reaction of AFB_1 with Tris buffer to form an AFB_1–8,9-epoxide-Tris adduct can also be exploited to monitor AFBO formation in microsomal incubations containing this buffer.[56]

The activation of AFB_1 to AFBO in human liver microsomes appears to be a complex process which is mediated by multiple P450 enzymes exhibiting different kinetic characteristics. At least five different human liver P450s, including 1A2, 2A6, 2B7, 3A3 and 3A4 are capable of activating AFB_1 to mutagenic metabolites and DNA-bound derivatives.[57,58] Crespi et al.[57] showed that human cell lines selectively expressing CYP 1A2 were 3 to 6-fold more effective than those expressing CYP 3A4, and 40 to 50-fold more effective than CYP 2A3, at activating AFB_1 to mutagenic metabolites at low substrate concentrations. Because CYP 3A4 is expressed in relatively high quantities in human liver it is the dominant CYP450 enzyme responsible for the in vitro activation of AFB_1 to AFBO in human liver microsomes at high substrate concentrations.[47,59] However, human CYP3A4-mediated activation of AFB_1 to AFBO follows non-linear Hill kinetics (typical of allosteric activation), rather than Michalelis-Menten kinetics, and aflatoxin itself is apparently capable of activating its own oxidation.[60] Thus, at the low AFB concentrations typical of human dietary exposure, the enzyme likely exhibits relatively low catalytic activity toward AFBO in vivo. Furthermore, human CYP3A4 forms both AFBO and AFQ_1 in a ratio of approximately 1:10 (AFBO:AFQ_1),[21,47] and thus is predominantly a detoxification pathway.

In vitro, human CYP1A2 forms both AFM_1 and AFBO in a ratio of approximately 2.5:1 (AFBO:AFM_1), and is thus primarily an activation pathway. However, in contrast to CYP3A4, which forms exclusively the DNA-binding exo-stereoisomer, human CYP1A2 forms both endo and exo isomers in approximately a 1:1 ratio.[21,60] From a series of studies using human liver microsomes, specific cDNA expressed CYP enzymes, and a highly effective and specific inhibitor of CYP1A2 (furafylline), Eaton and co-workers[47,60] have demonstrated that CYP1A2 is likely to be the predominant human CYP enzyme involved in activation of AFB_1 to carcinogenic metabolites in vivo at the hepatic AFB_1 concentration that would result from typical dietary intake of AFB_1. CYP1A2 expression exhibits considerable genetic variation in human liver - perhaps as high as 70 -fold.[61-63] CYP1A2 is also inducible by polycyclic aromatic hydrocarbons such as those found in cigarette smoke[64] and charbroiled beef.[65] Taken together, the differential expression of CYP1A2 and the role of CYP1A2 in AFB_1 activation in human liver suggest that there may be large interindividual differences in susceptibility to AFB-induced hepatocarcinogenicity.

3.2. Oxidation of AFB_1 by Non-CYP-Mediated Pathways

Although CYP450-mediated oxidation of AFB_1 is considered to be the dominant route for AFB_1 epoxidation, CYP450-independent pathways for AFB_1 activation have also been demonstrated. Battista et al.[66] reported that prostaglandin H synthase (PHS)-dependent epoxidation of AFB_1 can co-occur with CYP450-mediated AFB_1 epoxidation. Lipoxygenases from guinea pig liver and kidney are also capable of activating AFB_1 to DNA-bound derivatives.[67]

3.3. Phase II (Conjugation and Hydrolysis) Reactions with AFB Metabolites

Several of the products of the oxidative metabolism of AFB_1 serve as substrates for phase II detoxification enzymes. Extensive interspecies variation exists in the functional importance of the different phase II enzymes in AFB_1 detoxification, in particular with the glutathione S-transferases.

3.3.1. Conjugation with Glutathione. Detoxification of AFBO in rodents occurs mainly by conjugation with glutathione (GSH) via glutathione S-transferase (GST)[68-71] (Figure 3). In rodents, the amount of GST activity towards AFBO is inversely related to species susceptibility to AFB_1-induced hepatocarcinogenesis. Ultimately, the amount of AFB_1 that will bind to DNA is determined by the proportion activated to the epoxide and the fraction of the epoxide that is enzymatically conjugated with GSH.

The selectivity of GST isozymes towards AFBO serves as a critical determinant of mammalian species susceptibility to AFB_1-induced hepatocarcinogenesis.[72-78] Mouse liver cytosolic fractions have 50 to 100 fold greater AFBO conjugating activity than rat even though both species have comparable amounts of GST activity towards 1-chloro- 2,4-dinitrobenzene (CDNB).[79] Accordingly, mice are resistant to the hepatocarcinogenic effects of AFB_1 when compared to rats,[6] a difference reflected by 50 to 100- fold less AFB_1-DNA adduct formation by mice after in vivo AFB_1 exposure.[70] The efficient detoxification of AFBO by mouse GSTs appears to be a characteristic common to most mouse strains, as nine different strains exhibited similar high specific GST activities towards AFBO.[80]

Mice contain at least three alpha class GST subunits, one of which is constitutively expressed while the other two are inducible by chemoprotective agents.[81] Two alpha class GSTs from mouse liver have been cloned and sequenced.[82-84]

Human liver cytosolic fractions are very poor at conjugating either stereoisomer of AFBO, although the human GST mu form, GST M1a-M1a, had significant conjugating activity toward the *endo*-epoxide. Human alpha class GSTs had only marginally detectable activity toward either epoxide.[85] GST M1a-M1a is inherited in an autosomal dominant fashion in human liver and is therefore absent in approximately 50% of the Caucasian population. Thus, it is possible that AFB_1-exposed individuals who lack this enzyme and may be at increased risk for AFB_1 hepatocarcinogenesis. As noted above, however, it is questionable whether the *endo*-epoxide is of biological relevance, as it binds poorly to DNA and may not be mutagenic.[52]

A recent human molecular epidemiology study reported a relationship between susceptibility to aflatoxin-related hepatocellular carcinoma (HCC) and the presence of the null genotype for GST-M1.[86] The GST M1 null genotype was not significantly associated with increased risk for HCC, although small sample size hindered the power of this study to detect an association. Although the human GST M1–1 enzyme has relatively low activity catalytic activity toward the carcinogenic AFB-exo-8,9-epoxide[85] when compared with the high activity rodent alpha class GSTs (mGSTA3–3, rGST A5–5), given the absence of any high activity alpha form in human liver, the low GSTM1–1 activity may still play an important role in detoxification of AFB-exo-epoxide in humans. Recent studies in our laboratory with non-human primates suggest that the protective effects of dietary oltipraz could result in part from induction of a mu-class GST with low but significant catalytic activity toward AFB-8,9-exo-epoxide (Bammler and Eaton, unpublished observations).

3.3.2. Glucuronide and Sulfate Conjugates. The glucuronides of AFP_1 and 4,9a-dihydroxyaflatoxin B_1 have been identified as biliary metabolites in rats treated with AFB_1.[50] Studies with isolated hepatocytes suggest that AFP_1 is a better substrate for glucuronidation than are AFM_1 and AFQ_1.[87,88] However, as the enzymatic hydrolysis of non-aromatic glucuronide conjugates with bacterial ß-glucuronidase preparations can be inefficient and may underestimate the amount of conjugate present,[88] the extent and significance of glucuronidation of AFM_1 and AFQ_1 remains uncertain. AFL-glucuronide and AFL-M_1-glucuronide are the major AFB_1 conjugates in rainbow trout fed control diets.[89] Sulfate conjugates were not detected in rainbow trout exposed to AFB_1,[89] consistent with other studies indicating that rainbow trout excrete xenobiotics as glucuronides, as opposed to sulfates.[90] Treatment of aqueous urinary AFB_1 metabolites from rhesus monkeys with arylsulfatase released AFM_1, indicating the in vivo formation of the sulfate conjugate of AFM_1 in this species.[91]

3.3.3. Hydrolysis via Microsomal Epoxide Hydrolase. The significance of microsomal epoxide hydrolase (mEH) in AFBO hydrolysis has not been unequivocally established. Studies using isolated hepatocytes from rats and mice suggest that microsomal EH actively converts AFBO to AFB_1-dihydrodiol.[92,93] Shayiq & Avadhani[94] reported a reduction in AFB_1-DNA binding when EH was added to a reconstituted system containing purified P450s. However, other work using specific EH inhibitors in rat and mouse hepatic subcellular fractions indicates that EH does not play a significant role in the inactivation of AFBO.[70,78,95] This is apparently because either AFBO is not a substrate or because the K_m of EH for AFBO is relatively high as compared to that of GST. However, AFBO hydrolysis does occur, since the dihydrodiol is rapidly formed under in vitro conditions in the absence of GST.

The human molecular epidemiology study reported by McGlynn et al.[86] examined the relationship between susceptibility to aflatoxin-related hepatocellular carcinoma (HCC) and the presence of a variant form of the human microsomal epoxide hydrolase gene.[86] This study showed a significant association (OR = 3.3) between the variant EH allele and the incidence of HCC in an AFB-exposed population.

Recent studies from our laboratory have shown that co-expression of human microsomal epoxide hydrolase with human CYP1A1 results in a decrease in mutagenicity of AFB1 in the Ames salmonella mutagenicity assay when compared to CYP1A1 expression alone, lending further support to the hypothesis that human mEH could be an important detoxification pathway for AFB_1.

3.4. Reduction of AFB_1

Several species, including rabbit, chicken, and trout may rapidly convert AFB_1 to aflatoxicol (AFL) by reduction of the 1-keto-group through a cytosolic NADPH-dependent reductase.[96] AFL can be further metabolized by undergoing a 9a-hydroxylation to form AFL-M_1.[97] Although AFL is not an important reductive metabolite of AFB_1 in mammalian liver, it has been identified as the major in vivo AFB metabolite in the plasma of rats administered AFB_1 either orally or i.v.[98] The formation of AFL does not appear to be an important detoxification pathway for AFB_1, as AFL may be rapidly converted back to AFB_1 by a microsomal dehydrogenase,[96] thereby increasing the physiological half-life of AFB_1.

Hydrolysis of AFBO to the dihydrodiol occurs rapidly, but at physiological pH this product may rearrange to form AFB_1–8,9-aldehyde, which in turn can react with free amines via a Schiff base formation to yield AFB-protein (lysine) adducts. This reaction may be responsible for some of the acute toxic effects of AFB_1. Recently, a new aldo-keto reductase involved in the reduction of AFB_1–8,9-aldehyde (AFBA), has been described.[99,100] This enzyme, termed aflatoxin B_1-aldehyde reductase is dissimilar from any previously described reductase. It is inducible by ethoxyquin, and is expressed in greater quantities in preneoplastic foci.[99] Although the significance of this enzyme in the detoxification of AFB-aldehyde has yet to be demonstrated definitively in vivo, it may be of importance in protecting against acute and chronic hepatotoxicity of AFB_1. As cytotoxic effects of AFB may play a significant promotional role in tumor development,[101] it is also possible that this enzyme pathway may offer some protection against AFB_1 carcinogenicity, but it is unlikely to influence genotoxicity.

4. QUANTITATIVE RISK ASSESSMENT FOR DIETARY AFLATOXIN

Numerous studies have attempted to extrapolate laboratory animal data and/or human epidemiological data on aflatoxin exposure to human liver cancer risk.[102–107] This subject has been reviewed recently.[108] Using the standard US regulatory approach of extrapolating tumor dose response data from the most sensitive species with the Linearized Multi-Stage (LMS) model, a "Virtually Safe Dose" (VSD) of 0.016 ng/kg/day aflatoxin B_1 was obtained for a risk level of 1×10^{-5}, based on the rat tumor data of Wogan et al.[7] When expressed as a potency value with units of $[\text{ng/kg/day}]^{-1}$, the value is 6.25×10^{-4}. Using this estimate of potency, and an estimated average daily intake of aflatoxin in the southeast United States of 110 ng/kg/day,[109] Bruce[102] estimated an excess lifetime cancer risk of 6.9×10^{-2}, which yields an annual incidence of liver cancer of 98/100,000. This is about 20 times greater than the actual incidence of liver cancer in the United States from *all* causes. Clearly, either the potency estimate is wrong, or the dietary exposure estimates in the US were substantially overestimated.

Using human epidemiologic data, Bruce, Hoseyni and Gorelick[102,105,106,108] estimated a potency factor for aflatoxin of 4.8×10^{-5}, based on data from aflatoxin-endemic areas in Africa and Thailand, and a value of 8.2×10^{-5} using data from China (see[102] for details). These values are 13 and 7.6 times less than the rat potency estimate of 6.25×10^{-4}, respectively. As noted by these authors, these values are likely overestimates of the true potency for aflatoxin alone, as the tumor response data for both epidemiological estimates were derived from areas with endemic hepatitis B virus. Hoseyni[106] attempted to correct for the influence of hepatitis B virus on the potency estimate for aflatoxin by using an exponential-multiplicative relative risk function applied to Chinese liver cancer incidence data. This model projected potency values of 1.6×10^{-6} and 2.8×10^{-6} for the best estimate and upper 95% CL, respectively, for human cancer risk from lifetime exposure to aflatoxin in the absence of hepatitis B virus. Using the upper confidence limit value of the corrected risk (2.8×10^{-6}), this estimate is about 30 times less than the estimate for aflatoxin-related liver cancer risk in the presence of hepatitis B virus, and 223 times less than the estimate based on LMS extrapolation of rat data. This difference is remarkably similar to the estimated magnitude of effect of hepatitis B virus on aflatoxin carcinogenicity from the recent prospective epidemiological data of Groopman and coworkers.[14,15] As discussed previously, these authors found a relative risk for liver cancer in an aflatoxin-exposed population (Guangxi, China) of about 2 in the absence of HBV, but slightly over 60 when both hepatitis B virus and aflatoxin exposure were present.

REFERENCES

1. D. M. Wilson and G. A. Payne, 1994, Factors affecting *Aspergillus flavus* group infection and aflatoxin contamination of crops, In: *The Toxicology of Aflatoxins: Human Health, Veterinary and Agricultural Significance*, (D. L. Eaton and J. D. Groopman, eds.) pp. 309–326, Academic Press, New York.

2. J. D. Groopman, Z. Jiaqi, P. R. Donahue, A. Pikul, L. Zhang, J.-S. Chen and G. N. Wogan, 1992, Molecular dosimetry of urinary aflatoxin-DNA adducts in people living in Guangxi Autonomous Region, People's Republic of China, *Cancer Res.* **52**: 45–52.

3. J. D. Groopman, 1994, Molecular dosimetry methods for assessing human aflatoxin exposure, In: *The Toxicology of Aflatoxins: Human Health, Veterinary and Agricultural Significance*, (D. L. Eaton and J. D. Groopman, eds.) pp. 259–280, Academic Press, New York.

4. J. M. Cullen and P. M. Newberne, 1994, Acute hepatotoxicity of aflatoxins, In: *The Toxicology of Aflatoxins: Human Health, Veterinary and Agricultural Significance*, (D. L. Eaton and J. D. Groopman, eds.) pp. 3–26, Academic Press, New York.

5. G. Wogan, 1973, Aflatoxin carcinogenesis, In: *Methods in Cancer Research*, (H. Busch, eds.) pp. 309–344, Academic Press, New York, London.

6. G. N. Wogan and P. M. Newberne, 1967, Dose-reponse characteristics of aflatoxin B_1 carcinogenesis in the rat., *Cancer Res.* **27**: 2730-.

7. G. N. Wogan, S. Paglialunga and P. M. Newberne, 1974, Carcinogenic effects of low dietary levels of aflatoxin B_1 in rats, *Fd. Cosmet. Toxicol.* **12**: 681–685.

8. S. D. Vesselinovitch, N. Mihailovich, G. N. Wogan, L. S. Lombard and K. V. N. Rao, 1972, Aflatoxin B_1, a hepatocarcinogen in the infant mouse, *Cancer Res.* **32**: 2289–2291.

9. R. C. Shank, P. Siddhichai, B. Subhamani, N. Bhamarapravati, J. E. Gordon and G. N. Wogan, 1972, Dietary aflatoxins and human liver cancer. V. Duration of primary liver cancer and prevalence of hepatomegaly in Thailand, *Food Cosmet Toxicol.* **10**: 181–91.

10. F. G. Peers and C. A. Linsell, 1973, Dietary aflatoxins and liver cancer--A population based study in Kenya., *Br. J. Cancer.* **27**: 473–483.

11. S. J. Van Rensburg, P. Cook Mozaffari, D. J. Van Schalkwyk, J. J. Van der Watt, T. J. Vincent and I. F. Purchase, 1985, Hepatocellular carcinoma and dietary aflatoxin in Mozambique and Transkei, *Br J Cancer.* **51**: 713–26.

12. F. Peers, X. Bosch, J. Kaldor, A. Linsell and M. Pluumen, 1987, Aflatoxin exposure, hepatitis B virus infection and liver cancer in Swaziland, *Int. J. Cancer.* **39**: 545–53.

13. F. S. Yeh, M. C. Yu, C. C. Mo, S. Luo, M. J. Tong and B. E. Henderson, 1989, Hepatitis B virus, aflatoxins, and hepatocellular carcinoma in southern Guangxi, China, *Cancer Res.* **49**: 2506–9.

14. R. K. Ross, J.-M. Yuan, M. C. Yu, G. N. Wogan, G.-S. Qian, J.-T. Tu, J. D. Groopman, Y.-T. Gao and B. E. Henderson, 1992, Urinary aflatoxin biomarkers and risk of hepatocellular carcinoma, *Lancet.* **339**: 9343–946.

15. G. S. Qian, R. K. Ross, M. C. Yu, J. M. Yuan, Y. T. Gao, B. E. Henderson, G. N. Wogan and J. D. Groopman, 1994, A follow-up study of urinary markers of aflatoxin exposure and liver cancer risk in Shanghai, People's Republic of China, *Cancer Epidemiol Biomarkers Prev.* **3**: 3–10.

16. J. M. Essigmann, R. G. Croy, A. M. Nadzan, W. F. Busby, V. N. Reinhold, G. Buchi and G. N. Wogan, 1977, Structural identification of the major DNA adduct formed by aflatoxin B_1 *in vitro*, *Proc. Natl. Acad. Sci. USA*. **74**: 1870–1874.

17. J.-K. Lin, J. A. Miller and E. C. Miller, 1977, 2,3-dihydro-2-(guan-7-yl)-3-hydroxy-aflatoxin B_1, a major acid hydrolysis product of aflatoxin B_1-DNA or -ribosomal RNA adducts formed in hepatic microsome-mediated reactions and in rat liver *in vivo*, *Cancer Res.* **37**: 4430–4438.

18. C. N. Martin and R. C. Garner, 1977, Aflatoxin B_1-oxide generated by chemical or enzymatic oxidation of aflatoxin B_1 causes guanine substitution in nucleic acids, *Nature (London).* **267**: 863–865.

19. D. H. Swenson, J.-K. Lin, E. C. Miller and J. A. Miller, 1977, Aflatoxin B_1–2,3-oxide as a probable intermediate in the covalent binding of aflatoxins B_1 and B_2 to rat liver DNA and ribosomal RNA in vivo, *Cancer Res.* **37**: 172–181.

20. V. M. Raney, T. M. Harris and M. P. Stone, 1993, DNA conformation mediates aflatoxin B_1-DNA binding and the formation of guanine N^7 adducts by aflatoxin B_1 8,9-*exo*-epoxide, *Chem. Res. Toxicol.* **6**: 64–68.

21. K. D. Raney, T. Shimada, D.-H. Kim, J. D. Groopman, T. M. Harris and F. P. Guengerich, 1992, Oxidation of aflatoxins and sterigmatocystin by human liver microsomes: Significance of aflatoxin Q_1 as a detoxication product of aflatoxin B_1, *Chem. Res. Toxicol.* **5**: 202–210.

22. G. S. Bailey, 1994, Role of aflatoxin-DNA adducts in the cancer process, In: *The Toxicology of Aflatoxins: Human Health, Veterinary and Agricultural Significance,* (D. L. Eaton and J. D. Groopman, eds.) pp. 137–148, Academic Press, New York.

23. A. D. D'Andrea and W. A. Haseltine, 1978, Modification of DNA by aflatoxin B_1 creates alkali-labile lesions in DNA at postions of guanine and adenine, *Proc. Natl. Acad. Sci. USA.* **75**: 4120–4124.

24. F.-L. Yu, J.-X. Huang, W. Bender, Z. Wu and J. C. S. Chang, 1991, Evidence for the covalent biding of aflatoxin B_1-dichloride to cytosine in DNA, *Carcinogenesis.* **12**: 997–1002.

25. D. H. Bechtel, 1989, Molecular dosimetry of hepatic aflatoxin B_1-DNA adducts: Linear correlation with hepatic cancer risk, *Reg. Toxicol. Pharmacol.* **10**: 74–81.

26. P. Buss, M. Caviezel and W. K. Lutz, 1990, Linear dose-response relationship for DNA adducts in rat liver from chronic exposure to aflatoxin B_1, *Carcinogenesis.* **11**: 2133–2135.

27. R. H. Dashwood, D. N. Arbogast, A. T. Fong, J. D. Hendricks and G. S. Bailey, 1988, Mechanisms of anti-carcinogenesis by indole-3-carbinol: detailed in vivo DNA binding dose-response studies after dietary administration with aflatoxin B_1, *Carcinogenesis.* **9**: 427–432.

28. D. Li, Y. Cao, L. He, N. J. Wang and J. R. Gu, 1993, Aberrations of *p53* gene in human hepatocellular carcinoma from China, *Carcinogenesis.* **14**: 169–73.

29. P. Coursaget, N. Depril, M. Chabud, R. Nandi, V. Mayelo, P. LeCann and B. Yvonnet, 1993, High prevalence of mutations at codon 249 of the p53 gene in hepatocellular carcinomas from Senegal, *Br. J. Cancer Res.* **67**: 1395–1397.

30. M. Hollstein, D. Sidransky, B. Vogelstein and C. C. Harris, 1991, *p53* mutations in human cancers, *Science.* **253**: 49–53.

31. Y. Murakami, 1993, Co-inactivation of the p53 and RB genes in human hepatocellular carcinoma, *Nippon Rinsho.* **51**: 375–9.

32. G. J. Walker, N. K. Hayward, S. Falvey and G. E. Cooksley, 1991, Loss of heterozygosity in hepatocellular carcinoma, *Cancer Res.* **51**: 4367–4370.

33. G. McMahon, E. Davis and G. N. Wogan, 1987, Characterization of c-Ki-ras oncogene alleles by direct sequencing of enzymatically amplified DNA from carcinogen-induced tumors, *Proc Natl Acad Sci U S A.* **84**: 4974–8.

34. G. McMahon, E. F. Davis, L. J. Huber, Y.-S. Kim and G. N. Wogan, 1990, Characterization of c-Ki-*ras* and N-*ras* oncogenes in aflatoxin B_1-induced rat liver tumors, *Proc. Natl. Acad. Sci., USA.* **87**: 1104–1108.

35. M. Corominas, H. Kamino, J. Leon and A. Pellicer, 1989, Oncogene activation in human benign tumors of the skin (keratoacanthomas): Is H-*ras* involved in differentiation as well as proliferation?, *Proc. Natl. Acad. Sci. USA.* **86**: 6372–6377.

36. D. L. Eaton and E. P. Gallagher, 1994, Mechanisms of aflatoxin carcinogenesis, *Annu. Rev. Pharmacol. Toxicol.* **34**: 135–172.

37. Y. Fujimoto, L. L. Hampton, L. D. Luo, P. J. Wirth and S. S. Thorgeirsson, 1992, Low frequency of *p53* gene mutation in tumors induced by aflatoxin B_1 in nonhuman primates, *Cancer Res.* **52**: 1044–6.

38. S. L. Lilleberg, M. A. Cabonce, N. R. Raju, L. M. Wagner and L. D. Kier, 1992, Alterations in the structural gene and the expression of *p53* in rat liver tumors induced by aflatoxin-B_1, *Mol. Carcinogenesis.* **6**: 159–172.

39. J. E. Hulla, Z. Y. Chen and D. L. Eaton, 1993, Aflatoxin B_1-induced rat hepatic hyperplastic nodules do not exhibit a site-specific mutation within the *p53* gene, *Cancer Res.* **53**: 9–11.

40. L. Stoloff, M. J. Verrett, J. Dantzman and E. F. Reynaldo, 1972, Toxicological study of aflatoxin P_1 using the fertile chicken egg, *Toxicol. Appl. Pharmacol.* **23**: 528–531.

41. D. P. H. Hsieh, A. S. Salhab, J. J. Wong and S. L. Yang, 1974, Toxicity of aflatoxin Q_1 as evaluated with the chicken embryo and bacterial auxotrophs, *Toxicol. Appl. Pharmacol.* **30**: 237–242.

42. R. A. Coulombe, D. W. Wilson and D. P. H. Hsieh, 1984, Metabolism, DNA binding, and cytotoxicity of aflatoxin B_1 in tracheal explants from syrian hamster., *Toxicology.* **32**: 117–130.

43. R. A. Coulombe, D. W. Shelton, R. O. Sinnhuber and J. E. Nixon, 1982, Comparative mutagenicity of aflatoxins using a Salmonella/trout hepatic enzyme activation system., *Carcinogenesis.* **3**: 1261–1264.

44. D. P. H. Hsieh, J. M. Cullen and B. H. Ruebner, 1984, Comparative hepatocarcinogenicity of aflatoxins B_1 and M_1 in the rat, *Food Chem. Toxicol.* **22**: 1027–1033.

45. Z. A. Wong and D. P. H. Hsieh, 1980, The comparative metabolism and toxicokinetics of aflatoxin B_1 in the monkey, rat and mouse., *Toxicol. Appl Pharmacol.* **55**: 115–125.

46. H. S. Ramsdell, A. Parkinson, A. C. Eddy and D. L. Eaton, 1991, Bioactivation of aflatoxin B_1 by human liver microsomes: Role of cytochrome P450 IIIA enzymes, *Toxicol. Appl. Pharmacol.* **108**: 436–447.

47. E. P. Gallagher, L. C. Wienkers, P. J. Stapleton, K. L. Kunze and D. L. Eaton, 1994, Role of human microsomal and human complementary DNA-expressed cytochromes P4501A2 and P4503A4 in the bioactivation of aflatoxin B_1, *Cancer. Res.* **54**: 101–108.

48. G. Kirby, C. R. Wolf, G. Neal, P. Srivantanakul and C. Wild, 1993, Metabolism of aflatoxin B_1 by human liver tissue from Thailand, American Association for Cancer Research. **34**: 162.

49. J. T. Heinonen, R. Fisher, K. Brendel and D. L. Eaton, 1996, Determination of aflatoxin B_1 biotransformation and binding to hepatic macromolecules in human precision liver slices, *Toxicol. Appl. Pharmacol.* **136**: 1–7.

50. D. L. Eaton, D. H. Monroe, G. Bellamy and D. A. Kalman, 1988, Identification of a novel dihydroxy metabolite of aflatoxin B_1 produced in vitro and in vivo in rats and mice, *Chem. Res. Toxicol.* **1**: 108–114.

51. K. D. Raney, B. Coles, F. P. Guengerich and T. M. Harris, 1992, The endo-8,9-Epoxide of Aflatoxin-B_1: A New Metabolite, *Chem. Res. Toxicol.* **5**: 333–335.

52. R. S. Iyer, B. F. Coles, K. D. Raney, T. Ricarda, F. P. Guengerich and T. M. Harris, 1994, DNA adduction by the potent carcinogen aflatoxin B_1: mechanistic studies., *J. Am. Chem. Soc.* **116**: 1603–1609.

53. S. W. Baertschi, K. D. Roney, M. P. Stone and T. M. Harris, 1988, Preparation of the 8,9-epoxide of the mycotoxin aflatoxin B_1: the ultimate carcinogenic species., *J. Amer. Chem. Soc.* **110**: 7929-.

54. R. S. Iyer and T. M. Harris, 1993, Preparation of aflatoxin B1-epoxide using m-chloroperbenzoic acid., *Chem. Res. Toxicol.* **6**: 313–316.

55. G. Sabbioni, P. L. Skipper, G. Buchi and S. R. Tannenbaum, 1987, Isolation and characterization of the major serum albumin adduct formed by aflatoxin B_1 in vivo in rats, *Carcinogenesis.* **8**: 819–24.

56. G. E. Neal and P. J. Colley, 1979, The formation of 2,3-dihydro-2,3-dihydroxy aflatoxin B_1 by the metabolism of aflatoxin B_1 in vitro by rat liver microsomes., *FEBS Lett.* **101**: 382–386.

57. C. L. Crespi, B. W. Penman, D. T. Steimel, H. V. Gelboin and F. J. Gonzales, 1991, The development of a human cell line stably expressing human CYP3A3: Role in the metabolic activation of aflatoxin B_1 and comparison to CYP1A2 and CYP2A3, *Carcinogenesis.* **12**: 255–259.

58. L. M. Forrester, G. E. Neal, D. J. Judah, M. J. Glancey and C. R. Wolf, 1990, Evidence for involvement of multiple forms of cytochrome P-450 in aflatoxin B_1 metabolism in human liver, *Proc. Natl. Acad. Sci. USA.* **87**: 8306–8310.

59. T. Shimada and F. P. Guengerich, 1989, Evidence for cytochrome P-450NF, the nifedipine oxidase, being the principal enzyme involved in the bioactivation of aflatoxins in human liver, *Proc. Natl. Acad. Sci. USA.* **86**: 462–465.

60. E. P. Gallagher, K. L. Kunze, P. J. Stapleton and D. L. Eaton, 1996, The kinetics of aflatoxin B1 oxidation by human cDNA-expressed and human liver microsomal cytochromes P450 1A2 and 3A4., *Toxicol. Appl. Pharmacol.* **141**: 595–606.

61. S. A. Wrighton, C. Campanile, P. E. Thomas, S. L. Maines, P. B. Watkins, G. Parker, P. G. Mendez, M. Haniu, J. E. Shively, W. Levin and a. l. et, 1986, Identification of a human liver cytochrome P-450 homologous to the major isosafrole-inducible cytochrome P-450 in the rat, *Mol Pharmacol.* **29**: 405–10.

62. F. P. Guengerich and C. G. Turvy, 1991, Comparisons of levels of several human microsomal cytochrome P450 enzymes and epoxide hydrolase in normal and disease states using immunochemical analysis of surgical liver samples, *J. Pharmacol. Exp. Ther.* **256**: 1189–1194.

63. H. Schweikl, J. A. Taylor, S. Kitareewan, P. Linko, D. Nagorney and J. A. Goldstein, 1993, Expression of CYP1A1 and CYP1A2 genes in human liver, *Pharmacogenetics.* **3**: 239–49.

64. D. Sesardic, A. R. Boobis, R. J. Edwards and D. S. Davies, 1988, A form of cytochrome P450 in man, or-
    thologous to form d in the rat, catalyses the O-deethylation of phenacetin and is inducible by cigarette
    smoking, *Br J Clin Pharmacol*. **26**: 363–72.

65. A. H. Conney, P. J. Pantuck, K. C. Hsaio, W. A. Garland, K. E. Anderson, A. P. Alvares and A. Kappas,
    1976, Enhanced phenacetin metabolism in human subjects fed charcoal-broiled beef, *Clin. Pharmacol.
    Ther.* **20**: 633–642.

66. J. R. Battista and L. J. Marnett, 1985, Prostaglandin H synthase-dependent epoxidation of aflatoxin B$_1$,
    Carcinogenesis. **1**: 1227-.

67. L. Liu and T. E. Massey, 1992, Bioactivation of aflatoxin B-$_1$ by lipoxygenases, prostaglandin-H synthase
    and cytochrome-P450 monooxygenase in guinea-pig tissues, *Carcinogenesis*. **13**: 533–539.

68. G. O. Emerole, N. Neskovic and R. L. Dixon, 1979, The detoxication of aflatoxin B$_1$ with glutathione in
    the rat, *Xenobiotica*. **9**: 737–43.

69. G. H. Degen and H. G. Neumann, 1978, The major metabolite of aflatoxin B$_1$ in the rat is a glutathione
    conjugate, *Chem. Biol. Interact*. **22**: 239–255.

70. D. H. Monroe and D. L. Eaton, 1987, Comparative effects of butylated hydroxyanisole on hepatic in vivo
    DNA binding and in vitro biotransformation of aflatoxin B$_1$ in the rat and mouse, *Toxicol. Appl. Pharma-
    col.* **90**: 401–09.

71. H. S. Ramsdell and D. L. Eaton, 1990, Mouse liver glutathione S-transferase isoenzyme activity toward
    aflatoxin B$_1$–8,9-epoxide and benzo[a]pyrene-7,8-dihydrodiol-9,10-epoxide, *Toxicol. Appl. Pharmacol.*
    105: 216–225.

72. B. D. Roebuck and G. N. Wogan, 1977, Species comparison of in vitro metabolism of aflatoxin B$_1$., *Cancer
    Res.* **37**: 1649–1656.

73. K. O'Brien, E. Moss, D. Judah and G. Neal, 1983, Metabolic basis of the species difference to aflatoxin B$_1$
    induced hepatotoxicity., *Biochem Biophys Res Commun*. **114**: 813–821.

74. G. H. Degen and H.-G. Neumann, 1981, Differences in aflatoxin B$_1$-susceptibility of rat and mouse are
    correlated with the capability in vitro to inactivate aflatoxin B$_1$-epoxide., *Carcinogenesis*. **2**: 299–306.

75. D. L. Eaton, H. Ramsdell and D. H. Monroe, 1990, Biotransformation as a determinant of species suscepti-
    bility to aflatoxin B$_1$: *In vitro* studies in rat, mouse, monkey and human liver, In: *Microbial Toxins in Foods
    and Feeds: Cellular and Molecular Modes of Action*, (A. E. Pohland, V. R. J. Dowell and J. L. Richard,
    eds.) pp. 275–290, Plenum Press, New York and London.

76. D. L. Eaton, H. S. Ramsdell and G. E. Neal, 1994, Biotransformation of aflatoxins, In: *The Toxicology of
    Aflatoxins: Human Health, Veterinary and Agricultural Significance*, (D. L. Eaton and J. D. Groopman,
    eds.) pp. 45–72, Academic Press, New York.

77. D. L. Eaton and H. S. Ramsdell, 1992, Species and diet related differences in aflatoxin biotransformation,
    In: *Handbook of Applied Mycology*, 5, (D. Bhatnagar, eds.) pp. 157–182, Marcel Dekker, New York.

78. P. D. Lotlikar, E. C. Jhee, S. M. Insetta and M. S. Clearfield, 1984, Modulation of microsome-mediated
    aflatoxin B$_1$ binding to exogenous and endogenous DNA by cytosolic glutathione S-transferases in rat and
    hamster livers., *Carcinogenesis*. **5**: 269–276.

79. D. H. Monroe and D. L. Eaton, 1988, Effects of modulation of hepatic glutathione on biotransformation
    and covalent binding of aflatoxin B$_1$ to DNA in the mouse, *Toxicol. Appl. Pharmacol*. **94**: 118–127.

80. K. I. Borroz, H. R. Ramsdell and D. L. Eaton, 1991, Mouse strain differences in glutathione S-transferase
    activity and aflatoxin B$_1$ biotransformation, *Toxicol. Letters*. **58**: 97–105.

81. J. D. Hayes, D. J. Judah, L. I. McLellan and G. E. Neal, 1991, Contribution of the glutathione S-trans-
    ferases to the mechanisms of resistance to aflatoxin B$_1$, *Pharmacol. Ther*. **50**: 443–472.

82. T. Buetler and D. L. Eaton, 1992, Complimentary DNA cloning, messenger RNA expression, and induction
    of alpha class glutathione S-transferases in mouse tissues, *Cancer Res*. **52**: 314–318.

83. J. D. Hayes, D. J. Judah, G. E. Neal and T. Nguyen, 1992, Molecular cloning and heterologous expression
    of a cDNA encoding a mouse glutathione S-transferase Yc subunit possessing high catalytic activity for
    aflatoxin B$_1$–8,9-epoxide, *Biochem. J*. **285**: 173–180.

84. W. R. Pearson, J. Reinhart, S. C. Sisk, K. S. Anderson and P. Adler, 1988, Tissue-specific induction of mur-
    ine glutathione transferase mRNAs by butylated hydroxyanisole, *J. Biol. Chem*. **263**: 13324–13332.

85. K. D. Raney, D. J. Meyer, B. Ketterer, T. M. Harris and F. P. Guengerich, 1992, Glutathione conjugation of
    aflatoxin B$_1$ *exo*-epoxides and *endo*-epoxides by rat and human glutathione S-transferases, *Chem. Res.
    Toxicol.* **5**: 470–478.

86. K. A. McGlynn, E. A. Rosvold, E. D. Lustbader, Y. Hu, M. L. Clapper, T. Zhou, C. P. Wild, X.-L. Xia, A.
    Baffoe-Bonnie, D. Ofori-Adjei, G.-C. Chen, W. T. London, F.-M. Shen and K. H. Buetow, 1995, Suscepti-
    bility to hepatocellular carcinoma is associated with genetic variation in the detoxification of aflatoxin B$_1$.,
    Proc. Natl. Acad. Sci. USA. **92**: 2384–2387.

87. J. J. Ch'ih, T. Lin and T. M. Devlin, 1983, Activation and deactivation of aflatoxin B_1 in isolated rat hepatocytes., *Biochem. Biophys. Res. Commun.* **110**: 668–674.

88. S. A. Metcalfe and G. E. Neal, 1983, The metabolism of aflatoxin B_1 by hepatocytes isolated from rats following the in vivo administration of some xenobiotics., *Carcinogenesis.* **4**: 1007–1012.

89. P. M. Loveland, J. E. Nixon and G. S. Bailey, 1984, Glucuronides in bile of rainbow trout (*Salmo gairdneri*) injected with [^{3}H]aflatoxin B_1 and the effects of dietary β-naphthoflavone., *Comp. Biochem. Physiol.* **78C**: 13–19.

90. G. S. Bailey, J. D. Hendricks, J. E. Nixon and N. E. Pawlowski, 1984, The sensitivity of rainbow trout and other fish to carcinogens, *Drug Metab. Disp.* **15**: 725–750.

91. A. A. Wong, C. Wei, D. W. Rice and D. P. H. Hsieh, 1981, Effects of phenobarbital pretreatment on the metabolism and toxicokinetics of aflatoxin B_1 in the rhesus monkey., *Toxicol Appl Pharmacol.* **600**: 387–397.

92. J. J. Ch'ih, T. Lin and T. M. Devlin, 1983, Effect of inhibitors of microsomal enzymes on aflatoxin B_1-induced cytotoxicity and inhibition of RNA synthesis in isolated rat hepatocytes., *Biochem. Biophys. Res. Commun.* **115**: 15–21.

93. G. M. Decad, K. K. Dougherty, D. P. H. Hsieh and J. L. Byard, 1979, Metabolism of aflatoxin B_1 in cultured mouse hepatocytes: Comparison with rat and effects of cyclohexene oxide and diethylmaleate., *Toxicol. Appl. Pharmacol.* **50**: 429–436.

94. R. M. Shayiq and N. G. Avadhani, 1989, Purification and characterization of a hepatic mitochondrial cytochrome P450 active in aflatoxin B1 metabolism, *Biochem.* **28**: 7546–7554.

95. G. E. Neal and P. J. Colley, 1978, Some high-performance liquid- chromatographic studies of the metabolism of aflatoxins by rat liver microsomal preparations., *Biochem. J.* **174**: 839–851.

96. A. S. Salhab and G. S. Edwards, 1977, Comparative in vitro metabolism of aflatoxicol by liver preparations from animals and humans, *Cancer Research.* **37**: 1016–1021.

97. P. M. Loveland, J. S. Wilcox, J. D. Hendricks and G. S. Bailey, 1988, Comparative metabolism and DNA binding of aflatoxin B_1, aflatoxin M_1, aflatoxicol and aflatoxicol-M_1 in hepatocytes from rainbow trout (*Salmo gairdneri*), *Carcinogenesis.* **9**: 441–446.

98. Z. A. Wong and D. P. H. Hsieh, 1978, Aflatoxicol: aflatoxin B_1 metabolite in rat plasma, *Science.* 325.

99. D. J. Judah, J. D. Hayes, J.-C. Yang, L.-Y. Lian, G. C. K. Roberts, P. B. Farmer, J. H. Lamb and G. E. Neal, 1993, A novel aldehyde reductase with activity towards a metabolite of aflatoxin B_1 is expressed in rat liver during carcinogenesis and following the administration of an anti-oxidant, *Biochem. J.* **292**: 13–18.

100. J. D. Hayes, D. J. Judah and G. E. Neal, 1993, Resistance to aflatoxin B_1 is associated with the expression of a novel aldo-keto reductase which has catalytic activity towards a cytotoxic aldehyde-containing metabolite of the toxin, *Cancer Res.* **53**: 3887–94.

101. Y. P. Dragan and H. C. Pitot, 1994, Aflatoxin carcinogenesis in the context of multistage nature of cancer, In: *The Toxicology of Aflatoxins: Human Health, Veterinary and Agricultural Significance,* (D. L. Eaton and J. D. Groopman, eds.) pp. 179–206, Academic Press, New York.

102. R. D. Bruce, 1990, Risk Assessment of aflatoxins II. Implications of human epidemiology data, *Risk Anal.* **10**: 561–569.

103. F. W. Carlborg, 1979, Cancer, mathematical models and aflatoxin, *Food. Cosmet. Toxicol.* **17**: 159–166.

104. C. R. Dichter, 1984, Risk estimates of liver cancer due to aflatoxin exposure from peanuts and peanut products., *Food Chem. Toxicol.* **22**: 431–437.

105. N. J. Gorelick, 1990, Risk assessment of aflatoxin: I. Metabolism of aflatoxin B_1 by different species, *Risk Analy.* **10**: 539–549.

106. M. S. Hoseyni, 1993, Risk Assessment for aflatoxin: III. Modelling the relative risk of hepatocellular carcinoma, *Risk Analy.* **12**: 123–128.

107. T. Kuiper-Goodman, 1990, Uncertainties in the risk assessment of three mycotoxins: Aflatoxin, ochratoxin and zearalenone, *Can. J. Physiol. Pharmacol.* **68**: 1017–1024.

108. N. J. Gorelick, R. D. Bruce and M. S. Hoseyni, 1994, Human risk assessment based on animal data: Inconsistencies and alternatives, In: *The Toxicology of Aflatoxins: Human Health, Veterinary and Agricultural Significance,* (D. L. Eaton and J. D. Groopman, eds.) pp. 493–511, Academic Press, New York.

109. L. Stoloff, 1983, Aflatoxin as a cause of primary liver-cell cancer in the United States: A probability study, *Nutr. Cancer.* **5**: 165–185.

15

ROLE OF INDIVIDUAL ENZYMES IN THE CONTROL OF GENOTOXIC METABOLITES

Franz Oesch and Michael Arand

Institute of Toxicology
University of Mainz
Obere Zahlbacher Str. 67
D-55131 Mainz, Germany

1. INTRODUCTION

Excretion of abundant compounds is an essential property of living organisms. Each day, a vast amount of material is taken up by the body that is of no nutritional value. Uptake of such xenobiotics occurs mainly with the food but also by inhalation or transdermally. Accumulation of these compounds would result in an enormous body burden. Thus, efficient mechanisms for the excretion of such compounds developed that have their roots very early in the evolution of life.

Two major elimination pathways exist in human, the excretion via bile and the excretion via urine. For volatile compounds, exhalation can become a significant mode of elimination. Both, renal as well as biliary elimination demand water soluble substrates and become increasingly efficient with rising polarity of the respective compound. Some xenobiotics are sufficiently water soluble for excretion, others are too lipophilic and cannot be directly excreted but would accumulate if not processed to more polar derivatives. This is supposed to be the major reason for the development of a complex network of xenobiotic-metabolizing enzymes. In higher animals, xenobiotic metabolism resides mainly in the liver, but a number of other organs, such as kidney, gut and lung, are also capable of metabolising foreign compounds.

Some properties are particularly desirable for this enzyme network. In addition to a sufficient water solubility, the metabolites arising from xenobiotic metabolism should be free of adverse biological activity, in order to avoid toxic effects, and the enzyme systems involved should have a broad substrate specificity, in order to be able to deal with any newly encountered compound. Obviously, these criteria cannot always be fulfilled at the same time. Many lipophilic compounds are themselves chemically almost inert. The first step of their transformation in the so-called phase I (see below) of xenobiotic metabolism leads to chemically more reactive metabolites, often electrophilic in nature. On the one

Molecular and Applied Aspects of Oxidative Drug Metabolizing Enzymes,
edited by Arınç *et al.* Kluwer Academic / Plenum Publishers, New York, 1999.

hand, these metabolites can be rapidly conjugated to the endogenous nucleophile glutathione by enzymatic or non-enzymatic conjugation, to form readily water soluble metabolites. On the other hand, these compounds also react fast with cellular macromolecules, in particular with proteins and DNA, which results in cytotoxic and genotoxic effects. In many cases, the efficacy of the secondary metabolism of these compounds by glutathione S-transferases and epoxide hydrolases sufficiently protects from the toxicity of the intermediary metabolites. However, there is quite a number of examples where xenobiotic metabolism eventually leads to toxification.

2. THE PHASE CONCEPT OF XENOBIOTIC METABOLISM

The metabolism of foreign compounds often proceeds in several sequential steps. A common strategy of the organism to improve the water solubility of xenobiotics is to conjugate lipophilic substrates with hydrophilic endogenous building blocks, such as glucuronic acid, glutathione or sulfate. This conjugation step is dependent upon the presence of a suitable functional group in the xenobiotic that, if not present, has to be introduced or uncovered by one or more steps. This functionalization step is called the phase I of xenobiotic metabolism while the subsequent conjugation is termed phase II. Depending on the nature of the functional groups introduced in phase I these can be classified as being either electrophilic or nucleophilic. Typical electrophilic structures are epoxide functions and α,β-unsaturated carbonyl groups. Important nucleophilic structures are alcoholic or phenolic hydroxyl groups, amino and sulfhydryl functions or carboxylic groups. Because of their individual chemical reactivity, electrophiles can have a significant cytotoxic and/or mutagenic potential, due to their ability to react with electron-rich partners, in particular proteins, RNA and DNA. In contrast, nucleophilic metabolites are usually not capable of attacking endogenous macromolecules by covalent interaction and are thus, in general, less dangerous. On the other hand, they are quite often the determinants of the biological activity of a given compound in terms of pharmacological or acutely toxic effects. Conjugation reactions frequently terminate the potential of electrophiles to react with proteins and DNA, and the ability of nucleophiles to interact with a biological receptor. Simultaneously, they strongly increase the water solubility of the respective compound in most instances. Thus, conjugation reactions can be regarded as the major detoxification step in the metabolism of xenobiotics. However, there are also some exceptions to this rule. Figure 1 exemplifies the general route of metabolism of a carcinogenic compound.

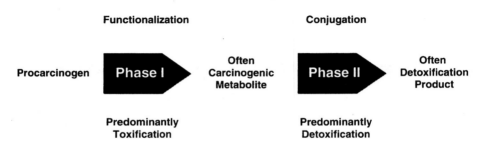

Figure 1. Phase model of xenobiotic metabolism. Many carcinogens require metabolic activation. If activation is necessary for a given compound to be carcinogenic, this occurs during oxidative phase I metabolism, in most instances. In the majority of cases, phase II metabolism terminates the genotoxic effect of a given carcinogen.

3. ENZYMES IMPLICATED IN XENOBIOTIC METABOLISM

The phase I/functionalization phase is carried out by two major groups of enzymes, the oxidoreductases and the hydrolases. Oxidoreductases with implication in the metabolism of xenobiotics are the cytochrome P450-dependent monooxygenases (CYP; the most prominent and quantitatively most important superfamily of xenobiotic-metabolizing enzymes) as well as flavin-containing monooxygenases (FMO), monoamine oxidases (MAO), and prostaglandin synthase (cyclooxygenase; Cox). These enymes introduce oxygen into or remove electrons from their substrates, with few exceptions. Dehydrogenases and reductases, that complete the large groupe of oxidoreductases, add or remove hydrogen to or from the target molecule. Among these, alcohol dehydrogenases, aldehyde dehydrogenases and carbonyl reductases play a role in xenobiotic metabolism. The hydrolases comprise families of enzymes specialized in the hydrolysis of either esters and amides, epoxides, or glucuronides, reactions of obvious importance for the metabolism of foreign compounds. The phase II/conjugation phase of the xenobiotic metabolism is carried out by transferases. Electrophilic substrates are taken over by the glutathione S-transferases. Nucleophilic substrates, i.e., those with hydroxy-, sulfhydryl-, amino- or carboxyl groups, are metabolized by UDP-glucuronosyltransferases (UGT), sulfotransferases (SULT), acetyltransferases (AT), acyl-CoA amino acid N-acyltransferase, and methyl transferases. An overview on these enzymes is given in Table 1.

Among the foreign compounds whose genotoxic effects are under the control of these enzymes, polycyclic aromatic hydrocarbons, aromatic amines, dialkyl nitrosamines and certain halogenated aliphatic compounds represent the largest groups (Table 2).

4. THE METABOLISM OF BENZO[A]PYRENE—A CLASSIC EXAMPLE

Benzo[a]pyrene is a polycyclic aromatic hydrocarbon that has been known to be a carcinogen since a long time. Over the decades, it has become evident that the compound requires metabolic activation to gain genotoxic potency.[1] An excerpt of its complex metabolism is shown in Figure 2. The presented pathways were selected either because of their particular relevance for the carcinogenicity of the compound or as illustrative examples for the role of many of the above listed enzymes in the metabolism of this carcinogen.

The benzo[a]pyrene-7,8-dihydrodiol 9,10-oxide, and in particular the 7R,8S,9S,10R isomer of this compound, is regarded as the most important metabolite of benzo[a]pyrene

Table 1. Enzymes involved in xenobiotic metabolism

Phase I Enzymes	Phase II Enzymes
Oxidoreductases	Transferases
Cytochromes P450	Glutathione S-transferases
Flavin Monooxygenases	UDP-Glucuronosyltransferases
Prostaglandin Synthases	Sulfotransferases
Dihydrodiol Dehydrogenases	Acetyltransferases
DT-Diaphorases	Methyltransferases
Alcohol Dehydrogenases	Aminoacyltransferases
Hydrolases	
Epoxide Hydrolases	
Esterases/Amidases	
Glucuronidases	

Table 2. Important groups of carcinogens that require enzymatic activation

Procarcinogens	Reactive intermediates	Enzymes involved in control
Aromatic Hydrocarbons	Epoxides	Cytochromes P450
		Glutathione S-transferases
		Epoxide Hydrolases
Aromatic Amines	Reactive esters	Cytochromes P450
		Sulfotransferases
		Acetyltransferases
		UDP-Glucuronosyltransferases
		Glutathione S-transferases
Dialkylnitrosamines	Carbonium ions/	Cytochromes P450
	electron-deficient alkyl groups	
Vicinal Dihaloalkanes	Episulfonium ions	Glutathione S-transferases

in terms of its carcinogenicity. The formation of this genotoxic species requires three enzymatic steps: (i) an epoxidation of the parent compound in the 7,8-position carried out by cytochrome P-450 enzymes of the family 1, best documented for the isoenzyme CYP1A1, leads to a reactive 7,8-oxide that can by itself react with DNA but rapidly undergoes isomerization to the corresponding phenol(s); (ii) alternatively, this short-lived metabolite can be further processed by the microsomal epoxide hydrolase (mEH) to form the corresponding 7,8-dihydrodiol, a compound of no inherent genotoxicity and thus, in principle, a detoxification product, that can, however, (iii) be epoxidized a second time, either by a number of different CYP isoenzymes or by Cox to the ultimate carcinogen, the 7,8-dihydrodiol 9,10-oxide. A general lesson to be learnt from this metabolism pathway is that CYP-mediated oxidation of aromatic compounds often leads to the formation of more or less genotoxic epoxides, bearing the risk for the organism to increase the carcinogenicity of a given compound. On the other hand, this reaction cascade is one of the very few examples where an epoxide hydrolase reaction eventually results in a toxification. It has, indeed, been experimentally proven that benzo[a]pyrene is more genotoxic in the presence than in the absence of mEH when metabolized by CYP1A1, which also indicates that the 7,8-epoxide is not as efficiently genotoxic as the 7,8-dihydrodiol 9,10-oxide.

Possible mechanisms of the organism to reduce the amount of diol epoxide formed lie in the sequestration of the intermediate of the pathway, the 7,8-dihydrodiol. On the one hand, this has been shown to be a moderate substrate for conjugation reactions carried out by UGT and SULT, however, the quantitative importance of these conjugations remains to be established. On the other hand, dihydrodiol dehydrogenase can metabolize the dihydrodiol to the corresponding catechol, that is further converted non-enzymatically (autoxidation) and possibly also enzymatically to the quinone. This reaction is not exclusively detoxifying, since quinones themeselves are cytotoxic, but in the present case the reaction has a higher potential to be beneficial.

Despite its high proven *in vivo* genotoxicity and carcinogenicity, the diol epoxide itself can be detoxified by xenobiotic metabolizing enzymes. Several glutathione S-transferase isoenzymes are capable of conjugating the 7,8-dihydrodiol 9,10-oxide to glutathione. The most proficient of these, the GSTP1, is an essentially extrahepatic isoenzyme, but GSTM1, one of the major GST in rat liver, also displays a high turnover number with the benzo[a]pyrene diol epoxide. In contrast, the two xenobiotic epoxide hydrolases, mEH and soluble epoxide hydrolase (sEH), do not show any enzymatic activity towards this compound. Surprisingly, it has been shown that the dihydrodiol dehydrogenase sig-

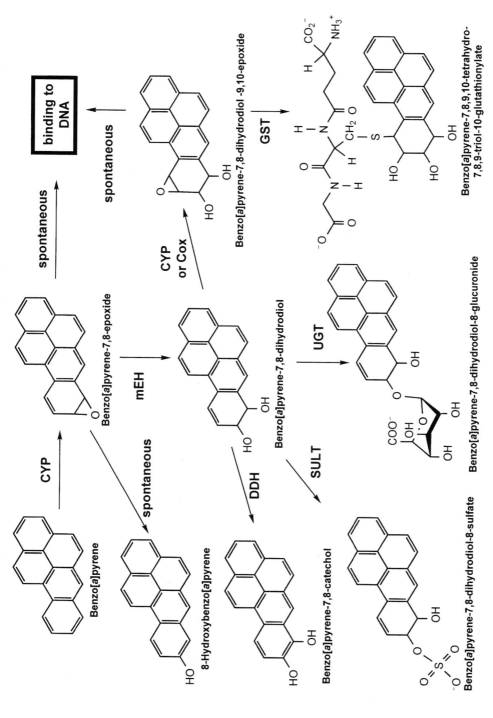

Figure 2. Selected metabolic pathways for the activation, inactivation, and sequestration of the procarcinogen benzo[*a*]pyrene. See main text for further explanation.

nificantly inhibits the mutagenicity of benzanthracene-8,9-diol 10,11-oxide[2] and possibly also reduces benzo[a]pyrene 7,8-dihydrodiol 9,10-oxide mutagenicity, since it was reported that the enzyme inhibits the mutagenicity of benzo[a]pyrene activated by methyl-cholanthrene-inducible murine liver enzymes.[3]

5. METABOLISM OF CHRYSENE DIOL EPOXIDE ISOMERS BY MAMMALIAN EPOXIDE HYDROLASES

Epoxide hydrolases[4] cleave the oxirane ring of epoxides by nucleophilic substitution under formation of an enzyme-substrate ester intermediate that is subsequently hydrolyzed.[5-9] Attempts to demonstrate a protective effect of epoxide hydrolases on the mutagenicity of diol epoxides have failed for many years,[2] and it is generally believed that the neighbourhood of the two hydroxy groups to the epoxide efficiently blocks the interaction between enzyme and diol epoxide, since corresponding tetrahydro epoxide compounds that only differ from the diol epoxide by the lack of the two hydroxy groups are good substrates for the mEH.[10] Recently, however, a protective effect of mEH on the bacterial mutagenicity of the reverse diol epoxides but not the diol bay region epoxides of chrysene has been reported[11] (see Figure 3A and B). We therefore investigated the potential of mEH and sEH to hydrolyze diol epoxides from several polycyclic aromatic compounds.

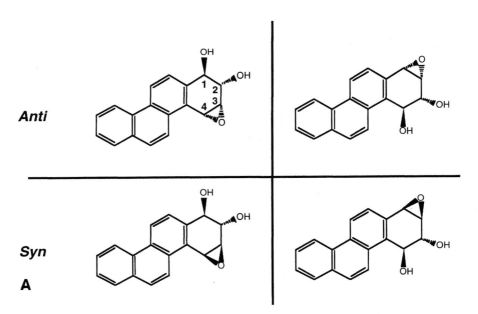

Figure 3. Metabolism of regio- and stereoisomeric chrysene diol epoxides by microsomal and soluble epoxide hydrolase. A. The racemic syn- and anti-chrysene-1,2-dihydrodiol 3,4-oxides (diol bay region epoxides left) and chrysene-3,4-dihydrodiol 1,2-oxides (reverse diol epoxides right) were analyzed. B. Previous results by Glatt et al.[10] had demonstrated a protective effect of rat mEH against the bacterial mutagenicity of the reverse diol epoxides. C. Hydrolysis of the reverse anti-diol epoxide (right) is significantly enhanced over the spontaneous hydrolysis in the presence of rat mEH. Under the same conditions, a 100-fold excess of the enzyme does not increase hydrolysis rate of the corresponding diol bay region epoxide (left). D. Hydrolysis of the bay region syn-diol epoxide is significantly enhanced over the spontaneous hydrolysis in the presence of human sEH. In contrast, the enzyme does not significantly increase hydrolysis rate of the corresponding reverse region diol epoxide (right).

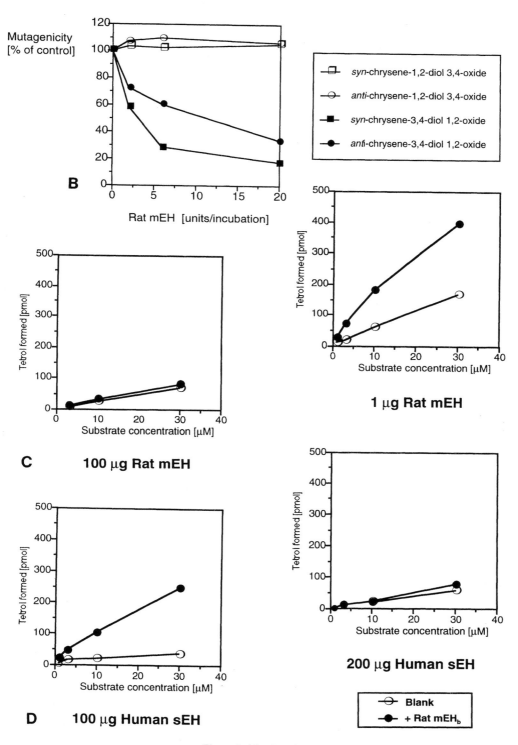

Figure 3. (*Continued*)

In this study, both enzymes did not detectably act on the diol bay region epoxides from benzo[a]pyrene, benzo[c]phenanthrene and benz[a]anthracene. In contrast, both reverse diol epoxides from chrysene were hydrolyzed to a significant extent by rat mEH, as implied by the above effect on the mutagenicity of these compounds. Of the chrysene diol bay region epoxides, the *syn*-diastereomer was also hydrolyzed by mEH yet about one order of magnitude slower than the reverse diol epoxides. The *anti*-diastereomer was not efficiently used as a substrate by mEH (Figure 3C). The sEH showed the best turnover with the *syn*-diol bay region epoxide (Figure 3D) that was, however, not much higher than the mEH activity towards this compound. The enzyme also hydrolyzed the *anti*-diastereomers of the bay region and the reverse diol epoxide. No catalytic activity was detectable against the *syn*-reverse diol epoxide. These results demonstrate that mammalian epoxide hydrolases can hydrolyze diol epoxides, yet at a slow rate, which together with the previous data from mutagenicity testings also indicates some relevance for the *in vivo* situation. Further details of this study will be presented elsewhere.

6. ACTIVATION OF ALIPHATIC CARCINOGENS

Prominent examples among the metabolically activated non-aromatic carcinogens are the alkyl nitrosamines and vicinal dihaloalkanes. Both will be briefly discussed in the following.

Alkyl nitrosamines can be strong carcinogens.[12] Important examples are the 4-(methyl nitrosamino)-1-(3-pyridyl)-1-butanone (NNK) that occurs in significant concentrations in tobacco smoke, or occupationally relevant compounds such as dimethyl nitrosamine, N-nitroso morpholine (NMOR) and N-nitrosodiethanolamine (NDELA).[13] All these compounds require metabolic activation. The underlying mechanism (see Figure 4) is a CYP-mediated α-C hydroxylation, usually carried out by either CYP2E1 or CYP2A6. The resulting unstable metabolite decomposes in a first step to form the corresponding aldehyde and alkyl diazohydroxide. Releases of a hydroxide anion from the latter intermediate leads to an alkyl diazonium, a strongly alkylating compound with high DNA damaging potential.

Glutathione S-transferases are highly important safeguards of the organism against chemically reactive electrophilic compounds, that often are genotoxic. One of the few ex-

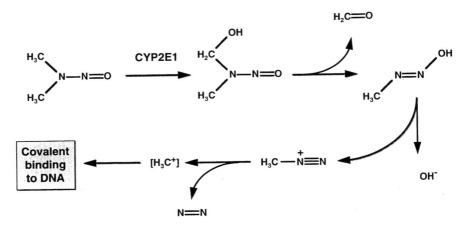

Figure 4. Metabolic activation of dialkylnitrosamines by CYP-mediated α-C hydroxylation. The presented example is dimethyl nitrosamine. See main text for further explanation.

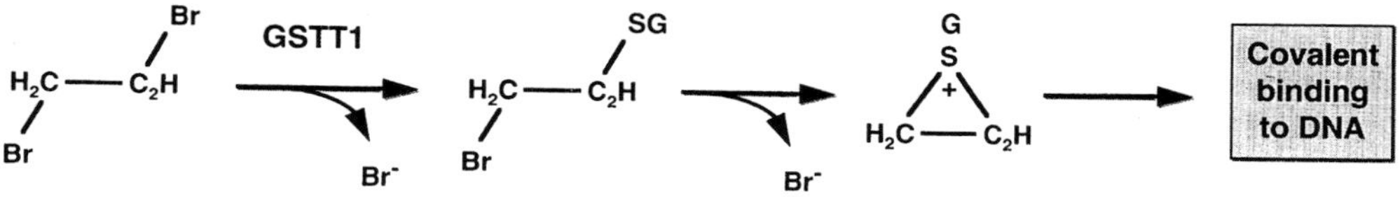

Figure 5. Glutathione S-transferase T1-mediated activation of 1,2-Dibromoethane. See main text for further explanation.

amples for the opposite role in the control of carcinogenic compounds of the otherwise so beneficial GST is the metabolism of vicinal dihaloalkanes.[14] 1,2-Dibromoethane, for instance, is debrominated by the rat isoenzyme GSTT1[15] (the former GST 5–5) to form the corresponding glutathione conjugate (see Figure 5). The resulting metabolite can undergo a spontaneous intramolecular nucleophilic substitution reaction, resulting in the release of the second bromine as a bromide anion and the formation of a highly reactive thiiranium ion (also called episulfonium ion). The thiiranium can be regarded as the thio-analogon of an epoxide. Its reactivity is strongly enhanced by the positive charge of the heteroatom and respective compounds are powerful genotoxins.

The relevance of this metabolic pathway has been elegantly proven by Guengerich and colleagues: In *Salmonella* strains that express recombinant rat GSTT1, 1,2-dibromoethane is a potent mutagen as revealed by the Ames test, while it is not mutagenic in wild type tester strains, even in the presence of purified GSTT1.[16,17] Obviously, the highly polar active species is not able to pass the bacterial cell membranes and has to be generated inside the cell to exert its genotoxic effects.

REFERENCES

1. C. Heidelberger, 1975, Chemical carcinogenesis, *Annu. Rev. Biochem.* **44**: 79–121.
2. H.R. Glatt, C.S. Cooper, P.L. Grover, P. Sims, P. Bentley, M. Merdes, F. Waechter, K. Vogel, T.M. Guenthner, F. Oesch, 1982, Inactivation of a diol-epoxide by dihydrodiol dehydrogenase, but not by two epoxide hydrolases., *Science* **215**: 1507–1509.
3. H.R. Glatt, K. Vogel, P. Bentley, F. Oesch, 1979, Reduction of benzo*(a)*pyrene mutagenicity by dihydrodiol dehydrogenase. *Nature* **277**: 319–320.
4. F. Oesch, 1973, Mammalian epoxide hydrases: Inducible enzymes catalyzing the inactivation of carcinogenic and cytotoxic metabolites derived from aromatic and olefinic compounds, *Xenobiotica* **3**: 305–340.
5. G.M. Lacourciere, R.N. Armstrong, 1993, The catalytic mechanism of microsomal epoxide hydrolase involves an ester intermediate, *J. Am. Chem. Soc.* **115**: 10466–10467.
6. M. Arand, D.F. Grant, J.K. Beetham, T. Friedberg, F. Oesch, B.D. Hammock, 1994, Sequence similarity of mammalian epoxide hydrolases to the bacterial haloalkane dehalogenase and other related proteins - Implication for the potential catalytic mechanism of enzymatic epoxide hydrolysis, *FEBS Lett.* **338**: 251–256.
7. B.D. Hammock, F. Pinot, J.K. Beetham, D.F. Grant, M.E. Arand, F. Oesch, 1994, Isolation of a putative hydroxyacyl enzyme intermediate of an epoxide hydrolase, *Biochem. Biophys. Res. Commun.* **198**: 850–856.
8. M. Arand, H. Wagner, F. Oesch, 1996, Asp_{333}, Asp_{495}, and His_{523} form the catalytic triad of rat soluble epoxide hydrolase, *J. Biol. Chem.* **271**: 4223–4229.
9. F. Müller, M. Arand, H. Frank, A. Seidel, W. Hinz, L. Winkler, K. Hänel, E. Blee, J.K. Beetham, B.D. Hammock, F. Oesch, 1997, Visualization of a covalent intermediate between microsomal epoxide hydrolase, but not cholesterol epoxide hydrolase, and their substrates, *Eur. J. Biochem.* **245**: 490–496.
10. A.W. Wood, R.L. Chang, W. Levin, R.E. Lehr, M. Schaefer-Ridder, J.M. Karle, D.M. Jerina, A.H. Conney, 1977, Mutagenicity and cytotoxicity of benz[a]anthracene diol epoxides and tetrahydro-epoxides: exceptional activity of the bay region 1,2-epoxides, *Proc. Natl. Acad. Sci. U. S. A.* **74**: 2746–2750.
11. H.R. Glatt, C. Wameling, S. Elsberg, H. Thomas, H. Marquardt, A. Hewer, D.H. Phillips, F. Oesch, A. Seidel, 1993, Genotoxicity characteristics of reverse diol-epoxides of chrysene., *Carcinogenesis* **14**: 11–19.

12. P.N. Magee, J.M. Barnes, 1956, The production of malignant hepatic tumours in the rat feeding dimethylni-trosamine, *Brit. J. Cancer* **10:** 114–119.

13. R. Preussmann, 1990, Carcinogenicity and structure-activity relationships of N-nitrosocompounds: a review, in: *The Significance of N-Nitrosation of Drugs*, (G. Eisenbrand, G. Bozler and H. von Nicolai, eds.), pp. 3–17, Gustav Fischer Verlag, Stuttgart, Germany.

14. P.B. Inskeep, F.P. Guengerich, 1984, Glutathione-mediated binding of dibromoalkanes to DNA: specificity of rat glutathione S-transferases and dibromoalkane structure, *Carcinogenesis* **5:** 805–808.

15. D.J. Meyer, B. Coles, S.E. Pemble, K.S. Gilmore, G.M. Fraser, B. Ketterer, 1991, Theta, a new class of glutathione transferases purified from rat and man, *Biochem. J.* **274:** 409–414.

16. R. Thier, J.B. Taylor, S.E. Pemble, W.G. Humphreys, M. Persmark, B. Ketterer, F.P. Guengerich, 1993, Expression of mammalian glutathione S-transferase 5–5 in *Salmonella typhimurium* TA1535 leads to base-pair mutations upon exposure to dihalomethanes, *Proc. Natl. Acad. Sci. U.S.A.* **90:** 8576–8580.

17. R. Thier, M. Müller, J.B. Taylor, S.E. Pemble, B. Ketterer, F.P. Guengerich, 1995, Enhancement of bacterial mutagenicity of bifunctional alkylating agents by expression of mammalian glutathione S-transferase, *Chem. Res. Toxicol.* **8:** 465–472.

16

ENHANCEMENT OF THE MUTAGENICITY OF ETHYLENE OXIDE AND SEVERAL DIRECTLY ACTING MUTAGENS BY HUMAN ERYTHROCYTES AND ITS REDUCTION BY XENOBIOTIC INTERACTION

Jan G. Hengstler, Jochen Walz, Rani Kübel, Tri Truong, Jürgen Fuchs, Albrecht Seidel, Michael Arand, and Franz Oesch

Institute of Toxicology
University of Mainz, Germany

ABSTRACT

According to the present state of knowledge mutagenicity or genotoxicity of the ultimate genotoxic agents ethylene oxide or styrene oxide cannot be increased by further metabolism. However, in the present study we demonstrate that mutagenicity of several ultimate genotoxic substances is increased by human erythrocytes. For instance mutagenicity of mafosfamide, N-nitroso-N-methylurea, ethylene oxide, and styrene oxide to *Salmonella typhimurium* TA 1535 was increased 5.5-, 5.1-, 2.7-, and 2.3-fold, respectively, by addition of human erythrocyte homogenate to the preincubation mixture in the *Ames* test. On the other hand, the mutagenicity of cumene hydroperoxide, benzo[a]pyrene-4,5-oxide, and 1-methyl-3-nitro-1-nitrosoguanidine was decreased approximately 3-, 5-, and 1000-fold by human erythrocytes, respectively.

Interindividual differences in enhancement of ethylene oxide mutagenicity by human erythrocytes were relatively small. However, drinking of 1.5 l of a standardized black tea preparation significantly decreased the mutagenicity-enhancing effect of human erythrocytes of 16 individuals (p=0.001). Preparation of subcellular fractions from human erythrocytes demonstrated that factors, which enhance ethylene oxide mutagenicity, are located predominantly in the cytosol and have a molecular weight >5000. Nevertheless, to a minor degree also cytosolic low molecular weight (<5000) constituents enhanced the mutagenicity of ethylene oxide. At least 3 single constituents of erythrocytes, namely hemoglobin, glutathione, and L-cysteine were able to enhance significantly the mutagenicity of ethylene oxide. At physiological concentrations hemoglobin (14 g/dl) could explain

Molecular and Applied Aspects of Oxidative Drug Metabolizing Enzymes,
edited by Arinç *et al.* Kluwer Academic / Plenum Publishers, New York, 1999.

~54% and glutathione (1 mM) ~24 % of the total effect of erythrocyte homogenate, whereas the contribution of L-cysteine was negligible, due to its low physiological concentration.

Since all tissues of origin for cancer are in close contact to erythrocytes, modification of the potency of mutagenic substances by human erythrocytes seems to be important for risk assessment.

1. INTRODUCTION

Nearly 50 years ago the observation was made that many carcinogenic chemicals are not mutagenic or genotoxic by themselves. In all these cases metabolic activation catalyzed mainly by hepatic enzymes (but also by non-hepatic enzymes) was recognized as necessary for generation of reactive intermediates, also termed ultimate genotoxic agents, that are capable of covalently binding to DNA.[1] According to the present state of knowledge mutagenicity or genotoxicity of the ultimate genotoxic agents ethylene oxide, styrene oxide, or phosphoramide mustard cannot be increased by further metabolism (but is abolished for many ultimate genotoxic substances by phase II-metabolism).[2,3] However, in the present study evidence is reported that mutagenicity of some ultimate genotoxic substances can be increased (up to ~6-fold in the case of mafosfamide) by human erythrocytes. Since all tissues of origin for cancer are in close contact to erythrocytes, modification of the effect of genotoxic substances by human erythrocytes might significantly influence the extent of DNA damage in cells of origin for cancer. In this study modification of mutagenicity of directly acting genotoxic substances (with special consideration of ethylene oxide) by human erythrocytes was examined. Evidence is given that increased mutagenicity is mediated mostly by hemoglobin and to a minor degree by glutathione, but does not require the action of "classical" erythrocytic drug-metabolizing enzymes, such as glutathione S-transferase theta or pi.

2. INCREASE IN MUTAGENICITY OF ETHYLENE OXIDE BY INTACT HUMAN ERYTHROCYTES AND ERYTHROCYTE HOMOGENATE

Mutagenicity of ethylene oxide was tested in standard medium (150 mM KCl; 10 mM sodium phosphate; pH 7.4) with and without addition of intact human erythrocytes and erythrocyte homogenate (Figure 1). The *Salmonella typhimurium* reverse mutation assay (1 h preincubation at 37 °C) using TA 1535 was applied.[4,5] *Salmonella* tester strain TA 1535 carries the *his G46* mutation, which is located in the *his G* gene coding for the first enzyme of histidine biosynthesis.[4] This mutation substitutes –GGG- (proline) for –GAG- (leucine) in the wild-type organism.[5] Thus, TA 1535 detects mutagens that cause base-pair substitutions, primarily at G-C pairs. TA 1535 has been shown to detect mutagenicity of ethylene oxide with high sensitivity[6] and, thus, was used for the experiments in the present study.

In standard medium ethylene oxide caused a dose-dependent increase in revertants per plate in a dose range between 5 and 60 mM ethylene oxide (Fig. 1). At higher concentrations a decrease in revertants was observed, due to toxicity of ethylene oxide. Addition of human erythrocytes to the standard medium in a concentration similar to the *in vivo*

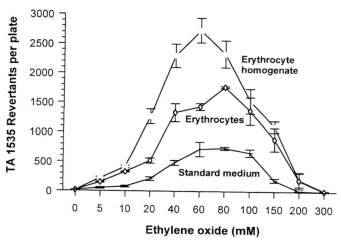

Figure 1. Increase in mutagenicity of ethylene oxide by intact human erythrocytes and erythrocyte homogenate in the Ames test (1 h preincubation at 37 °C) using *Salmonella typhimurium* TA 1535. Standard medium consisted of 150 mM KCl, 10 mM sodium phosphate, pH 7.4. Erythrocytes were prepared by centrifugation of heparinized venous blood (400 × g; 15 min). The blood plasma and buffy coat were decanted and the erythrocytes were washed twice with standard medium. (Washing with standard medium was performed to remove the blood plasma. Removal of blood plasma was important for reproducibility of the experiments since plasma interferred with growth of *Salmonella typhimurium* TA 1535.) Subsequently, standard medium was added to the erythrocytes up to the original volume of venous blood and erythrocytes were carefully resuspended. Bacteria (TA 1535 in 100 µl of standard medium) and ethylene oxide solution (in 300 µl of standard medium) were added to 500 µl of erythrocyte suspension and incubated for 1 h at 37 °C in a shaking water-bath. Solutions with ethylene oxide were prepared as described by Hengstler et al.[6] For incubations with erythrocyte homogenate, the suspension with erythrocytes was homogenized by sonication prior to addition of ethylene oxide. Erythrocytes of a healthy male non-smoking subject, not taking medicine or vitamin pills were used. Data shown are mean values ± standard deviations of three independent incubations of a representative experiment.

situation (5×10^6/µl) caused an approximately 2-fold increase in the mutagenicity of ethylene oxide. If erythrocytes were homogenized prior to incubation with ethylene oxide an even higher, approximately 4-fold increase in mutagenicity compared to standard medium was observed. The difference in mutagenicity between intact and homogenized erythrocytes might be due to difficulties in penetration through the erythrocytic cell membrane, since a putative highly mutagenic metabolite of ethylene oxide formed in erythrocytes has to pass the cell membrane of intact erythrocytes, before it can bind to DNA of *Salmonella typhimurium* TA 1535. No significant difference in revertants per plate was observed between the standard medium, intact erythrocytes and erythrocyte homogenate, if no ethylene oxide was added.

3. INTERINDIVIDUAL DIFFERENCES IN ENHANCEMENT OF ETHYLENE OXIDE MUTAGENICITY BY HUMAN ERYTHROCYTES

Since in the previous section erythrocytes of only one individual were used, we next examined the extent of interindividual differences in enhancement of ethylene oxide mu-

Table 1. Influence of black tea and coffee consumption on the ability of the erythrocytes of two individuals to enhance ethylene oxide mutagenicity

Individual no.	Ethylene oxide			
	0 mM	20 mM	40 mM	60 mM
First experiment with individuals no. 8 and 13 (with high consumption of coffee or tea for at least one month)				
8	10 ± 3	288 ± 23	590 ± 20	803 ± 15
13	7 ± 2	298 ± 6	633 ± 49	797 ± 49
Second experiment (after 3 days without consumption of coffee and tea)				
8	13 ± 1	423 ± 8	727 ± 15	1115 ± 13
13	10 ± 1	447 ± 25	760 ± 30	1140 ± 27
Third experiment (Erythrocytes taken 90 min after drinking of 1.5 l of black tea immediately after the second experiment)				
8	10 ± 1	305 ± 13	607 ± 15	857 ± 31
13	10 ± 1	368 ± 24	690 ± 10	967 ± 25

Data shown are mean numbers of revertants per plate $\pm$ standard deviations of three independent incubations.
Salmonella typhimurium TA 1535 was incubated with 0, 20, 40, and 60 mM ethylene oxide (1 h at 37 °C) with addition of erythrocytes of individuals no. 8 and 13.
Erythrocytes were prepared as described for Figure 1.

tagenicity by human erythrocytes. Intact erythrocytes of 21 individuals (non-smokers, without medication, not taking vitamin pills) were examined as described for Fig. 1. No significant influence of erythrocytes on the number of revertants per plate was observed, if no ethylene oxide was added (Figure 2a; 0 mM ethylene oxide). However, for almost all individuals an approximately 2-fold increase in mutagenicity of ethylene oxide was caused by addition of erythrocytes to the preincubation mixture with 20 (Fig. 2a), 40, and 60 (Fig. 2b) mM ethylene oxide. No significant difference in enhancement of ethylene oxide mutagenicity between erythrocytes of male and female individuals was observed (data not shown). Interindividual differences were relatively small with the exception of the erythrocytes of two individuals (no. 8 and no. 13) that obviously exerted smaller effects (Fig. 2). For these individuals an extremely high consumption of black tea (individual no. 8: 3 to 4 liters of black tea per day since more than one month) and coffee (individual no. 13: at least 10 cups with approximately 200 ml of coffee per day since more than 2 months) was known. To examine, whether consumption of tea and coffee might interact with the mutagenicity-enhancing effect of erythrocytes, individuals no. 8 and 13 were asked not to drink tea and coffee for 3 days. Thereafter, blood was taken and the effects of the erythrocytes on ethylene oxide mutagenicity was examined as described for Fig. 2. An increase in the effect caused by erythrocytes was observed for both individuals, who had stopped drinking of tea or coffee for 3 days (Table 1). Individuals no. 8 and 13 were then asked to drink 1.5 l of a standardized black tea preparation within 45 min and another blood sample (heparinized venous blood) was taken 90 min after the beginning of drinking. After drinking of black tea the erythrocytes of both individuals caused a smaller enhancement in ethylene oxide mutagenicity compared to the value before tea drinking (Table 1).

To examine systematically whether drinking of black tea influences the ability of erythrocytes to enhance ethylene oxide mutagenicity, 18 healthy individuals (non-smokers,

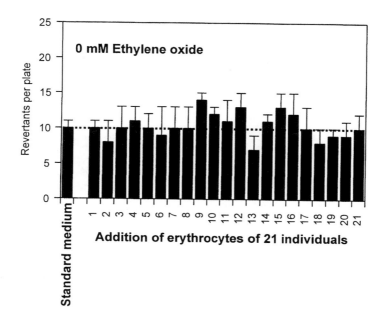

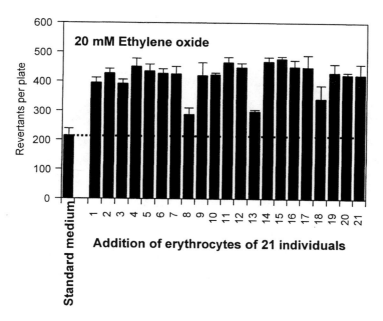

Figure 2. Enhancement of ethylene oxide mutagenicity by intact erythrocytes of 21 individuals (non-smokers, without medication, not taking vitamin pills). Erythrocytes were prepared as described for Fig. 1 and incubated with *Salmonella typhimurium* TA 1535 after addition of 0 and 20 mM (Fig. 2a), as well as 40 and 60 mM (Fig. 2b) ethylene oxide (1h preincubaton at 37 °C). Data shown are mean values ± standard deviations of three independent incubations.

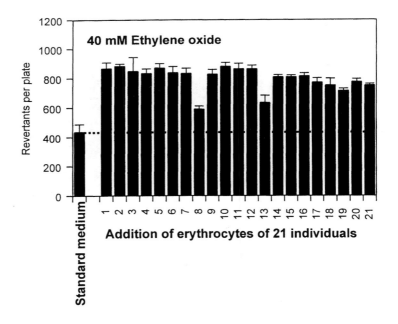

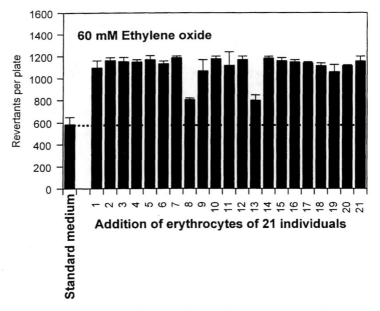

Figure 2. (*Continued*)

without medication, not taking vitamin pills) were asked to drink 1.5 l of a standardized black tea preparation within 60 min. (Two individuals had to be excluded from the study because of vomiting.) Before and 90 min after the beginning of tea drinking blood samples were taken. Eight individuals, who drank 1.5 l of water instead of black tea served as controls. Erythrocytes taken before and after tea drinking were added to the preincubation mixture with TA 1535 as described for Fig. 1. No significant difference in the number of revertants between erythrocytes taken before and after tea drinking was observed if no ethylene oxide was added (Figure 3). The mean number of TA 1535 revertants was 11 ± 3 ($\pm$ standard deviation) if erythrocytes taken before tea drinking were added, versus 10 ± 3 for addition of erythrocytes taken 90 min after tea drinking (Fig. 3). However, enhancement of ethylene oxide mutagenicity was significantly smaller ($p < 0.001$; Wilcoxon-test for paired data; two-tailed test) for erythrocytes taken 90 min after tea drinking compared to erythrocytes taken before drinking. Although the difference was very small (mean no. of revertants before and after tea drinking: 1170 ± 263 and 1010 ± 229, respectively), it appeared significant in the Wilcoxon test for paired data, since a decrease was observed for 14 of 16 individuals examined. Drinking of water (1.5 l) did not significantly influence enhancement of ethylene oxide mutagenicity by erythrocytes (Figure 4).

In conclusion, human erythrocytes and erythrocyte homogenates enhanced ethylene oxide mutagenicity approximately 2- and 4-fold, showing relatively small interindividual differences. Consumption of a relatively high (but in every-day life not unrealistic) amount of black tea caused a small (in the average ~15 %, in some individuals up to ~30%), but statistically significant decrease in the mutagenicity-enhancing effect of human erythrocytes.

4. SUBCELLULAR LOCALIZATION OF AN ERYTHROCYTIC FACTOR, WHICH INCREASES MUTAGENICITY OF ETHYLENE OXIDE

In this experiment subcellular fractions of human erythrocytes were prepared, which were added to the preincubation mixture with *Salmonella typhimurium* TA 1535 together with ethylene oxide solution. Human erythrocytes were homogenized by sonication and microsomes (100 000 × g sediment of the 9 000 × g supernatant), cytosol (100 000 × g supernatant of the 9000 × g supernatant) and a fraction with mitochondria and lysosomes (9000 × g sediment) were separated by ultracentrifugation. Standard medium was added to the respective fractions up to the original volume of the erythrocytes and 500 µl of these suspensions were added to the respective preincubation mixtures together with 100 µl bacterial suspension (TA 1535) and 300 µl of ethylene oxide solution (Figure 5). No significant modification of the number of revertants by the subcellular fractions was observed for the controls (Fig. 5; 0 mM ethylene oxide). If ethylene oxide (20 mM) was added, erythrocyte homogenate caused a 3.1-fold increase in the number of revertants compared to standard medium. A 2.4-fold increase was obtained if cytosol was added, whereas microsomes as well as the fraction with mitochondria and lysosomes (9000 × g sediment) caused only a minor 1.5-fold and 1.6-fold enhancement of ethylene oxide mutagenicity. Thus, most of the factors responsible for enhancement of ethylene oxide mutagenicity seem to be located in the cytosol of erythrocytes, although—to a minor degree—also microsomes as well as the fraction with mitochondria and lysosomes possess the ability to enhance the mutagenicity of ethylene oxide.

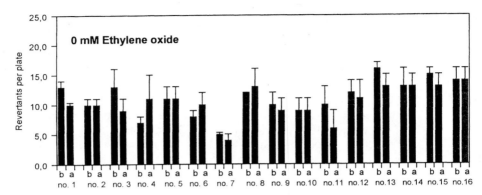

Erythrocytes of 16 individuals before (b) and 90 min after (a) drinking of 1.5 l of a standardized black tea preparation

Number of revertants (mean ± standard deviation)

Before tea drinking:	11 ± 3
After tea drinking:	10 ± 3

Wilcoxon test for paired data: not significant

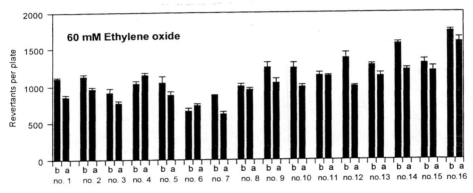

Erythrocytes of 16 individuals before (b) and 90 min after (a) drinking of 1.5 l of a standardized black tea preparation

P < 0.001 (Wilcoxon-test for paired data)

Number of revertants (mean ± standard deviation)

Before tea drinking:	1170 ± 263
After tea drinking:	1010 ± 229

Wilcoxon test for paired data: p < 0.001

Figure 3. Influence of consumption of black tea on the ability of erythrocytes to enhance ethylene oxide mutagenicity. Erythrocytes were taken from 16 individuals (non-smokers, without medication, not taking vitamin pills) before and 90 min after drinking of 1.5 l of a standardized black tea preparation and added to the preincubation mixture. Data shown are mean numbers of TA 1535 revertants per plate ± standard deviations of three independent incubations.

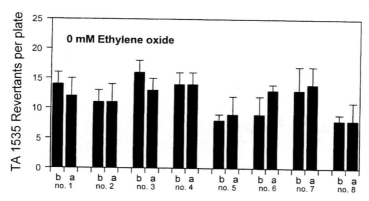

Erythrocytes of 8 individuals before (b) and 90 min after (a) drinking of 1.5 l of water

Number of revertants (mean ± standard deviation)

Before drinking of water:	12 ± 3
After drinking of water	12 ± 3

Wilcoxon test for paired data: not significant

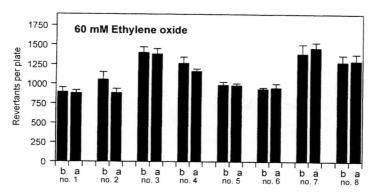

Erythrocytes of 16 individuals before (b) and 90 min after (a) drinking of 1.5 l of water

Number of revertants (mean ± standard deviation)

Before drinking of water:	1150 ± 206
After drinking of water:	1124 ± 233

Wilcoxon test for paired data: not significant

Figure 4. Influence of consumption of water on the ability of erythrocytes to enhance ethylene oxide mutagenicity (control experiment to the study shown in Fig. 3). Instead of black tea individuals drank 1.5 l of water and erythrocytes were taken before and 90 min after drinking. Data shown are mean numbers of TA 1535 revertants per plate ± standard deviations of three independent incubations.

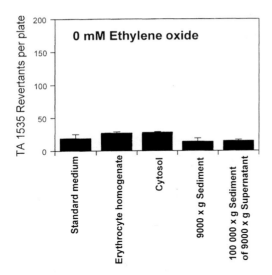

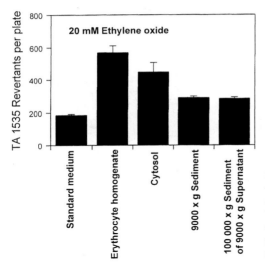

Figure 5. Influence of erythrocyte homogenate, cytosol, microsomes (100 000 × g sediment of 9000 × g supernatant) and a fraction with mitochondria and lysosomes (9000 × g sediment) on the mutagenicity of ethylene oxide. Data shown are mean values ± standard deviations of three independent incubations of a representative experiment.

The cytosol of erythrocytes was further separated into subfractions with constituents of relatively large (>5000) and relatively low (<5000) molecular weight using Sephadex-columns (Pharmacia). These subfractions (500 µl each) were added to the preincubation mixture with 100 µl bacterial suspension (TA 1535) and 300 µl ethylene oxide solution (Figure 6). The cytosolic subfraction with constituents of high molecular weight (>5000) increased the mutagenicity of ethylene oxide to a similar extent as the original cytosol (Fig. 6). Compared to the fraction with high molecular weight the number of revertants was markedly smaller if the low molecular weight fraction was added. Nevertheless, the number of revertants in the samples with the low molecular weight fraction was still significantly higher compared to the samples with standard medium.

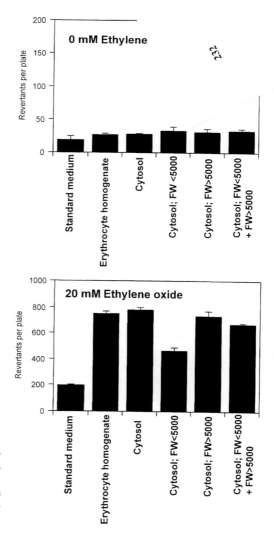

Figure 6. Influence of high (FW>5000) and low (FW<5000) molecular weight constituents of erythrocytic cytosol on the mutagenicity of ethylene oxide. Data shown are mean numbers of TA 1535 revertants per plate ± standard deviations of three independent incubations of a representative experiment.

In conclusion, factors which enhance ethylene oxide mutagenicity have been shown to be located predominantly in the cytosol of erythrocytes and to have a molecular weight >5000. Nevertheless, to a minor degree also cytosolic low molecular weight (<5000) constituents as well as erythrocytic microsomes are able to enhance the mutagenicity of ethylene oxide.

5. INFLUENCE OF GST T1, GST M1, AND HIGH TEMPERATURE

Recently, Thier et al.[8] expressed human GST T1 in *Salmonella typhimurium* either in sense or in antisense orientation, the latter serving as a control. Expression of GST T1 enhanced mutageniciy of several substrates of GST T1, including 1,2,3,4-diepoxybutane, 1,3-dichloroacetone, epibromohydrin and 1,2-dibromoethane.[8] Bioactivation of haloalkanes

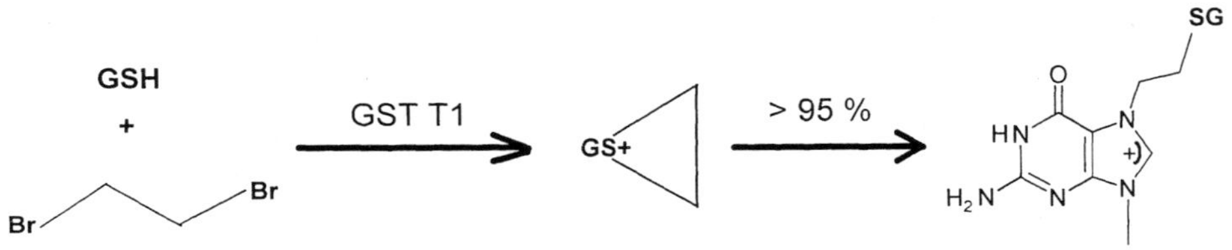

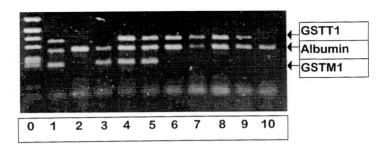

Figure 7. Glutathione-dependent bioactivation of dibromoethane via formation of an episulfonium ion.[1]

mediated by glutathione S-transferases has also been described by Guengerich.[1] Conjugation of haloalkanes with glutathione generates the highly reactive episulfonium ion, which predominantly binds to the N7-position of guanine (≥ 95 %), but also to other sites including O^6-guanine (Figure 7). GST T1 is known to be polymorphic in humans, due to a large deletion in its structural gene[2,9,10] and is known to be expressed in human erythrocytes.[11] The polymorphism of GST T1 offers a good opportunity to examine, whether this enzyme contributes to the observed enhancement of ethylene oxide mutagenicity. Thus, we established a multiplex polymerase chain reaction (PCR) assay, which allows determination of GST T1 (and also GST M1) in a single PCR reaction, including primers for the albumin gene as an amplification control[12,13] (primers and reaction conditions: Table 2; representative gel with amplification products: Figure 8). Erythrocytes from 9 GST T1 positive as well as 5 GST T1 negative individuals (500 µl per individual) were added to the preincubation mixture in the *Ames* test together with 300 µl of ethylene oxide solution and 100 µl of TA 1535 bacterial suspension. Enhancement in mutagenicity of ethylene oxide by erythrocytes of GST T1 negative individuals was very similar to erythrocytes of GST T1 positive individuals (Figure 9). Individuals with the deletion of the GST T1 structural gene (GST T1⁻) necessarily lack GST T1 enzyme activity, but their erythrocytes nevertheless increased the number of revertants. Therefore, GST T1 does not seem to play an important role for the observed enhancement of ethylene oxide mutagenicity. Similarly, enhancement of ethylene oxide mutagenicity did not depend on the GST M1 genotype (Fig. 9).

To examine, whether the factor(s) in erythrocytes responsible for enhancement of ethylene oxide mutagenicity can be inactivated at high temperature, erythrocyte homogen-

Figure 8. Multiplex PCR analysis of the GST polymorphisms. The PCR reaction was performed on 10 randomly selected samples as described[12,13] under conditions shown in table 2. The resulting DNA fragments were separated by electrophoresis on a 1.5% agarose gel containing 0.4 µg ethidium bromide / ml, and were visualized by UV detection. The analysis resulted in assignment of the following genotypes: GSTT1⁻/GSTM1⁻: no. 2 and 10; GSTT1⁺/GSTM1⁺: no. 1, 4, 5; GSTT1⁻/GSTM1⁺: no. 3; GSTT1⁺/GSTM1⁻: No. 6, 7, 8, 9; lane 0: AluI-restricted pBR322 DNA size marker.

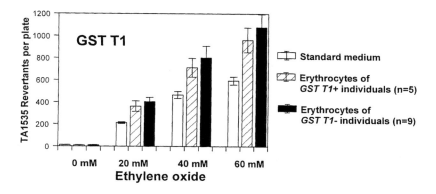

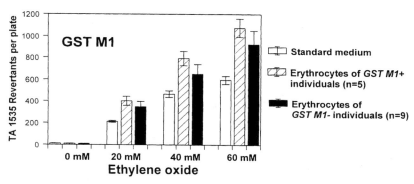

Figure 9. Enhancement of ethylene oxide mutagenicity of erythrocytes of GST T1 and GST M1 negative as well as positive individuals.

ate was heated at 70 °C for 30 min before it was added to the preincubation mixture (Figure 10). These are conditions, which cause a complete inactivation of the majority of drug-metabolizing enzymes. However, erythrocyte homogenates, which have been warmed up, still caused a significant enhancement of ethylene oxide mutagenicity compared to the samples with standard medium, although the extent of enhancement appeared smaller (esp. for individual no. 2).

In conclusion, enhancement of ethylene oxide mutagenicity by human erythrocytes did not depend on the polymorphism of GST T1 or GST M1 and is mediated by relatively heat-insensitive factors.

6. INVESTIGATION OF A POSSIBLE STABILIZING EFFECT BY HUMAN ERYTHROCYTES

A possible explanation for the enhancement of ethylene oxide mutagenicity by human erythrocytes might be stabilization of the mutagenic properties of ethylene oxide (e.g.

Table 2. Primer sequences and reaction conditions for the multiplex polymerase
chain reaction protocol for the simultaneous analysis of the GSTT1
and GSTM1 polymorphisms

Primer	Sequence	Product length
GSTT1		
(forward primer)	TTCCTTACTG GTCCTCACAT CTC	480 bp
(reverse primer)	TCACCGGATC ATGGCCAGCA	
GSTM1		
(forward primer)	GAACTCCCTG AAAAGCTAAA GC	215 bp
(reverse primer)	GTTGGGCTCA AATATACGGT GG	
Albumin		
(forward primer)	GCCCTCTGCT AACAAGTCCT AC	350 bp
(reverse primer)	GCCCTAAAAA GAAAATCGCC	
	AATC	

Step	Temperature	Duration
Primary denaturation	95 °C	2 min
Denaturation	94 °C	1 min
Annealing	62 °C	1 min
Extension	72 °C	1 min
Final elongation	72 °C	5 min

by decreasing the rate of non-enzymatic hydrolysis). Thus, we incubated ethylene oxide
(20 mM) with erythrocytes as well as with standard medium for various time periods between 15 and 120 min before bacteria (*Salmonalla typhimurium* TA 1535) were added
(Figure 11). As expected, mutagenicity of ethylene oxide decreased with increasing time
periods before addition of bacteria. However, the decrease in mutagenicity was not re-

Table 3. Modification of mutagenicity of various directly acting genotoxic substances by
human erythrocyte homogenate

Test substance	Modification factor	Test substance/ plate	*Salmonella typhimurium* strain
Mafosfamide	5.5 ± 1.71	333 μg	TA 1535
N-Nitroso-N-methyl-urea	5.1 ± 0.24	50 μg	TA 102
Ethylene oxide	2.7 ± 0.35	1.6 mg	TA 1535
Styrene oxide	2.3 ± 0.41	888 μg	TA 1535
Methylmethane-sulfonate	1.5 ± 0.39	500 μg	TA 1535
1,2-Epoxybutane	1.4 ± 0.035	12 mg	TA 1535
2,3-Epoxy-1-propanol	1.3 ± 0.080	600 μg	TA 1535
(+)-*anti*-Benzo[*c*]phen-anthrene-3, 4-diol-1,2-epoxide	0.69 ± 0.21	106 ng	TA 100
(±)-*anti*-Benzo[*a*]pyrene-7,8-dihydrodiol- 9,10-epoxide	0.45 ± 0.19	33 ng	TA 100
Cumene hydroperoxide	0.34 ± 0.17	10 μg	TA 102
Benzo[*a*]pyrene-4,5-oxide	0.20 ± 0.060	1 μg	TA 98
1-Methyl-3-nitro-1-nitrosoguanidine	0.001 ± 0.0008	3 μg	TA 1535

Mutagenicity was tested in the *Ames* test (preincubation assay) with and without addition of human erythrocyte homogenate. The modification factor gives the ratio of induced revertants with and without addition of erythrocyte homogenate at non-toxic concentrations where the ratio was maximal. Mean value ± standard deviation of three independent
experiments (using erythrocyte homogenate of 3 different individuals) are given.

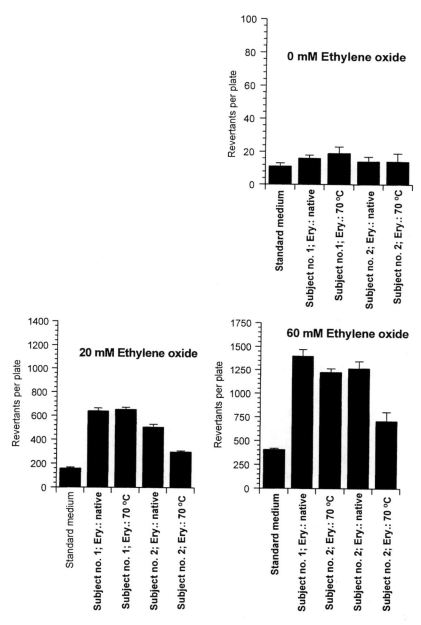

Figure 10. Enhancement of ethylene oxide mutagenicity by native erythrocyte homogenate and by erythrocyte homogenate, which has been warmed up to 70°C for 30 min.

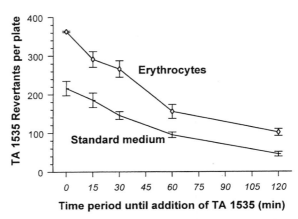

tarded by human erythrocytes (Fig. 11). Thus, a stabilization of ethylene oxide by human erythrozytes does not account for the observed increase in mutagenicity.

7. INFLUENCE OF HUMAN HEMOGLOBIN, GLUTATHIONE, AND L-CYSTEINE *IN VITRO* ON THE MUTAGENICITY OF ETHYLENE OXIDE

Hemoglobin is the main constituent of the erythrocytic cytosol. To examine, whether hemoglobin influences the mutageniciy of ethylene oxide, commercially available human hemoglobin (Sigma) was added to the preincubation mixture in various concentrations up to 20 mg/ml (Figure 12). (At higher concentrations hemoglobin precipitated.) A relatively small, but statistically significant enhancement of ethylene oxide mutagenicity was caused by hemoglobin. It should be considered that a concentration of 20 mg hemoglobin/ml is small compared to the hemoglobin content of human blood, which is approximately 140 mg/ml. Thus, although (the commercial) hemoglobin precipitated at concentrations >20 mg/ml we performed incubations with higher concentrations of hemoglobin, which was kept in suspension in a shaking water bath. Although hemoglobin was not dissolved, a further enhancement of ethylene oxide mutagenicity up to the highest hemoglobin concentration tested (160 mg/ml) was observed (Figure 13). The ability of hemoglobin to enhance ethylene oxide mutagenicity was not significantly altered by infusion of O_2, CO_2 or CO into the preincubation mixture (data not shown). In addition the iron chelator desferoxamine at concentrations up to 1 mM did not inhibit the ability of hemoglobin to enhance ethylene oxide mutagenicity.

Glutathione is another organic molecule present in the cytosol of erythrocytes at relatively high levels. The mean concentration of glutathione present in total venous blood of 44 individuals was 1.28 ± 0.11 mM ($\pm$ 95 %-confidence interval; range: 0.73–2.49 mM), determined by the assay with 5,5-dithiobis(2-nitrobenzoic)acid under conditions as described by Oesch et al.[14] Glutathione was added to the preincubation mixture in concentrations between 0.4 and 39.5 mM together with *Salmonella typhimurium* TA 1535 and ethylene oxide solution (Figure 14). An enhancement of ethylene oxide mutagenicity was observed at glutathione concentrations ranging from ~0.7 and ~7 mM at 60 mM ethylene

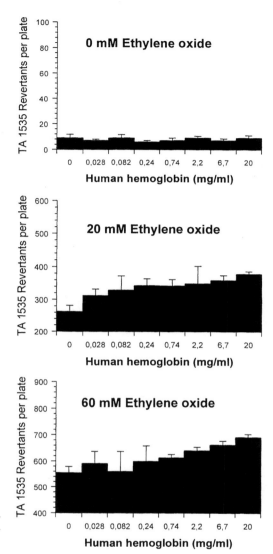

Figure 12. Influence of human hemoglobin (dissolved in standard medium) on ethylene oxide mutagenicity.

oxide. The decrease in revertants per plate at higher concentrations of glutathione was due to toxicity of the tripeptide (dat not shown).

Since the sulfhydryl group of glutathione may be involved in toxification of ethylene oxide, possibly via the formation of an episulfonium ion (Fig. 7), we also examined the influence of the amino acid L-cysteine. In concentrations between 1.0 and 3.2 mM L-cysteine enhanced the mutagenicity of 60 mM ethylene oxide. At a concentration of 20 mM ethylene oxide the influence of L-cysteine was very small, but a slight trend towards an enhancement of ethylene oxide mutagenicity was still observable (Figure 15).

Since at least 3 single constituents of human erythrocytes (hemoglobin, glutathione and L-cysteine) were able to enhance the mutagenicity of ethylene oxide, we tested hemoglobin, glutathione and L-cysteine in physiological concentrations present in blood, to examine whether these single constituents can quantitatively explain the effect of

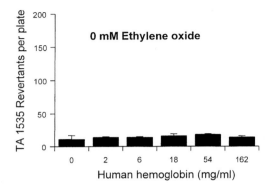

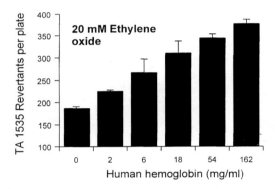

Figure 13. Influence of human hemoglobin, which was kept in suspension in a shaking water bath at concentrations of 54 and 162 mg/ml, on ethylene oxide mutagenicity.

erythrocytes (Figure 16). At physiological concentrations hemoglobin (14 g/dl) exerted the largest effect, which was 54 % of the total effect of erythrocyte homogenate. Glutathione (1 mM) caused 24% of the total effect of erythrocyte homogenate, whereas the contribution of L-cysteine seems to be negligible, due to its low physiological concentration (Fig. 16). Thus, based on the assumption that the effect of single constituents behaves additive, glutathione and hemoglobin might explain ~78% of the effect of erythrocyte homogenate.

To examine, whether the ability to enhance the mutagenicity of ethylene oxide are relatively unique properties of hemoglobin and glutathione, or represent a general feature of many proteins, we examined the effects of two ubiquitous proteins namely serum albumin and γ-globulins (Figure 17). Serum albumin and and γ-globulins were added to the preincubation mixture at the relatively high concentration of 140 mg/ml, together with bacterial suspension (TA 1535) and ethylene oxide solution. However, neither serum albumin nor γ-globulins caused a significant enhancement of ethylene oxide mutagenicity.

8. MODIFICATION OF MUTAGENICITY OF VARIOUS DIRECTLY ACTING GENOTOXIC SUBSTANCES BY HUMAN ERYTHROCYTES

In section 1–5 we examined modification of the mutagenicity of ethylene oxide by human erythrocytes. To examine whether erythrocytes modify also the mutagenicity of various directly acting genotoxic substances, several known mutagens were examined in

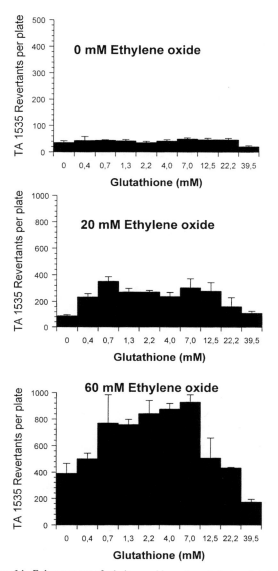

Figure 14. Enhancement of ethylene oxide mutagenicity by glutathione.

the Ames test with and without addition of erythrocyte homogenate to the preincubation mixture (Table 3). *Salmonella typhimurium* tester strains which have been reported to be most sensitive for determination of mutagenicity of the individual substances have been used. Mutagenicity of a wide range of substances was modified by human erythrocytes, ranging from a 5.5-fold increase in mutagenicity of mafosfamide to an approximately 1000-fold decrease in mutagenicity of 1-methyl-3-nitro-1-nitrosoguanidine (MNNG) (Table 3). Enhancement of the mutagenicity of ethylene oxide (2.7-fold) was only intermediate in this range of substances. Modification of mutagenicity by human erythrocytes was not limited to base substitution mutations (detected by TA 1535 and TA 100) but included also frameshift mutations (detected by TA 98). In addition the number of mutations to G-C

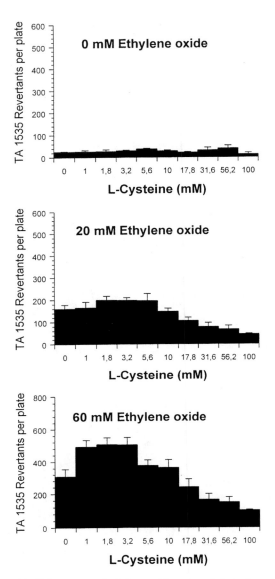

Figure 15. Enhancement of ethylene oxide mutagenicity by L-cysteine.

base pairs (detected by TA 1535 and TA 98, which contain G-C at the site of the mutation), as well as mutations to A-T base pairs (detected by TA 102, which carries A-T at the site of the mutation) was increased by human erythrocytes. Although the number of substances examined is still relatively small, there seem to be some general features of directly acting mutagens, which are further activated by human erythrocytes, such as a highly strained, three-membered ring. Addition of aliphatic (see: 1,2-epoxybutane; 2,3-epoxy-1-propanol in Table 3) or aromatic (see: styrene oxide) substituents to the "model substance" ethylene oxide decreased the extent of further toxification by erythrocytes. Also mafosfamide (the substance with the highest modification factor in Table 3), a self-activating derivative of

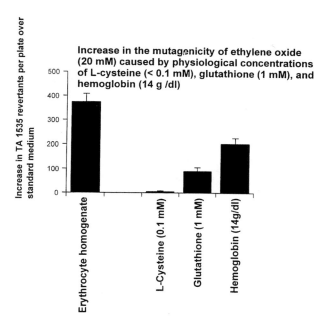

Figure 16. Influence of physiological concentrations of hemoglobin (14 g/dl), glutathione (1 mM), and L-cysteine (0.1 mM) on mutagenicity of ethylene oxide.

cyclophosphamide (which releases 4-hydroxycyclophosphamide in aqueous solution) forms a highly strained, three-membered ring (the nitrenium ion) after non-enzymatic formation of phosphoramide mustard. On the other hand mutagenicity of arene oxides (benzo[*a*]pyrene-4,5-oxide) and dihydrodiol epoxides ((±)-*anti*-benzo[*a*]pyrene-7,8-dihydrodiol-9,10-epoxide and (+)-*anti*-benzo[*c*]phenanthrene-3,4-diol-1,2-epoxide) of polycyclic aromatic hydrocarbons was decreased by human erythrocytes (Table 3).

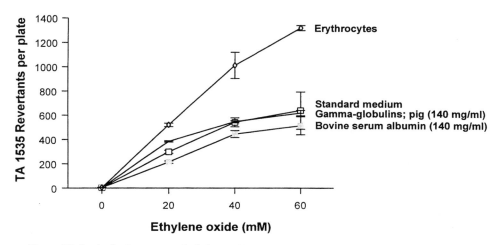

Figure 17. Lack of enhancement of ethylene oxide mutagenicity by serum albumin and γ-globulins.

Since differences in modification factors by erythrocytes were larger than 5000-fold for the substances shown in Table 3, modification of the effect of genotoxic substances by erythrocytes might be important for genotoxic risk assessment.

9. MODIFICATION OF THE EXTENT OF DNA SINGLE STRAND BREAK INDUCTION BY HUMAN ERYTHROCYTES

In sections 1–6 mutagenicity in *Salmonella typhimurium* was examined. In addition we examined, whether human erythrocytes also modify the extent DNA strand lesions (DNA single strand breaks and alkali labile sites) using the alkaline elution technique as described[15] with minor modifications.[16] Human mononuclear blood cells were isolated from heparinized venous blood by centrifugation with Ficoll-Metrizoate[15] and incubated with ethylene oxide (for 1 h at 37 °C) either in phosphate buffered saline (PBS) or in PBS with addition of erythrocytes (in a physiological concentration: 5×10^6 erythrocytes/μl). Addition of human erythrocytes significantly decreased the extent of DNA strand lesions induced by ethylene oxide to approximately 50% (Figure 18). Thus, erythrocytes influenced the extent of DNA strand lesions and the number of mutations in opposite directions, whereby DNA strand lesions were decreased and mutagenicity was enhanced.

10. A CONCEPT, WHICH MAY EXPLAIN ENHANCEMENT OF MUTAGENICITY IN CONTRAST TO THE DECREASE IN DNA STRAND BREAKS CAUSED BY HUMAN ERYTHROCYTES

At first glance it might be surprising that human erythrocytes enhance ethylene oxide mutagenicity, but on the other hand decrease induction of DNA strand lesions (DNA single strand breaks and alkali labile sites). However, it should be considered that strand lesions

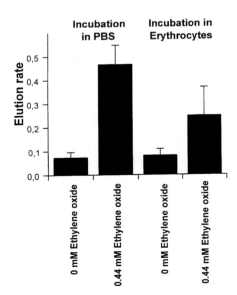

Figure 18. Influence of intact human erythrocytes on the extent of DNA strand lesions induced by ethylene oxide. Mononuclear blood cells of 10 individuals (non-smokers, without medication, not taking vitamin pills) were incubated with 0.44 mM ethylene oxide either in phosphate buffered saline (PBS) or in PBS with addition of erythrocytes (37 °C for 1 h). DNA strand lesions were examied by alkaline elution.[15]

and mutations are different end points, which require different mechanisms and also different initial DNA base alterations. The initial event, which is responsible for the majority of DNA strand lesions, is the formation of a 2-hydroxyethyl adduct at the N7 position of guanine by ethylene oxide (Figure 19). Binding to N7 of guanine labilizes the N-glycosidic bond between N9 of guanine and C1 of deoxyribose. Thus, the guanine adduct may be cleaved off, leaving an apurinic site (Fig. 19). At the apurinic site an equilibrium between the closed furanose and the open aldehyde form may exist. Subsequent proton-abstraction from the aldehyde form of deoxyribose (at C2) may occur in the presence of OH⁻, causing a strand break at the 3'-site of the apurinic site known as β-elimination (Figure 20). Since proton-abstraction from C2 of deoxyribose depends on the concentration of OH⁻, β-elimination is a relatively rare event at physiological pH, but occurs almost at 100 % of all apurinic / apyrimidinic sites at pH 12.6, the pH at which alkaline elution is performed.

It is well known that point mutations are efficiently induced by alkylation of the O^6-position of guanine. Since 30–50% of O^6-methylguanine adducts pair with thymine instead of cytosine, a high percentage of GC→AT transitions will be caused.[17,18] Alkylation at the O^6-position of guanine does not induce DNA single strand breaks and no single strand breaks occur during repair of O^6-methylguanine by O^6-methylguanine-DNA-alkyltransferase. Alkylation at N7 of guanine may—to a relatively small extent—also induce point mutations (e.g. by incorporation of adenine opposite of an apurinic site during DNA replication), however, the percentage of apurinic sites causing point mutations is much lower than the percentage for O^6-methylguanine.[17,18] The ratio of N7 to O^6-alkylation of guanine varies widely between different alkylating substances. Reaction through an S_N2 mechanism results predominantly in N7-alkylated guanine, whereas more ionized sub-

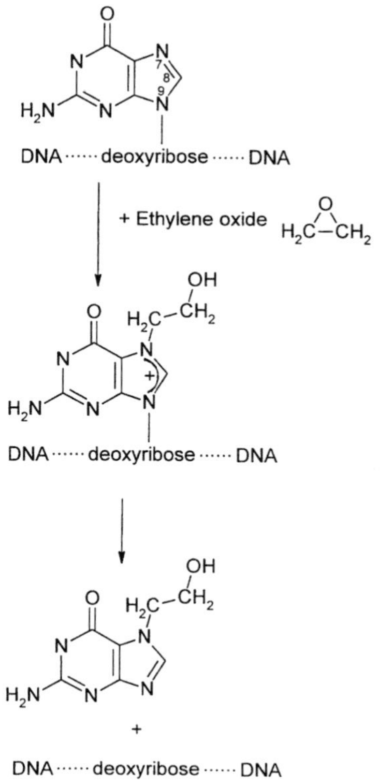

Figure 19. Formation of an apurinic site after binding of ethylene oxide to the N7-position of guanine.

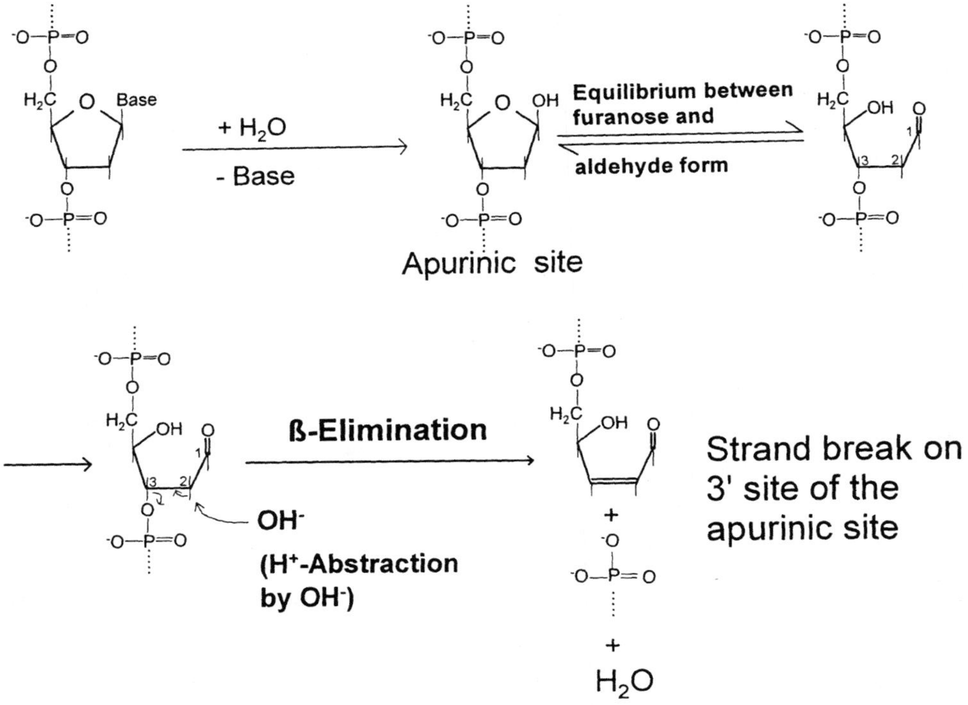

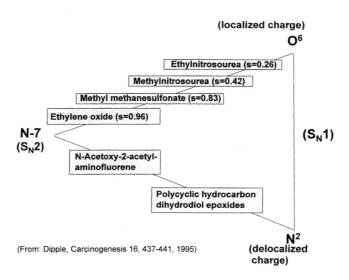

Figure 20. Mechanism of DNA strand breakage by β-elimination due to an apurinic site.

Figure 21. Reaction of various genotoxic substances on guanine residues in DNA.

stances, which react through an S_N1 mechanism give rise to alkylation of the O^6-position of guanine[18] (if the charge is not delocalized, resulting in N^2 substitution) (Figure 21). For instance the ratio of N7 to O^6 alkylation of guanine is 330 for methyl methanesulfonate, but only 1.7 for ethylnitrosourea.[19] The ratio of N7 to O^6-alkylation for ethylene oxide is even higher compared to methyl methanesulfonate (Fig. 21). Assuming that ethylene oxide might be metabolized to a more ionized substance by erythrocytic constituents an increase in alkylation at O^6 (and a decreased alkylation at N7) of guanine would be expected. This shift in alkylation pattern could explain the simultaneous enhancement of mutations and decrease in induction of DNA strand lesions due to human erythrocytes.

REFERENCES

1. F.P. Guengerich, 1996, Metabolic Control of Carcinogens, in: *Control Mechanisms of Carcinogenesis*, J.G. Hengstler and F. Oesch (eds.), Mainz, Publishing House of the Editors, 12–35.
2. J.G. Hengstler, M. Arand, M.E. Herrero, F. Oesch, 1998, Polymorphisms of N-acetyltransferases, glutathione S-transferases, microsomal epoxide hydrolase and sulfotransferases: Influence on cancer susceptibility, in: Genes and Environment in Cancer (Rabes, ed.), Karger-publishing house, in press.
3. J.D. Hayes and D.J. Pulford, 1995, The glutathione S-transferase supergene family: Regulation of GST and the contribution of the isoenzymes to cancer chemoprotection and drug resistance, *Critical Reviews in Biochemistry and Molecular Biology* 30: 445–600.
4. B.N. Ames, 1971, The detection of chemical mutagens with enteric bacteria, In: *Chemical Mutagens, Principles and Methods for their Detection*, Plenum, New York, Vol. 1, pp. 267–282.
5. D.M. Maron and B.N. Ames, 1983, Revised methods for the Salmonella mutagenicity test, *Mut. Res.* 113: 173–215.
6. J.G. Hengstler, J. Fuchs, S. Gebhard, F. Oesch, 1994, Glycolaldehyde causes DNA-protein crosslinks: a new aspect of ethylene oxide genotoxicity, *Mut. Res.:* 304, 229–234.
7. V.L. Dellarco, W.M. Generoso, G.A. Sega, J.R. Fowle, D. Jacobson-Kram, 1990, Review of the mutagenicity of ethylene oxide, *Environmental and Molecular Mutagenesis* 16, 85–103.
8. R. Thier, S.E. Pemble, H. Kramer, J.B. Taylor, F.P. Guengerich, B. Ketterer, 1996, Human glutathione S-transferase T1–1 enhances mutagenicity of 1,2-dibromoethane, dibromomethane and 1,2,3,4-diepoxybutane in Salmonella typhimurium, *Carcinogenesis* 17: 163–166.
9. S. Pemble, K.R. Schroeder, S.R. Spencer, D.J. Meyer, E. Hallier, H.M. Bolt, B. Ketterer, J.B. Taylor, 1994, Human glutathione S-transferase theta (GSTT1): cDNA cloning and the characterization of a genetic polymorphism. *Biochem. J.* 300: 271–276.
10. U. Fost, M. Tornqvist, M. Leutbrecher, F. Granath, E. Hallier, L. Ehrenberg, 1995, Effects of variation in detoxification rate on dose monitoring through adducts. *Hum. Exp. Toxicol.* 14: 201–203.
11. K.R. Schröder, E. Hallier, D.J. Meyer, F.A. Wiebel, A.M. Muller, H.M. Bolt, 1996, Purification and characterization of a new glutathione S-transferase, class theta, from human erythrocytes, *Arch. Toxicol.* 70: 559–566.
12. M. Arand, R. Mühlbauer, J.G. Hengstler, E. Jäger, J. Fuchs, L. Winkler, F. Oesch, 1996, A multiplex polymerase chain reaction protocol for the simultaneous analysis of the glutathione S-transferase GSTM1 and GSTT1 polymorphisms, *Anal. Biochem.* 236: 184–186.
13. J.G. Hengstler, A. Kett, M. Arand, B. Oesch, F. Oesch, P.G. Knapstein, B. Tanner, 1998, Glutathione S-transferase T1 and M1 gene defects in ovarian carcinoma. *Cancer Letters*, in press.
14. F. Oesch, J.G. Hengstler, J. Fuchs, 1994, Cigarette smoking protects mononuclear blood cells of carcinogen exposed workers from additional work-exposure induced DNA single strand breaks, *Mutat. Res.* 321: 175–185.
15. J.G. Hengstler, J. Fuchs, F. Oesch, 1992, DNA strand breaks and cross-links in peripheral mononuclear blood cells of human ovarian cancer patients during chemotherapy with cyclophosphamide/carboplatin, *Cancer Res.* 52: 5622–5626.
16. J.G. Hengstler, J. Fuchs, B. Tanner, B. Oesch-Bartlomowicz, C. Hölz, F. Oesch, 1997, Analysis of DNA single strand breaks in human venous blood: A technique which does not require isolation of white blood cells, *Environmental and Molecular Mutagenesis* 29: 58–62.
17. F. Oesch, J. Fuchs, M. Arand, S. Gebhard, A. Hallier, B. Oesch-Bartlomowicz, D. Jung, B. Tanner, U. Bolm-Audorff, G. Hiltl, G. Bienfait, J. Konietzko, J.G. Hengstler, 1997, Möglichkeiten und Grenzen der

alkalischen Filterelution zum Biomonitoring gentoxischer Belastungen, in: *Molekulare Marker bei Beruflich Verursachten Tumoren*, Ed.: Bundesanstalt für Arbeitsschutz und Arbeitsmedizin.

18. A. Dipple, 1995, DNA adducts of chemical carcinogens, *Carcinogenesis* 16: 437–441.
19. J.V. Frei, D.H. Swenson, D.H. Warren, P.D. Lawley, 1978, Alkylation of deoxyribonucleic acid in vivo in various organs of C57BL mice by the carcinogens N-methyl-N-nitrosourea, N-ethyl-N-nitrosourea and ethyls of high-pressure liquid chromatography, *Biochem. J.* 174: 1031–1044.

17

CYTOCHROME P4501A1 (CYP 1A1) AND ASSOCIATED MFO ACTIVITIES IN FISH AS AN INDICATOR OF POLLUTION WITH SPECIAL REFERENCE TO IZMIR BAY

Emel Arınç and Alaattin Sen

Joint Graduate Program in Biochemistry
Department of Biological Sciences
Middle East Technical University
06531 Ankara, Turkey

1. INTRODUCTION

Biotransformation of relatively insoluble organic chemicals to more water-soluble compounds is a prerequisite for their detoxification and excretion. The first step in biotransformation is usually the oxidative step, catalyzed by the microsomal cytochrome P450 dependent mixed function oxidase (MFO) system. This "Phase I" metabolism is usually followed by "Phase II" in which oxygenated groups of xenobiotics are conjugated with glucuronate, sulfate, acetyl, or glutathione by different families of transferase enzymes. Thus, resulting polar and water-soluble end product can be excreted from the organism through bile or urine.[1]

In addition, cytochrome P450 dependent mixed-function oxidation is also responsible for the activation of foreign chemicals to the reactive intermediates that ultimately results in toxicity, carcinogenicity and mutagenicity.[2,3] The degree of detoxification versus toxication of these compounds depends on which metabolic pathway predominates. Thus, it would be important to identify the factors (homeostatic and environmental) influencing the direction of metabolism of these compounds. Diverse fields including biochemistry, molecular biology, endocrinology, pharmacology, and therapeutics, genetics and chemical carcinogenesis are unified to deal cytochrome P450 dependent MFO systems.

2. FISH MFO SYSTEM

Although, Brodie and Maickel[4] in 1962 suggested that fish lacked the required enzymes for the metabolism of xenobiotics, it is now well established that a liver MFO sys-

Molecular and Applied Aspects of Oxidative Drug Metabolizing Enzymes,
edited by Arınç *et al*. Kluwer Academic / Plenum Publishers, New York, 1999.

tem, with the capability of metabolizing a variety of chemicals exists in both freshwater and marine fish.[5-18] Resolution of the fish liver microsomal MFO system was first described by Arinç *et al.*[7] in 1976. Little skate liver microsomal MFO system was resolved into three components: cytochrome P450, NADPH cytochrome P450 reductase and lipid.[7,19,20] Thus, MFO systems in fish and invertebrates appear to be multicomponent systems similar to the microsomal cytochrome P450 dependent electron transport systems in mammals. The major MFO system in fish, as in mammals, is associated with microsomes from hepatic tissues, although it has been observed in virtually all tissues examined. A number of reviews have appeared on these topics.[16,21-24]

Defining the functions of MFO system and linking a given characteristic of MFO system to the action of environmental chemicals require knowing the features associated with the varied biological or physiological conditions. In mammals, sex, diet and age are among the factors known to influence MFO systems. Marked sex differences also occur in cytochrome P450 and/or MFO activity of fish. Moreover, in fish, there are changes in MFO activity associated with season and gonadal status (for reviews, see[21,22]). The implication is that MFO of marine species might be regulated by hormonal factors, as well as that MFO system participates in hormone metabolism. As in the mammalian system, multiple forms of cytochrome P450 belonging to the families of CYP1A, CYP2B, CYP2E, CYP2M, CYP2K, CYP3A, CYP11A, CYP17, and CYP19 are found in fish. There is about 50 cytochrome P450 forms reported to be purified, partially purified or cloned from aquatic species.[16,18,21-27]

Among the purified fish cytochromeP450 forms, P4501A1 homologues hold the priority due to its role in metabolism of carcinogens, mutagens and environmental pollutants. So the occurrence and functions of cytochrome1A forms in diverse organisms are being investigated vigorously.

3. CYTOCHROME P4501A

Two cytochrome P4501A proteins, P4501A1 and P4501A2 have been identified from in all mammalian species studied up to now, including humans. These forms are previously called as cytochromes P448. A number of planar compounds including 3-methylcholanthrene (3-MC), benzo(a)pyrene, B(a)P, 2,3,7,8-tetrachlorodibenzodioxin (TCDD), dibenzofurans, polychlorinated biphenyls (PCBs) are known to be potent common inducers of these two forms of cytochrome P450. In each group, this appears to proceed via Ah receptor mediated mechanism. These two forms of cytochrome P4501A subfamily share 75% of identical amino acid sequences, a very similar gene organization and chemical properties, but show clear differences in their catalytic properties for carcinogenic chemicals and in mechanisms of transcriptional regulation of two genes. In animals, such as rat, rabbit and mouse cytochrome P4501A1 had a markedly high specific B(a)P hydroxylase activity while P4501A2 had a limited activity. In contrast, rat and rabbit hepatic cytochrome P4501A2 has a high arylamine and/or heterocyclic amine N-oxidation activity resulting in carcinogenicity.[28,29] Expression of cytochrome P4501A1 has been correlated with development of polycyclic aromatic hydrocarbon-associated cancers and other disorders in rodents.[30] The formation of the B(a)P 7,8 diol 9,10 epoxide and BPDE adduct of deoxyguanosine monophosphate by cytochrome P4501A1 was demonstrated.

In animals, *CYP1A1* gene is inducible expressed in liver and in extrahepatic tissues such as kidney, lung and skin while inducible expression of cytochrome P4501A2 gene is exclusively limited to the liver.[28,29] In humans, cytochrome P450 gene is not appreciably expressed in liver, is but inducible expressed in extrahepatic tissues and in placenta. Many cell lines derived from hepatocytes and other tissues such as Hepa-1 and He La cells exhibit inducible expression of *CYP1A1* in response to the inducer, similar to the those in livers of experimental animals, while cell lines with the ability of inducible expression of *CYP1A2* have not been reported. Primary cultures of hepatocytes also rapidly loose the inducibility of *CYP1A2*, whereas that of *CYP1A1* is retained long time.[31]

Induction of cytochrome P4501A1 by cigarette smoke in human lung has been reported. A restriction fragment length polymorphism near P4501A1 gene was found to be associated with increased lung cancer risk in Japanese smokers.[32,33]

4. CYTOCHROME P4501A1 IN FISH

Several fold induction of cytochrome P4501A1 content and P4501A1 associated arylhydrocarbon hydroxylase (AHH) activities upon treatment of fishes by 3-MC and related compounds has been demonstrated (for reviews see[22,34–36]). About 8- to 18- fold induction of cytochrome P4501A1 and/or its associated EROD and B(a)P hydroxylase activities are detected in little skate[19] and in gilthead seabream[25] liver microsomes in response to 3-MC, TCDD or β-NF treatment. It is remarkable to observe that the degree of induction of AHH activities in fish has been very similar to those obtained in rabbit lung[37] in response to TCDD treatment[38] (Figure 1).

A single gene in trout[39] or a protein in other teleost species homologous to mammalian cytochrome P4501A1 has been established in fish species. Whether multiple cyto-

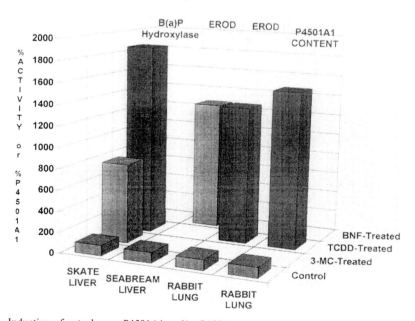

Figure 1. Induction of cytochrome P4501A1 and/or P4501A1 associated enzyme activities in little skate and gilthead seabream liver and in rabbit lung following the treatment of animals with 3-MC, TCDD or β-NF. Constructed from[19,25,37].

Table 1. General properties of mullet liver P4501A1

$Fe^2 \pm CO$ absorption maxima at 447 nm
Low spin heme iron
Molecular weight: $58\,000 \pm 500$
Primary activity: 7-ethoxyresorufin O-deethylase
EROD activity strongly inhibited by α-NF
Does not support oxidative hydroxylation reactions of benzphetamine
 or nitrosodimethylamine or aniline
Strongly induced by PAH-type inducers
Strong cross-reactivity with antibodies produced against other fish
 P4501A1 homologues

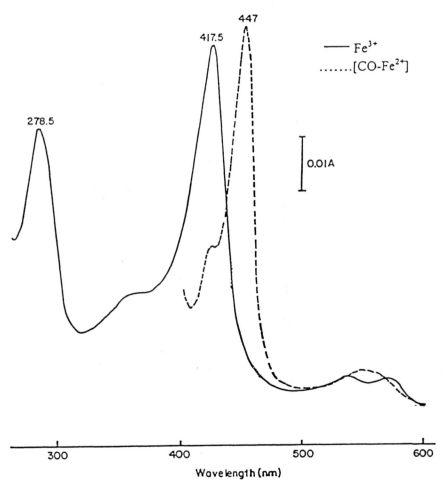

Figure 2. Absolute spectra obtained with the purified liver cytochrome P4501A1 from mullet. Oxidized (solid line); dithionite-reduced plus carbon monoxide (dashed line).

chrome P4501A genes occur in fish is uncertain.[16] The cytochrome P4501A genes of scup and toadfish appear to be closely related to trout and plaice *CYP1A1*.[40]

Rainbow trout (*Oncorhynchus mykiss*) P450LM4b,[41,42] scup (*Stenotomus chrysops*) P450E,[43] Atlantic cod (*Gadus morhua*) P450 c[44] and perch (*Perca fluviatis*) P450 V[45] are found to share the similar properties. These proteins are classified as cytochrome P4501A1 on the basis of sequence comparison with the mammalian counterparts and on those of biocatalytic and spectral properties.

It is well established that the distinct difference between cytochromes P4501A1 and 1A2 resides in their spin state. While P4501A1 is a low spin heme protein, P4501A2 exists as a high spin heme protein.

Cytochromes P4501A1 purified from feral fish or β-NF-, TCDD- or 3-MC- treated fish have showed the characteristics of low spin heme protein. Recently, we purified cytochrome P4501A1 in a homogenous form from leaping mullet (*Liza saliens*) caught from the most polluted part of Izmir Bay on the Aegean Coast and characterized by the analysis of its catalytic, spectral, electrophoretic and immunological properties.[26] As illustrated in Figure 2, the absolute absorption spectrum of the purified P4501A1 from the feral fish mullet shows two minor α and β peaks at 570 nm and 535 nm and one major Soret peak at 417.5 nm, which is a characteristic of low-spin cytochrome P4501A1.

Polyclonal antibodies (Pab anti-mullet P4501A1) raised against leaping mullet P4501A1 in this laboratory and Mab 1-12-3 against scup (*stenotomus chrysops*) P4501A1 (provided by J.J. Stegeman) and Pab against trout (*oncorynchus mykiss*) P4501A1 (provided by D.R. Buhler) showed strong cross-reactivity with the purified mullet liver

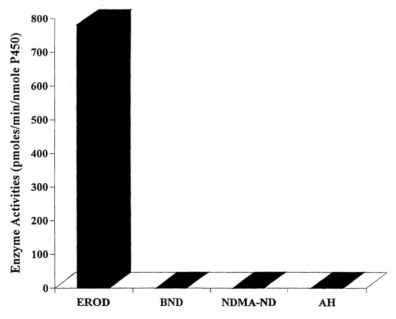

Figure 3. Biocatalytic activities of the purified mullet liver cytochrome P4501A1 with various substrates in the reconstituted systems containing purified fish liver reductase and synthetic lipid. EROD: 7-ethoxyresorufin O-deethylase; BND: benzphetamine N-demethylase; AH: aniline 4-hydoxylase. Reproduced from Sen and Arinç,[26] with permission.

P4501A1 and produced a single band with the mullet liver microsomes identical to the purified mullet P4501A1.

Biocatalytic properties of leaping mullet cytochrome P4501A1 determined according to its ability to catalyze the O-deethylation of 7-ethoxyresorufin, N-demethylation of benzphetamine and N-nitrosodimethylamine, and p-hydroxylation of aniline in the reconstituted system containing purified mullet cytochrome P450 reductase[46] and synthetic lipid, have demonstrated that mullet cytochrome P450 1A1 is very active in the O-deethylation of 7-ethoxyresorufin but showed no oxidative hydroxylation activity towards the other substrates, that is aniline, N-dinitrosodimethylamine and benzphetamine[26] (Figure 3).

5. INDUCTION OF HEPATIC CYTOCHROME P4501A1 IN FISH AS AN ENVIRONMENTAL BIOMARKER

Of particular interest from an environmental standpoint has been the finding that fishes exposed to petroleum products under the laboratory and field conditions show induced AHH activity. Payne and Penrose[47] have suggested the use of AHH activity as an environmental monitor for the first time in 1975 since not only does AHH activity respond to the presence of hydrocarbons but it also activates certain PAHs to carcinogenic metabolites. Subsequent studies have shown that fish caught in waters contaminated with petroleum oil products, paper pulp effluents, industrial and municipal wastes exhibit increased level of P450 and associated AHH and EROD activities.

Even tough the chemical analysis is able to measure a wide range of pollutants quantitatively and accurately, the complex mixture of chemical pollutants can hardly be assessed. Furthermore, it does not reveal the impact of chemical pollution on the aquatic environment. Utilization of biochemical factors to evaluate biological responses to pollutants, especially to carcinogenic compounds such as PAHs, PCBs, debenzodioxines and dibenzofurans has increased considerably over the past 15 years. Induction of cytochrome P4501A1 and of its associated MFO activities, namely AHH and EROD activities in fish by these environmental chemicals, are probably the most widely used biochemical measurements in biomonitoring environmental contaminants (for reviews see[22,34–36,48–50]). In addition, Antibodies produced against purified P4501A1 have been used to assess environmental induction of P4501A1 in deep-sea fish from Northern Atlantic;[51] in English sole from Puget Sound, USA;[52] in winter flounder and scup from Northeastern part of USA;[53,54] in flounder, plaice and dalp in Norway;[35,55] in a number of flatfish and other species samples following Exxon Valdez oil spill;[56] in perch, pike, dab and blenny from waters of Sweden.[36]

6. BIOMONITORING STUDIES ALONG THE IZMIR BAY ON THE AEGEAN COAST OF TURKEY

Izmir Bay is one of the most polluted areas in Turkey. The bay extends approximately 24 km in the East-West direction, with an average width of 5 km. The bay is usually considered to consist of three sections, according to topography and hydrology: Inner, Middle, and Outer Bays. The water depth at the Inner Bay changes between 0 and 20 m. Average water depths at the Middle and Outer Bays are 16 m and 49 m, respectively. The port of Izmir City and several industries are located at the Inner Bay. Urbanization, industrial activities, and agriculture impact heavily on the water quality of the Bay. Ship traffic is also very heavy in the Bay. As indicated by Balkas and Juhasz,[57] 25% of Turkey's export and 55% of her import pass through the port. In addition, defence activities are also present in the bay. The average number of commercial ships visiting the harbour each year is approximately 2000. Domestic and industrial wastes, urban and agricultural run off, discharges from ships, sediments and contaminated waters of rivers have cumulatively had a significant adverse effects on the water quality of the Bay. Industrial activities cover a large range of industries including food processing, tanneries, paint, paper and pulp factories, chemical and textile factories, vegetable oil and soap production, and a petroleum refinery. Industrial and domestic wastes as well as contaminated waters of several small rivers heavily pollute the Inner Bay. The Middle Bay is a transition zone with pollutant concentrations intermediate between Outer and Inner Bays, and the pollution in the Outer Bay is considered not significant.[58]

Here, we describe the results of some studies caried out in Izmir Bay. In these studies, the degree of induction of cytochrome P4501A1 associated EROD activity and immunochemical detection of cytochrome P4501A1 in leaping mullet were used as biomarker for assessment of PAH- and PCB- type organic pollutants along the Izmir Bay on the Aegean Sea coast.

All the sampling sites for the fish are given in Figure 4. Leaping mullet (n = 200) sampled at a site (site 1, Pasaport) known to contain high level of organic compounds had approximately 62 times more EROD activity than the feral fish sampled from an uncontaminated site in the Outer Bay (Figure 5). Hepatic microsomal EROD activity of the mullet caught from site 10 (reference site) was determined to be 25 ± 9, mean ± SE, n = 4 picomoles/min/mg protein. Mullet caught along the pollutant gradient at the three other sites in the Bay exhibited less but highly significant induced EROD activity (Figure 5). An

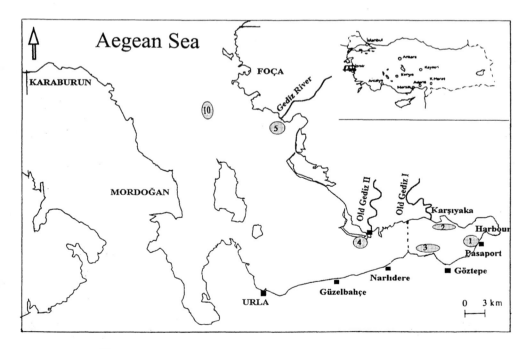

Figure 4. The map of Izmir Bay on the Aegean Sea and the sampling sites.

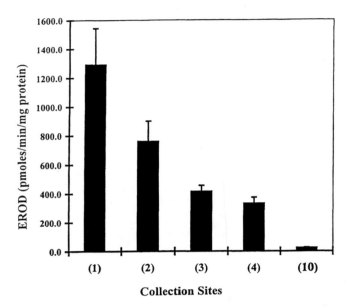

Figure 5. EROD activities of liver microsomes caught in Izmir Bay at five different sampling sites. The numbers in parentheses show collection site numbers. 1: Pasaport; 2: Karsiyaka; 3: Inciralti; 4: Tuzla; 10: Reference site in the Outer Bay.

inverse relationship was found between the EROD activity in the fish and the distance between the catch point and the discharge region of polluted rivers and creeks and of industrial and domestic wastes into the Harbour. In addition, studies with Western blot analysis using the polyclonal antibodies produced against purified mullet cytochrome P4501A1 in our laboratory[26] showed a similar trend with higher P4501A1 cross-reactivity at the contaminated regions, lower P4501A1 values at less polluted areas, and presence of no P4501A1 in the mullet caught from the reference site in the Outer Bay (Figure 5). These results indicated that Izmir Harbour is highly contaminated with PAH-, PCB-type toxic organic compounds.

REFERENCES

1. D. W. Nebert, and F. J. Gonzales, 1987, P-450 genes: Structure, evolution and regulation. *Ann. Rev. Biochem.* **56**: 945–993.
2. H. Conney, and J. J. Burns, 1972, Metabolic interactions among environmental chemicals and drugs. *Science* **178**: 576–586.
3. O. Pelkonen, and D. W. Nebert, 1982, Metabolism of polycyclic aromatic hydrocarbons; Etiologic role in carcinogenesis. *Pharmacol. Rev.* **34**: 189–222.
4. B. B. Brodie, and R. P. Maickel, 1962, Comparative biochemistry of drug metabolism. In *Proceedings of the First International Pharmacology Meeting* (B. B. Brodie, and E. G. Erdos, Eds.), vol. 6, pp. 299–324.
5. D.R. Buhler, and M. E. Rasmusson, 1968, The oxidation of drugs by fishes. *Comp. Biochem. Physiol.* **25**: 223–239.
6. R. J. Pohl, J. R., Bend, A. M. Guarino, J. R. Fouts, 1974, Hepatic microsomal mixed-function oxidase activity of several marine species from Coastal Maine. *Drug Metab. Disp.* **2**, 545–555.

7. E. Arinç, R. M. Philpot, and J. R. Fouts, 1976, Partial purification of hepatic microsomal cytochrome P450 and NADPH-cytochrome *c* reductase from little skate, *Raja erinacea*, and reconstitution of mixed-function oxidase activity. *Fedn. Proc.* **36**, 666.

8. J. R. Bend, and M. O. James, 1978, Xenobiotic metabolism in marine and freshwater species, in: *Biochemical and Biophysical Perspectives in Marine Biology* (D. C. Mallins, and J. R. Sargent, eds.), **Vol. 4**, pp. 125–188, Academic Press, London.

9. M. O. James, M. A. Q. Khan, and J. R. Bend, 1979, Hepatic microsomal mixed-function oxidase activities in several marine species common to coastal Florida. *Comp. Biochem. Physiol.* **62C**, 155–164.

10. J. J. Stegeman, and H. B. Kaplan, 1981, Mixed-function oxygenase activity and benzo(a)pyrene metabolism in the barnacle *Balanus eburneus* (Crustacea: Cirripedia). *Comp. Biochem. Physiol.* **68C**, 55–61.

11. J. J. Lech, M. J. Vodicnick, and C. R. Elcombe, 1982, Induction of monooxygenase activity in fish, in: *Aquatic Toxicology*, (L. J. Weber, ed), pp. 107–148. Raven Press, New York.

12. E. Arinç, and O. Adali, 1983, Solubilization and partial purification of two forms of cytochrome P-450 from trout liver microsomes. *Comp. Biochem. Physiol.* **76B**, 653–662.

13. G. Monod, A. Devaux, and J. L. Riviére, 1987, Characterization of some monooxygenase activities and solubilization of hepatic cytochrome P450 in two species of freshwater fish, the nase (*Chondrostoma nasus*) and the roach (*Rutilus rutilus*). *Comp. Biochem. Physiol.* **88C**, 83–89.

14. A. Goksoyr, T. Andersson, T. Hansson, J. Klungsoyr, Y. S. Zhang, and L. Förlin, 1987, Species characteristics of the hepatic xenobiotic and steroid biotransformation system of two teleost fish, Atlantic cod (*Gadus morhua*) and rainbow trout (*Salmo gairdneri*). *Toxicol Appl. Pharmacol.* **89**, 347–360.

15. M. Celander, and L. Förlin, 1991, Catalytic activity and immunochemical quantification of hepatic cytochrome P450 in β-naphthoflavone and isosafrole treated rainbow trout (Oncorhynchus mykiss). *Fish Physiol. Biochem.* **9**, 189–197.

16. J. J. Stegeman, 1993, Cytochrome P450 forms in fish, in: *Handbook of Experimental Pharmacology* (J. B. Schenkman, and H. Greim, eds.) **Vol. 105**, pp. 279–291, Springer-Verlag Press, Heidelberg.

17. E. Arinç, and A Sen, 1993, Characterization of cytochrome P450 dependent mixed-function oxidase system of gilthead seabream (*Sparus aurata* ; *Sparidae*) liver. *Comp. Biochem. Physiol.* **104B**, 133–139.

18. J. J. Stegeman, B. R. Woodin, H. Singh, M. F. Oleksiak, and M. Celander, 1997, Cytochromes P450 (CYP) in tropical fishes: Catalytic activities, expression of multiple CYP proteins and high levels of microsomal P450 in liver of fishes from Bermuda. *Comp. Biochem. Physiol.* **116 C**: 61–75.

19. J. R. Bend, R. J. Pohl, E. Arinç, and R. M. Philpot, 1977, Hepatic microsomal and solubilized mixed-function oxidase systems from the little skate, *Raja erinacea*, a marine elasmobranch, in: *Microsomes and Drug Oxidations* (V. Ullrich, I. Roots, A. G. Hildebrant, R. W. Estabrook, and A. H. Conney, eds), pp. 160–169, Pergamon Press, Oxford.

20. E. Arinç, R. J. Pohl, J. R. Bend, and R. M. Philpot, 1978, Biotransformation of benzo(a)pyrene by little skate hepatic microsomes – Stimulation and reconstitution of benzo(a)pyrene activity. *IVes Journess Etud. Pollutions*, pp. 273–276, C.I.E.S.M.

21. T. Andersson, and L. Förlin, 1992, Regulation of the cytochrome P450 enzyme system in fish. *Aquat. Toxicol.* **24**, 1–20.

22. T. D. Bucheli, and K. Fent, 1995, Inducion of cytochrome P450 as a biomarker for environmental contamination in aquatic ecosystems. *Critical Reviews in Sci. Tech.* **25**, 210–268.

23. R. D. Buhler, 1995, Cytochrome P450 expression in rainbow trout: An Overview, in: *Molecular Aspects of Oxidative Drug Metabolizing Enzymes: Their Significance in Environmental Toxicology, Chemical Carcinogenesis and Health* (E. Arinç J. B. Schenkman, and E. Hodgson, eds.), pp. 159–180, Springer-Verlag, Heidelberg.

24. J. J. Stegeman, 1995, Diversity and regulation of cytochrome P450 in aquatic species, in: *Molecular Aspects of Oxidative Drug Metabolizing Enzymes: Their Significance in environmental Toxicology, Chemical Carcinogenesis and Health* (E. Arinç, J. B. Schenkman, and E. Hodgson, eds.), pp. 135–158, Springer-Verlag, Heidelberg.

25. A. Sen, and E. Arinç, 1977, Separation of three P450 isozymes from liver microsomes of gilthead seabream treated with β-NF and partial purification of cytochrome P4501A1. *Biochem. Molec. Biol. Int.* **41**: 131–141.

26. A. Sen, and E. Arinç, 1998, Preparation of highly purified cytochrome P4501A1 from leaping mullet (*Liza saliens*) liver microsomes and its biocatalytic, molecular and immunological properties. *Comp. Biochem. Physiol.* In press.

27. I. Cok, J. L. Wang-Buhler, M. M. Kedzierski, C. L. Miranda, Y. H. Yang, and D. R. Buhler, 1988, Expression of CYP2M1, CYP2K1 and CYP3A27 in brain, blood, small intestine and other tissues of rainbow trout. *Biochem. Biophys. Res. Commun.* **244**: 790–795.

28. Y. Yamazoe, and R. Kato, 1993, Activation of Chemical Carcinogens, in: *Cytochrome P-450* (T. Omura, Y. Ishimura, Y. Fujii-Kuriyama, eds.), pp. 159–170, Kodansha Ltd., Tokyo.

29. C. Ioannidis, and D. V. Parke, 1993, Induction of cytochrome P4501A as an indicator of potential chemical carcinogenesis. *Drug Metab. Rev.* **25**: 485–501.

30. D. W. Nebert, 1989, The Ah Locus: genetic differences in toxicity, cancer, mutation and birth defects. *Cur. Rev. Tox.* **20**: 153–174.

31. E. G. Shuetz, D. Li, D. J. Omiecinski, U. Muller-Ebergard, H. K. Kelinman, B. Elswick, and P. S. Guzelian, 1988, Regulatin of gene expression in adult rat hepatocytes cutured on a basement membrane matrix. *J. Cell. Physiol.* **134**: 309–323.

32. K. Kawajiri, K. Nakachi, K. Imai, A. Yoshii, N. Shinoda, and J. Watanabe, 1990, Identification of genetically high risk individual to lung cancer by DNA polymorphisms of the P4501A1 gene. *FEBS Lett.* **263**: 131–133.

33. S. Hayashi, J. Watanabe, and K. Kawajiri, 1992, High susceptibility to lung cancer analyzed in terms of combined genotypes of P4501A1 and mu-class glutathione s-transferase gene. *Jpn. J. Cancer Res.* **83**: 866–870.

34. D. R. Buhler, and D. E. Williams, 1989, Enzymes involved in metabolism of PAHs by fishes and other aquatic animals, in: *Metabolism of Polynuclear Aromatic Hydrocarbons in the Aquatic environment* (U. Varanasi, ed.), pp. 151–184, CRC Press Inc., New York.

35. A. Goksoyr, and L. Förlin, 1992, The cytochrome P-450 system in fish, aquatic toxicology and environmental monitoring. *Aqua. Toxic.* **22**, 287–312.

36. L. Förlin, and M. Celander, 1993, Induction of cytochrome P4501A in teleosts: environmental monitoring in Swedish fresh, brackish and marine waters. *Aquat. Toxicol.* **26**: 41–56.

37. B. Domin, and R. M. Philpot, 1986, The effect of substrate on the expression of activity catalyzed by cytochrome P450: Metabolism mediated by rabbit isozyme 6 in pulmonary m,crosomal and reconstituted monooxygenase systems. *Arch. Biochem. Biophys.* **246**: 128–142.

38. E. Arinç, O. Adali, and A. Sen, 1995, Similar regulation and expression of cytochrome P4502B4 and P4501A1 homologues in mammalian lung and fish liver, in: *Book of Abstracts of 9th International Conference on Cytochrome P450: Biochemistry, Biophysics and Molecular Biology*, pp. 87, Zurich.

39. L. J. Heilmann, Y-Y. Sheen, S. W. Bigelow, and D. W. Nebert, 1988, Trout P4501A1: cDNA and deduced protein sequence, expression in liver, and evolutionary significance. *DNA* **7**, 379–387.

40. H. G. Morrison, M. J. Oleksiak, N. W. Cornell, M. L. Sogin, and J. J. Stegeman, 1995, Identification of cytochrome P-450 1A (CYP1A) genes from two teleost fish, toadfish (*Opsanus tau*) and scup (*Stenotomus dhaysops*), and phylogenetic analysis of CYP1A genes. *Biochem. J.* **308**: 97–104.

41. D. E. Williams, and D. R. Buhler, 1982, Purification of cytochrome P448 from β-naphthoflavone-treated rainbow trout. *Biochim. Biophys. Acta* **33**, 3743–3753.

42. D. E. Williams, and D. R. Buhler, 1984, Benzo(a)pyrene-hydroxylase catalyzed by purified isozymes of cytochrome P450 from β-naphthoflavone-fed rainbow trout. *Biochem. Pharmacol.* **33**, 3743–3753.

43. A. V. Klotz, J. J. Stegeman, and C. Walsh, 1983, An aryl hydrocarbon hydroxylating hepatic cytochrome P450 from the marine fish *Stenotomus chrysops*. *Arch. Biochem. Biophys.* **226**, 578–592.

44. A. Goksoyr, 1985, Purification of hepatic microsomal cytochrome P450 from β-naphthoflavone-treated Atlantic cod (*Gadus morhua*), a marine teleost fish. *Biochem. et Biophys. Acta* **840**, 409–417.

45. Y. S. Zhang, A. Goksoyr, T. Andersson, and L. Förlin, 1991, Initial purification and Characterization of hepatic microsomal cytochrome P450 from BNF-treated perch (*Perca fluviatis*). *Comp. Biochem. Physiol.* **98B**, 97–103.

46. A. Sen, and E. Arinç, 1998, Purification and characterization of cytochrome P450 reductase from feral leaping mullet (*Liza saliens*) liver microsomes. *J. Biochem. Molec. Toxicol.* **12**: 103–113.

47. J. F. Payne, and W. R. Penrose, 1975, Induction of aryl hydrocarbon benzo(a)pyrene hydroxylase in fish by petroleum. *Bull. Environ. Contam. Toxicol.* **14**, 112–116.

48. J. F. Payne, L. L. Fancey, A. D. Rahimtula, and E. L. Poretr, 1987. Review and perspective on the use of mixed-function oxygenase enzymes in biological monitoring. *Comp. Biochem. Physiol.* **86C**: 233–245.

49. K. M. Kleinow, M. J. Melancon, and J. J. Lech, 1987, Biotransformation and induction: Implications for toxicity, bioaccumulation and monitoring of environmental xenobiotics in fish. *Environ. Health Perspect.* **71**: 105–119.

50. R. F. Addison, 1995, Monooxygenase measurement as indicators of pollution in the field, in: *Molecular Aspects of Oxidative Drug Metabolizing Enzymes: Their Significance in Environmental Toxicology, and Health* (E. Arinç, J. B. Schenkman, E. Hodgson, eds.), pp. 554–565, Springer-Verlag, Heidelberg.

51. J. J. Stegeman, P. J. Klopper-Sams, and J. W. Farrington, 1986. Monooxygenase induction and chrobiphenyls in the deep-sea fish *Coryphaenooides armatus*. *Science* **231**: 1287–1289.

52. U. Varanasi, T. K. Collier, D. E. Williams, and R. D. Buhler, 1986, Hepatic cytochrome P450 isozymes and aryl hydrocarbon hydroxylase in English sole (*Parophrys vetulus*). *Biochem. Pharmacol.* **35**: 2967–2971.

53. J. J. Stegeman, F. Y. Teng, and E. A. Snowberger, 1987, Induced cytochrome P450 in winter flounder (*Pseudopleurononectes americanus*) from coastal Massachusetts evaluated by catalytic and monoclonal antibody probes. *Can. J. Fish. Aquat. Sci.* **44**: 1270–1277.

54. A. A. Elksus, J. J. Stegeman, L. C. Susani, D. Black, R. J. Pruell, and S. J. Fluck, 1989, Polychlorinated biphenyls concentration and cytochrome P-450 expression in winter flounder from contaminated environments. *Mar. Env. Res.* **28**: 25–30.

55. A. Goksoyr, A. –M. Husoy, H. E. Larsen, J. Klungsoyr, S. Wilhemsen, A. Maage, E. M. Brevik, T. Anderson, M. Celander, M. Pesonen, and L. Förlin, 1991, Environmental contaminants and biochemical responses in flatfish from the Hvaler Archipelago in Norway. *Arch. Environ. Contam. Toxicol.* **21**: 486–496.

56. T. K. Collier, D. Conor, B. -T. L. Eberthart, B. F. Anulacion, A. Goksoyr, and U. Varanasi, 1992, Using cytochrome P-450 to monitor the aquatic environment: initial results from regional and national surveys. *Mar. Env. Res.* **34**: 193–198.

57. I. T. Balkas, and F. Juhasz, 1993, Costs and benefits of measures for the reduction of degradation of the environment from land-based sources of pollution in coastal area: A case study of Izmir Bay. *MAP technical report series*, **No:72**, UNEP, Athens.

58. I. T. Balkas, F. Juhasz, Ü. Yetis, and G. Tuncel, 1992, The Izmir Bay wastewater management project-economical considerations. *Wat. Sci. Tech.* **26**: 2613–2616.

18

CYP 1A CONCENTRATIONS AS AN INDICATOR OF EXPOSURE OF FISH TO PULP-MILL EFFLUENTS

R. F. Addison and J. Y. Wilson*

Department of Fisheries and Oceans
Institute of Ocean Sciences
P.O. Box 6000
Sidney, British Columbia, Canada V8L 4B2

1. INTRODUCTION

The pulping of wood and subsequent steps in paper manufacture produce a large volume of waste water, which is usually discharged to lakes, rivers or the sea. This waste water contains a wide range of organic compounds which may have several lethal or sublethal effects on fish.[1] Polychlorinated dibenzo-*p*-dioxins and furans (PCDD/F) are among the chemicals often present in waste waters. These compounds have attracted considerable interest because of their potential toxicity and persistence. PCDD/F are effective inducers of cytochrome P-450 1A (CYP 1A) and its associated enzyme activity in fish, among other organisms.[2,3] PCDD/F are notoriously difficult (and therefore expensive) to analyse, and during the past decade or so interest has grown in using measurements of CYP 1A and its associated enzyme activity in resident fish species to complement traditional chemical analyses of these chemicals in receiving waters.

In this paper, we review recent studies of the response of fish hepatic CYP 1A to pulp mill effluents. We also describe results of some studies carried out in British Columbia (BC) to assess the suitability of field measurements of CYP 1A and its associated enzyme activity as an indicator of the impact of pulp mill effluents, or of their content of PCDD/F.

---

* Present address: Woods Hole Oceanographic Institution, Redfield Bldg. 3-42, Woods Hole, Massachusetts 02543.

Molecular and Applied Aspects of Oxidative Drug Metabolizing Enzymes,
edited by Arınç *et al*. Kluwer Academic / Plenum Publishers, New York, 1999.

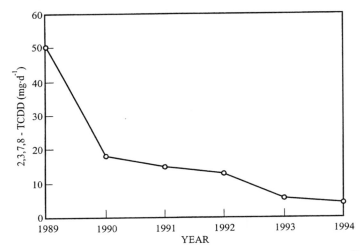

Figure 1. Decline in discharges of 2,3,7,8-tetrachlorodibenzodioxin (TCDD) from pulp mills, 1989–1994.

2. PCDD/F PRODUCTION DURING PULP AND PAPER MANUFACTURING PROCESSES

McCubbin[4] has reviewed the main processes used in the industry, and their potential environmental impacts. Briefly, wood is reduced to chips which are separated into fibres by various chemical and physical processes. The most common process used at present is the "kraft" or sulphate process in which chips are cooked with NaOH and Na_2S. The resulting pulp is eventually bleached, usually with Cl_2 and/or ClO_2. The combined effluents from

Table 1. Sources of PCDD/F congeners during wood pulp manufacture (simplified from Yunker and Cretney[8])

Source	Congener formed	TEF
Pentachlorophenol (wood treatment)	1,2,3,4,6,7,8-HpCDF	0.01
	Non 2,3,7,8-substituted PeCDF	
	Non 2,3,7,8-substituted HxCDF	
	Non 2,3,7,8-substituted HpCDF	
Polychlorophenol condensation (during cooking)	1,2,3,7,8-PeCDD	0.5
	1,2,3,6,7,8-HxCDD	0.1
	1,2,3,7,8,9-HxCDD	0.1
	Non 2,3,7,8-substituted HxCDD	
Chlorine bleaching	2,3,7,8-TCDF	0.1
	1,2,3,7,8-PeCDF	0.05
	2,3,4,7,8-PeCDF	0.5
	Non 2,3,7,8-substituted TCDF	

Abbreviations: TCDF: Tetrachlorodibenzofuran; PeCDF: Pentachlorodibenzofuran; HxCDF: Hexachlorodibenzofuran; HpCDF: Heptachlorodibenzofuran; PeCDD: Pentachlorodibenzodioxin; HxCDD: Hexachlorodibenzodioxin; TEF: Toxic Equivalent Factor (relative to 2,3,7,8-TCDD).

these processes are described as "bleached kraft mill effluent" (BKME). After bleaching, the pulp is blended, and various additives and fillers are added for paper production.

The chemistry of PCDD/F production during these processes has been studied in some detail.[5-7] Yunker and Cretney[8] have compared the PCDD/F congeners in sediments from near coastal pulp mills to the pathways of PCDD/F synthesis, and have identified three main sources in pulp mill emissions. These were (a) the pentachlorophenol (PCP) used in wood treatment or to preserve wood chips, (b) the effect of cooking on polychlorophenol components of PCP, and (c) the effects of bleaching using Cl_2 (Table 1).

In the late 1980's some marine invertebrates sampled from around BC coastal pulp mills were found to be contaminated with PCDD/F, and this led to closure of some areas to fishing. At that point, the use of PCP as a wood preservative was restricted and the bleaching processes used in the mills were changed to rely more on ClO_2. A result of these changes was that discharges of PCDD/F from BC mills declined appreciably during the early 1990's (Figure 1). This in turn led to a decline in the extent of PCDD/F contamination of receiving environments, and fisheries that were previously closed because of contamination are now being re-opened.

3. FIELD STUDIES OF THE IMPACT OF BKME ON FISH HEPATIC CYP 1A

The response of hepatic mono-oxygenases in several species of freshwater, marine or brackish water fish to pulp mill effluents has been studied extensively, mainly in Scandinavia and Canada.[9-27] Some results of this work are shown in Table 2. The general conclusions can be summarised as follows:

1. Pulp mill effluents (usually, but not always, BKME) induce concentrations of CYP 1A[19,27,28] and its associated enzyme activities. These are usually measured as ethoxyresorufin O-de-ethylase[9-14,16-26,28,34,35](EROD) but occasionally arylhydrocarbon hydroxylase[13,15] (AHH) or pentoxy-resorufin O-de-alkylase[14] (PROD) has been measured.
2. Phase II enzymes such as UDP glucuronosyl transferase (UDP-GT) and /or glutathione-S-transferases (GST) show no, or inconsistent, responses.[13,25,26]
3. Several other biochemical or physiological changes which are not obviously related directly to CYP 1A induction also occur; these include alterations in carbohydrate metabolism and in ion balance[10,12,15,22] and sometimes,[16,21] though not always,[15] in concentrations of plasma steroid hormones.
4. Although the CYP 1A system is consistently induced, there is no clear relationship with the pulp mill process: induction occurs whether or not Cl_2 bleaching is used[20,21] but secondary treatment of the effluent may slightly reduce the extent of induction or eliminate it completely.[17,21]

Considering the variety of physical and chemical processes used in pulp mills and the range of composition of feedstock used (softwoods and hardwoods) it is not surprising that variable biochemical responses of fish to effluents should be seen from mill to mill; indeed, it is remarkable that induction of the CYP 1A system is so consistent. This implies that the inducing agent is common to a wide variety of processes, and also that induction of the CYP 1A system in resident fish may be useful in assessing the zone of impact of pulp mill effluents, as a complement to chemical analyses. In the following sections, we describe some work carried out in BC during the last few years to address these topics.

Table 2. Examples of sub-lethal effects of bleached Kraft mill effluent (BKME) on fish

Site	Effluent	Species	Effect observed	Ref.
Coastal Sweden	BKME	*Perca fluviatilis; Myoxocephalus quadricornis*	EROD induced; UDP-GT induced occasionally.	9
Sweden (lab. studies)	BKME	*M. quadricornis*	EROD induced.	10
Coastal Sweden	BKME	*P. fluviatilis*	EROD induced; changes in blood variables and gonad size.	11
Coastal Sweden	BKME	*P. fluviatilis*	EROD induced; gonad development, ion balance, carbohydrate and vitamin metabolism changed; changes in blood variables.	12
Finland (lake)	BKME	Various freshwater spp	EROD etc. induced.	13
Finland (lake)	BKME	*Salmo gairdneri*	EROD etc. induced.	14
Canada (river)	BKME	*Catostomus commersoni*	AHH induced; changes in blood variables.	15
Canada (lake)	BKME	*Coregonus clupeaformis*	EROD induction; plasma steroid changes; delayed maturation; skin lesions.	16
Canada (freshwater)	BKME and others	*Oncorhynchus mykiss*	EROD induction; metallothionein induction.	17
Coastal Sweden	BKME	*P. fluviatilis*	EROD induction, haemaotcrit changes, gonado-somatic index changes.	18
Canada (river)	BKME	Various freshwater spp.	CYP 1A and EROD induction varying with spp; few other changes.	19
Canada (various freshwater sites)	BKME and others	*O. mykiss*	EROD induction.	20
Canada (various freshwater sites)	BKME and others	*C. commersoni*	EROD induction; not correlated with plasma steroid.	21
Sweden	BKME	*P. fluviatilis*	EROD induction; haematology, plasma ions, carbohydrate metabolism and gonad size changes.	22
Canada (river)	BKME	*C. commersoni*	EROD induction.	23
New Zealand (freshwater)	BKME and others	Various spp.	EROD induction.	24
Finland (lake)	BKME	*Coregonus lavaretus*	EROD induced; UDP-GT unchanged.	25
Finland (laboratory studies)	BKME	*C. lavaretus*	EROD induced; UDP-GT unchanged; various changes in blood variables.	26
USA (river)	BKME	*Ictalurus punctatus*	CYP 1A and EROD induction	27
Canada (river)	BKME	*Oncorhynchus tshawytscha*	EROD induction.	32
Canada (river)	BKME	*O. tshawytscha*	EROD induction; inter-renal nuclear diameters increased.	33

Abbreviations as in text.

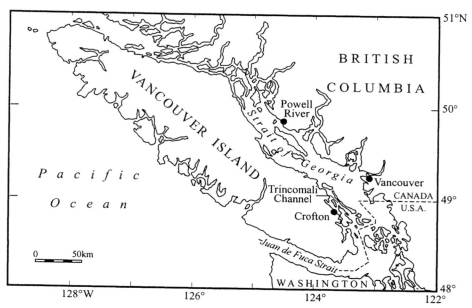

Figure 2. Map of the Strait of Georgia, BC, showing sampling sites at coastal pulp mills (Crofton and Powell River) and the reference site (Trincomali Channel).

3.1. Studies of BC Coastal Fish

We have carried out some preliminary studies[29] around coastal pulp mills in the Strait of Georgia (Figure 2). We used the benthic flatfish English sole (*Parophrys vetulus*) as a monitoring species because it does not migrate far, and because it is exposed to sediment-bound contaminants such as PCDD/F. We sampled male and female sole in June 1994 and 1995 from pulp mills at Crofton and Powell River and at a reference site in Trincomali Channel, and measured EROD aboard ship on $10,000 \times g$ supernatants (post-mitochondrial supernatants: PMS) prepared from fresh liver samples using standard methods.[30] We selected fish to fall within a specific range of lengths (20 - 30 cm) to reduce potential variability due to age.

As is usually the case with field studies on "wild" fish, variance was high. EROD activity was usually induced significantly at the mill sites over activities recorded at the reference site, but females usually showed induction more consistently than males (though this may have reflected the smaller sample size of males) (Table 3). This conclusion is a little surprising as EROD activity in females is usually considered to be more sensitive to natural variables associated with hormonal cycles, and in most field monitoring operations males are preferred. We measured CYP 1A concentrations (in 1995 samples only) by western blotting using a polyclonal antibody raised to amino acids 277–294 in rainbow trout CYP 1A.[31,32] CYP 1A concentrations were also elevated in fish from near the two mills (Crofton and Powell River) compared to those from the reference site; again, females gave a more consistent response than males. The EROD response was not very consistent with CYP 1A concentrations, though this again is not unusual in field monitoring.[33] However, although EROD activity and CYP 1A concentrations were usually elevated at the mill sites, they were not consistent with PCDD/F concentrations in Dungeness crab (*Cancer magister*, used as an indicator species in routine monitoring) collected at or near these sites (Table 3).

Table 3. Indices of hepatic mono-oxygenase induction (mean ± s.d. (no of samples)) in English sole (*Parophrys vetulus*) sampled near coastal pulp mills and at a reference site in BC, and PCDD/F-derived TEQ in crab from the same sites

	Site		
Index of induction	Trincomali (reference)	Crofton (mill)	Powell River (mill)
EROD (nmol·min⁻¹·PMS protein⁻¹)			
Females 1994	0.030 ± 0.034 (9)	0.215 ± 0.211* (4)	NA
Males 1994	0.038 ± 0.043 (6)	0.327 ± 0.236* (6)	NA
EROD (pmol·min⁻¹·microsomal protein⁻¹)			
Females 1995	166 ± 84.4 (6)	160 ± 225 (8)	972 ± 788* (3)
Males 1995	239 ± 260 (6)	758 ± 1280 (5)	933 ± 860 (3)
CYP 1A (Relative density to 0.5 pmol rat CYP 1A1/50 μg protein)			
Females 1995	0.95 ± 0.51 (6)	2.44 ± 0.89* (3)	2.68 ± 0.95* (8)
Males 1995	1.58 ± 0.97 (6)	1.82 ± 1.30 (5)	2.28 ± 0.57 (2)
Total PCDD/F TEQ in crab hepatopancreas (pg/g)	21.1, 21.7 (1990, 1992)	39.6, 48.5 (1994, 1995)	5.8, 7.8 (1994, 1995)

*Significantly different from reference site (P < 0.05). NA: Not analysed.

Taken together, these data show that effluents from coastal pulp mills induce hepatic CYP 1A and EROD activity in resident fish benthic flatfish. However, the absence of any consistency with environmental PCDD/F, as indicated by the crab hepatopancreas analyses, suggests that some other component(s) of pulp mill effluents may be the inducing agent.

3.2. Studies of Fraser River (BC) Freshwater Fish

The Fraser River system is a major salmon spawning and nursery ground in BC. It also represents the "receiving waters" for effluents from several pulp mills (Figure 3). The impact of these effluents on salmon biochemistry, physiology and behaviour has been studied for the last thirty years or so, and the recent focus of some of this work has been on the possible effects of PCDD/F on CYP 1A in resident juvenile chinook salmon (*Oncorhynchus tshawytscha*). Preliminary field observations[34] and lab exposures of BKME to chinook salmon[35] in the late 1980's showed a strong correlation between effluent exposure and EROD induction, and suggested that the responsible agent in the BKME may have been PCDD/F. These data stimulated a more detailed study, and in 1994 and 1995 we sampled juvenile chinook which over-winter in the Fraser River at various points downstream from pulp mills and at a reference site upstream from the mills. (Chinook spawn in the fall, and the fish which hatch the following spring usually spend the next year close to their hatching site, before moving downstream to the sea about 18 months after spawning. They therefore represent a useful monitoring species.) We measured EROD activities in the field on $10{,}000 \times g$ supernatants prepared from fresh homogenates as described above;

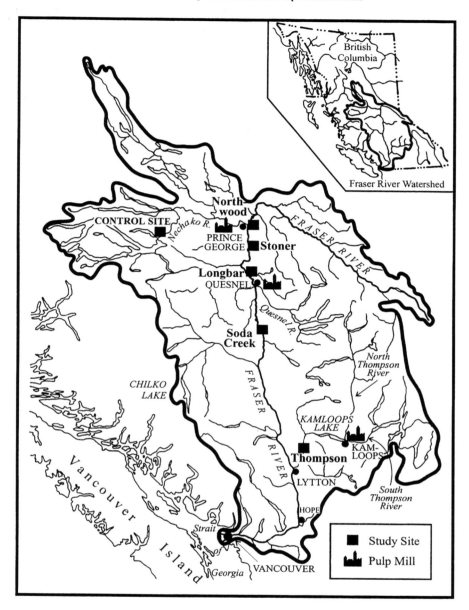

Figure 3. Map of the Fraser River watershed showing pulp mills and sampling and reference sites.

we froze other samples in liquid nitrogen for subsequent CYP 1A determination as described above, and for measurement of DNA adduct formation[36]. We stored carcasses for chemical analyses for PCDD/F and for potentially toxic non-*ortho* and mono-*ortho* substituted polychlorinated biphenyls (PCB).

We expected that fish from the reference site (Nechako River) would have low indices of CYP 1A induction; those closest to the mill (Northwood) would have highest indices, and fish from downstream sites (Stoner, Longbar, Soda Creek and Thompson River)

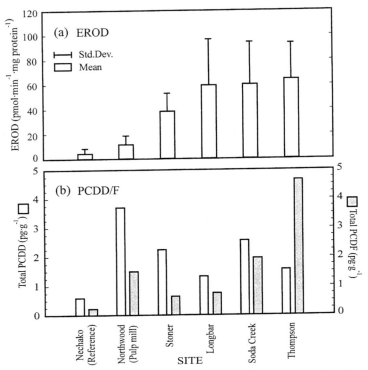

Figure 4. (a) Ethoxyresorufin O-de-ethylase (EROD) activity in juvenile chinook salmon in the Fraser River, 1994. (b) Total PCDD/F concentrations in juvenile chinook salmon from the Fraser River, 1994.

would have steadily declining indices of induction reflecting the dilution of the effluents (though the presence of mills at Quesnel and Kamloops could influence the Soda Creek and Thompson sites, respectively). However, the results showed the following:

1. EROD activity at the pulp mill site (Northwood) was higher than at the reference site (Nechako), as expected;

2. EROD activity tended to *increase* with distance downstream from the mills (Figure 4a: similar trends were seen in 1995: J.Y.Wilson, unpublished data) and CYP 1A was generally higher downstream than at the pulp mill sites (analysed in 1995 samples only);

3. PCDD/F concentrations were generally higher at the pulp mill site than at the reference site, but downstream residue concentrations showed no clear decrease with distance from the pulp mill (Figure 4b);

4. EROD activity was only poorly correlated with PCDD/F concentrations; none of the correlations was statistically significant ($P > 0.05$);

5. EROD activities and 2,3,7,8-tetrachlorodibenzofuran (TCDF) concentrations (an indicator of exposure to BKME) in chinook from the pulp mill and downstream sites had declined consistently since the late 1980's (Figure 5).

Our preliminary conclusions from these data were that:

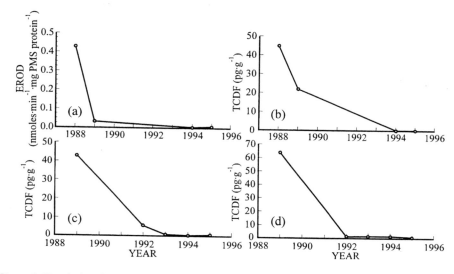

Figure 5. Trends in ethoxyresorufin O-de-ethylase (EROD) activity (a) at Northwood and 2,3,7,8-tetrachlorodibenzofuran (TCDF) concentrations in juvenile chinook salmon Northwood, Longbar and Soda Creek ((b), (c) and (d), respectively) 1988–1996.

1. Current (i.e., mid-1990's) and past (late 1980's) EROD induction and PCDD/F concentrations at the pulp mill site (Northwood) were both higher than those at the reference site, and suggest a possible cause-effect relationship;

2. The apparent EROD induction further downstream in the Fraser River could not be attributed to PCDD/F concentrations;

3. Neither EROD induction nor PCDD/F concentrations at the downstream sites could be attributed to the effects of BKME from the pulp mill site.

Other data argue against a cause-effect relationship between EROD induction and *current* PCDF concentrations. Total TEQs attributed to PCDD/F in these samples were in the range 0.1–0.7 pg • g^{-1} 2,3,7,8-tetrachlorodibenzodioxin (TCDD) equivalents. Servizi et al.[35] have shown for EROD activity to be doubled over reference values, body burdens of around 2–3 TEQs derived from BKME are required. At most downstream sites on the Fraser River, TEQs were about one-tenth of that value, from which it seems unlikely that the downstream increase in EROD activity could be related to PCDD/F burdens. However, the EROD induction observed[34,35] in the late 1980's was attributable to PCDD/F body burdens.

We have complemented these field studies with lab studies, in which juvenile chinook were exposed over a two-month period to increasing concentrations of BKME (0, 2, 4, 8, and 16% by volume) at the mill site. EROD activities and CYP 1A concentrations increased with BKME concentration, but total carcass contaminant burdens were independent of BKME exposure (J.Y. Wilson, unpublished data). In other words, the EROD and CYP 1A responses were not correlated with PCDD/F exposure in BKME. Furthermore, total PCDD/F burdens had TEQs around 0.1–0.2 pg/g 2,3,7,8-TCDD equivalents; even when TEQs derived from non-*ortho* and mono-*ortho*-substituted PCB were added, the total TEQ of around 1 pg/g was still well below the value of approx. 3 pg/g inferred as a threshold for EROD induction.[35]

Taken together, then, the field and laboratory data lead to the following conclusions.

1. BKME currently discharged into the Fraser River can induce CYP 1A and EROD activity, at least at sites close to pulp mills, but not necessarily further downstream;
2. CYP 1A induction (at the pulp mill site, or downstream) is not now attributable to the presence of PCDD/F or related chlorinated organics.
3. in the late 1980's, EROD induction at the mill site could be attributed to PCDD/F in BKME.

4. SUMMARY AND CONCLUSIONS

Both field and laboratory studies show that components of BKME induce fish hepatic CYP 1A and its associated enzyme activities. Data from the upper Fraser River show that in the late 1980's EROD induction was well correlated with the presence of PCDD/F in fish exposed to BKME, but the more recent data (1994 and 1995) show convincingly that induced CYP 1A and EROD cannot be attributed to the presence of PCDD/F or of PCBs. In the BC coastal environment, the induced CYP 1A concentrations and EROD observed in the mid-1990's cannot be ascribed convincingly to environmental PCDD/F burdens.

If PCDD/F are not responsible for the continuing evidence for CYP 1A induction in the Fraser River chinook (and possibly in English sole from the BC coastal sites) what component of pulp mill effluent is responsible? Analyses of DNA adducts in chinook from most of the downstream sites strongly suggest a hydrocarbon agent and that would be consistent with conclusions from other Canadian freshwater studies[19]. Furthermore, the induction of the CYP 1A system, and the DNA adduct analyses of chinook from the downstream Fraser sites show clearly that these responses need not be associated with pulp mill effluent sources.

We conclude that any field measurement of CYP 1A induction to assess the zone of influence of pulp mill effluents must be made with caution, taking into account the possibility of induction by other agents, and must be supported by laboratory studies (e.g., of dose-effect relationships) and appropriate chemical analyses.

ACKNOWLEDGMENTS

This work was supported in part by the Fraser River Action Plan (FRAP).

REFERENCES

1. T. Kovacs, 1986, Effects of bleached kraft mill effluent on freshwater fish: a Canadian perspective, *Water Poll. Cont. Res. J. Can.* **21**: 91–118.
2. D.C.G. Muir, A.L. Yarechewski, D.A. Metner, W.L. Lockhart, G.R.B. Webster, and K.J. Friesen, 1990, Dietary accumulation and sustained hepatic mixed function oxidase enzyme induction by 2,3,4,7,8-pentachlorodibenzofuran in rainbow trout, *Environ. Toxicol. Chem.* **9**: 1463–1472.
3. M.E.J. van der Weiden, J. van der Kolk, A.H. Penninks, W. Seinen and M. van den Berg, 1990, A dose/response study with 2,3,7,8-TCDD in the rainbow trout (*Oncorhynchus mykiss*), *Chemosphere* **20**: 1053–1058.
4. N. McCubbin, 1983, The basic technology of the pulp and paper industry and its environmental protection practices, *Environmental Protection Service Report EPS 6-EP 83–1*, Environment Canada, Ottawa, xii + 204 pp.
5. H. Hagenmaier and H. Brunner, 1987, Isomer specific analysis of pentachlorophenol and sodium pentachlorophenate for 2,3,7,8-substituted PCDD and PCDF at sub-ppb levels, *Chemosphere* **16**: 1759–1764.

6. G. Amendola, D. Barna, R. Blosser, A. McBride, F. Thomas, T. Tiernan and R. Whittemore, 1989, The occurrence and fate of PCDD's and PCDF's in five bleached kraft pulp and paper mills, *Chemosphere* **18**: 1181 - 1188.

7. R.E. Clement, C. Tashiro, S. Suter, E. Reiner and D. Hollinger, 1989, Chlorinated dibenzo-*p*-dioxins (CDDs) and dibenzofurans (CDFs) in effluents and sludges from pulp and paper mills, *Chemosphere* **18**: 1189–1196.

8. M.B.Yunker and W.J. Cretney, 1995, Chlorinated dioxin trends between 1987 and 1993 for samples of crab hepatopancreas from pulp and paper mill and harbour sites in British Columbia, *Can Tech. Rep. Fish. Aquat. Sci.* 2082; xii + 138 pp.

9. L. Förlin, T. Andersson, B. Bengtsson, J. Härdig and A. Larsson, 1985, Effects of pulp bleach plant effluents on hepatic xenobiotic transformation enzymes: laboratory and field studies, *Mar. Env. Res.* **17**: 109–112.

10. T. Andersson, B.-E. Bengtsson, L. Förlin, J. Härdig, and A. Larsson, 1987, Long-term effects of bleached kraft mill effluents on carbohydrate metabolism and hepatic xenobiotic biotransformation enzymes in fish, *Ecotoxicol. Environ. Saf.* **13**: 53–60.

11. T. Andersson, L. Förlin, J. Hardig, and A. Larsson, 1988, Physiological disturbances in fish living in coastal water polluted with bleached kraft pulp mill effluents, *Can. J Fish. Aquat. Sci.* **45**: 1525–1536.

12. A. Larsson, T. Andersson, L. Förlin, and J. Härdig, 1988, Physiological disturbances in fish exposed to bleached kraft mill effluents, *Water Sci. Technol.* **20**: 67–76.

13. P. Lindström-Seppä and A. Oikari, 1990, Biotransformation activities of feral fish in waters receiving bleached pulp mill effluents, *Environ. Toxicol. Chem.* **9**: 1415–1424.

14. P. L. Seppä and A. Oikari, 1990, Biotransformation and other toxicological and physiological responses in rainbow trout (*Salmo gairdneri* Richardson) caged in a lake receiving effluents of pulp and paper industry, *Aquat. Toxicol.* **16**: 187–204.

15. P. V. Hodson, M. McWhirter, K. Ralph, B. Gray, D. Thivierge, J.H. Carey, G. Van Der Kraak, D.M. Whittle, and M.C. Levesque, 1992, Effects of bleached kraft mill effluent on fish in the St. Maurice River, Quebec, *Environ. Toxicol. Chem.* **11**: 1635–1651.

16. K.R. Munkittrick, M.E. McMaster, C.B. Portt, G.J. Van der Kraak, I.R. Smith, and D.G. Dixon, 1992, Changes in maturity, plasma sex steroid levels, hepatic mixed-function oxygenase activity, and the presence of external lesions in lake whitefish (*Coregonus clupeaformis*) exposed to bleached kraft mill effluent, *Can. J. Fish. Aquat. Sci.* **49**:1560–1569.

17. F. Gagne and C. Blaise, 1993, Hepatic metallothionein level and mixed function oxidase activity in fingerling rainbow trout (*Oncorhynchus mykiss*) after acute exposure to pulp and paper mill effluents, *Water Res.* **27**: 1669–1682.

18. L. Balk, L. Förlin, M. Söderström and Å. Larsson, 1993, Indicators of regional and large-scale biological effects caused by bleached pulp mill effluents, *Chemosphere* **27**: 631–650.

19. P.J. Kloepper-Sams and E. Benton, 1994, Exposure of fish to biologically treated bleached-kraft effluent. 2. Induction of hepatic cytochrome P4501A in mountain whitefish (*Prosopium williamsoni*) and other species, *Environ. Toxicol. Chem.* **13**: 1483–1496.

20. P.H Martel, T.G. Kovacs, B.I. O'Connor, and R.H. Voss, 1994, A survey of pulp and paper mill effluents for their potential to induce mixed function oxidase enzyme activity in fish, *Water Res.* **28**: 1835–1844.

21. K.R. Munkittrick, G.J. van der Kraak, M.E. McMaster, C.B. Portt, M.R. van den Heuvel, and M.R. Servos, 1994, Survey of receiving-water environmental impacts associated with discharges from pulp mills. 2. Gonad size, liver size, hepatic EROD activity and plasma sex steroid levels in white sucker (*Catostomus commersoni*), *Environ. Toxicol. Chem.* **13**: 1089–1101.

22. L. Förlin, T. Andersson, L. Balk, and Å.Larsson, 1995, Biochemical and physiological effects in fish exposed to bleached kraft mill effluents, *Ecotoxicol. Environ. Saf.* **30**: 164–170.

23. M.M. Gagnon, D. Bussieres, J.J. Dodson, and P.V. Hodson, 1995, White sucker (*Catostomus commersoni*) growth and sexual maturation in pulp mill-contaminated and reference rivers, *Environ. Toxicol. Chem.* **14**: 317–327.

24. P.D. Jones, D.J. Hannah, S.J. Buckland, F.M. Power, A.R. Gardner, and C.J. Randall, 1995, The induction of EROD activity in New Zealand freshwater fish species as an indicator of environmental contamination, *Australas. J. Ecotoxicol.* **1**: 99–105.

25. R. Soimasuo, I. Jokinen, J. Kukkonen, T. Petänen, T. Ristola, and A. Oikari, 1995, Biomarker responses along a pollution gradient: Effects of pulp and paper mill effluents on caged whitefish, *Aquat. Toxicol.* **31**: 329–345.

26. R. Soimasuo, T. Aaltonen, M. Nikinmaa, J. Pellinen, T. Ristola, and A. Oikari, 1995, Physiological toxicity of low-chlorine bleached pulp and paper mill effluent on whitefish (*Coregonus lavaretus* L. s.l.): A laboratory exposure simulating lake pollution, *Ecotoxicol. Environ. Saf.* **31**: 228–237.

27. L.A. Banby, P.A. Van Veld, D.L. Borton, L. LaFleur and J.J. Stegeman, 1995, Responses of cytochrome P4501A in freshwater fish exposed to bleached kraft mill effluent in experimental stream channels, *Can. J. Fish. Aquat. Sci.* **52**: 434–447.

28. P.J. Kloepper-Sams, E. Benton, L. Förlin, and T. Andersson, 1995, Application of a sensitive chemiluminescent technique for comparison of cytochrome P4501A induction in hepatic and intestinal tissues of fish exposed to bleached kraft mill effluent, *Mar. Environ. Res.* 39: 213–218.

29. R.F. Addison and T.L. Fraser, 1996, Hepatic mono-oxygenase induction in benthic flatfish sampled near coastal pulp mills in British Columbia, *Mar Env. Res.* **42**: 273.

30. R.F. Addison and J.F. Payne, 1986, Assessment of hepatic mixed function oxidase induction in winter flounder (*Pseudopleuronectes americanus*) as a marine petroleum pollution monitoring technique, with an Appendix describing practical field measurements of MFO activity. *Can. Tech. Rept. Fish. Aquat. Sci.* no. 1505, 52 pp.

31. C.R. Myers, L.A. Sutherland, M.L. Haasch and J.J. Lech, 1993, Antibodies to a synthetic peptide that react specifically with rainbow trout hepatic cytochrome P450 1A1, *Environ. Toxicol. Chem.* **12**: 1619–1626.

32. S. Lin, P.L. Bullock, R.F. Addison and S.M. Bandiera, 1998, Detection of cytochrome P450 1A in several species using antibody against a synthetic peptide derived from rainbow trout cytochrome P450 1A1, *Environ. Toxicol. Chem.*, In press.

33. A. Goksøyr, A.-M. Husøy, H. Jarsen, J. Klungsøyr, S. Wihelmsen, A. Maage, E. Brevik, T. Andersson, M. Celander, P. Pesonen and L. Förlin, 1991, Environmental contaminants and biochemical responses in flatfish from the Hvåler Archipelago in Norway, *Arch. Env. Contam. Toxicol.* **21**: 486 - 496.

34. I.H. Rogers, C.D. Levings, W.L. Lockhart and R.J. Norstrom, 1989, Observations on overwintering juvenile chinook salmon (*Oncorhynchus tshawytscha*) exposed to bleached kraft mill effluent in the Upper Fraser River, British Columbia, *Chemosphere* **19**: 1853–1868.

35. J.A. Servizi, R.W. Gordon, D.W. Martens, W.L. Lockhart, D.A. Metner, I.H. Rogers, J.R. McBride and R.J. Norstrom, 1993, Effects of biotreated bleached kraft mill effluent on fingerling chinook salmon (*Oncorhynchus tshawytscha*), *Can. J. Fish. Aquat. Sci.* 50: 846–847.

36. W.L. Reichert and B. French, 1994, The ^{32}P-postlabeling protocols for assaying levels of hydrophobic DNA adducts in fish, *US Department of Commerce, NOAA Technical Memorandum NMFS-NWFSC-14*, 89 pp.

19

THE BIOCHEMISTRY AND PHYSIOLOGY OF PROSTACYCLIN- AND THROMBOXANE-SYNTHASE

Volker Ullrich[*]

Faculty of Biology
University of Konstanz
D 78457 Konstanz, Germany

1. INTRODUCTION

The prostaglandins form a family of signalling molecules that communicate between cells of a given organ (autacoids). They mediate key events in inflammation, autoimmune diseases, or shock syndromes but are also involved in physiological regulations associated with growth, differentiation, parturition or sleep (for reviews see[1-7]). Hence pharmacological research has focused on their pathways of biosynthesis and inhibition and also on their mode of action. Like steroid hormones, prostaglandins are not stored and secreted but synthesized and released on demand. They are formed and degraded much more rapidly than steroids, and since they act exclusively within organs, only their degradation products rather than active prostaglandins can be found in blood or serum. Because of this and their low concentrations in tissues, the analysis, the structural evaluation and the chemical synthesis of prostaglandins have been a challenge (for reviews see[8,9]).

The biosynthetic pathways for prostanoid formation in higher organisms start out from arachidonic acid (5Z, 8Z, 11Z, 14Z)-icosatetraenoic acid), which on account of its four double bonds and abundance of allylic hydrogens offers a large biosynthetic potential.[10] Arachidonate is readily available from biological membranes after activation of phospholipases. The second step is then the conversion of arachidonate into 15-hydroxy-9a-epidioxyprosta-5,13-dienic acid (PGH$_2$).[11] This requires two enzymatic activities, that of cyclooxygenase and of a hydroperoxidase, which are colocated on PGH-synthase.

The PGH$_2$ molecule contains the relatively weak epidioxy bond that can be opened for subsequent isomerization reactions leading to the five different prostaglandins. It is ob-

* Address correspondence to: Prof. Dr. V. Ullrich, Universität Konstanz, Fakultät für Biologie, Fach X910 - Sonnenbühl, D 78457 Konstanz, Germany. Tel.: +49 7531 882287; Fax: +49 7531 884084; e-mail: volker.ullrich@uni-konstanz.de.

Molecular and Applied Aspects of Oxidative Drug Metabolizing Enzymes,
edited by Arınç *et al*. Kluwer Academic / Plenum Publishers, New York, 1999.

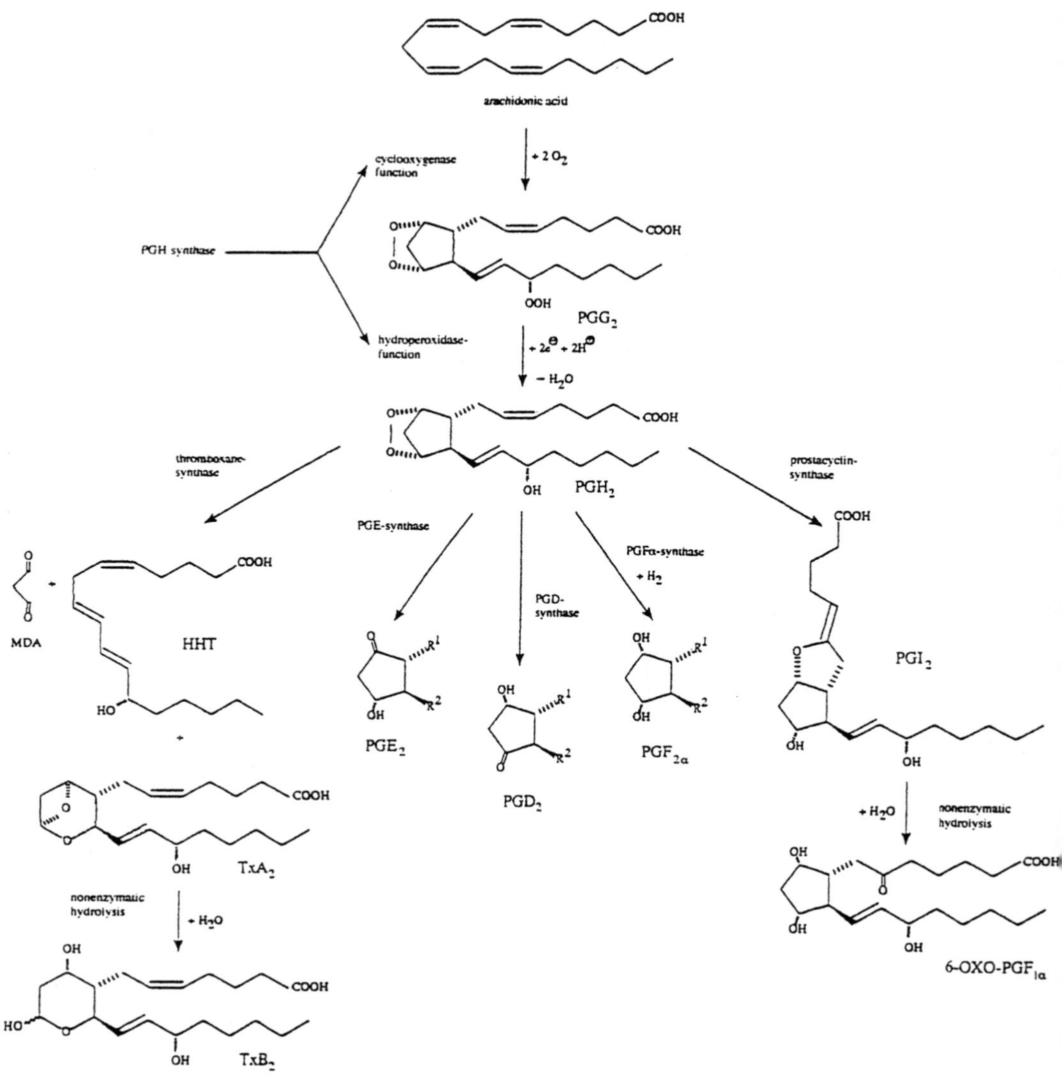

Scheme 1. Biosynthesis of prostanoids and thromboxane.

vious that prostaglandins D$_2$ and E$_2$ are formed after a heterolytic cleavage of the O-O bond by a nucleophilic mechanism. However, the biosynthesis of prostacyclin (PGI$_2$) and thromboxane (TxA$_2$) with the formation of the oxygen-containing heterocycles appears more complex. After ^{18}O-experiments had established the isomerase character of both reactions,[12,13] mechanisms of heterolytic cleavage of the epidioxide and rearrangements to thromboxane TxA$_2$[14] and PGI$_2$)[15] have been proposed. Such chemistry requires an efficient stabilizing catalyst to prevent carbonyl formation from the positively charged oxygen atom. This led us to suggest an involvement of heme since heme containing thiolate ligands in the cytochrome P450 monooxygenases can transfer oxygen atoms from hydroperoxides to organic substrates.[16,17] This working hypothesis could be verified by making use of the characteristic P450 spectra which allowed us to isolate the two isomerases for converting PGH_2 into TxA_2 and PGI_2, respectively.[18]

2. ISOLATION AND SPECTRAL CHARACTERIZATION

PGI_2 is mainly produced by the endothelium of blood vessels[19] which therefore served as the starting material for purification.[20] In accordance with literature the PGI_2-synthesizing enzymatic activity was associated with the endoplasmic reticulum obtained as the microsomal cell fraction.[21] Difference spectroscopy indeed revealed the presence of a P450 (= heme-thiolate) protein with its characteristic carbon monoxide difference spectrum at 450 nm after reduction with Na-dithionite. Enzymatic reduction by $NADPH_2$ could not be observed, which indicates that the P450 protein could not act as a monooxygenase. Solubilization of porcine aortic microsomes with detergents resulted in still active preparations, which could be further purified by classical methods or affinity chromatography. This purified fraction from aortae yielded a homogeneous protein on SDS-polyacrylamide gel electrophoresis with a molecular weight of about 52 kDa and a molecular activity of about 3 s^{-1}.[22]

Thromboxane A_2 is released from blood platelets, which are thus a suitable starting material for the isolation of the TxA_2-synthase. We identified a P450 protein in the same fraction, solubilized it with detergents, and further purified it to homogeneity.[23] On SDS gels it showed a band at 59 kD and converted PGH_2 into TxB_2, the stable hydrolysis product of TxA_2. TxA_2-synthase has a molecular activity of about 27 s^{-1} but in addition 12-hydroxy-heptadecatrienoic acid is formed at a similar rate and has to be considered as a second product derived from a common intermediate in the reaction cycle.

It is well established that the heme group in cytochrome P450 monooxygenases is linked to the thiolate group of a cysteine residue, which causes the typical characteristics in the optical and the EPR spectra.[24-28]

The oxidized iron(III) cytochrome P450 is low spin with an unusual narrow splitting of the g-tensor.[29] The Soret band of this species peaks at 417 nm and by a comparison with heme models[30] as well as from the X-ray structure of P450 proteins[31,32] it contains water at the sixth coordination site. Upon addition of their substrates, which are generally lipophilic, the spectra of P450 monooxygenases change dramatically: EPR shows that the blue shift of about 26 nm in the Soret band is due to the generation of a high spin five-coordinated iron center.[33] Upon reduction by sodium dithionite in the presence of carbon monoxide, the Soret band shifts to 450 nm. Denaturation of the protein or protonation of the thiolate ligand leads to inactive "P420" enzymes with Soret bands at 420 nm where they are located in most hemoproteins.

3. AMINO ACID SEQUENCES

According to their spectral data, PGI- and TxA_2-synthase clearly belong to the heme-thiolate superfamily of enzymes, which includes an ever-growing number of more than sevenhundred monooxygenases.[34,35] According to the proposed nomenclature, different families of hemethiolate proteins are defined when the amino acid sequence homology is less than 40%, an arbitrarily chosen value.

The sequences of these synthases were recently established in cooperation with the group of T. Tanabe in Osaka.[36] Human TxA_2-synthase has 533 amino acids, with homolo-

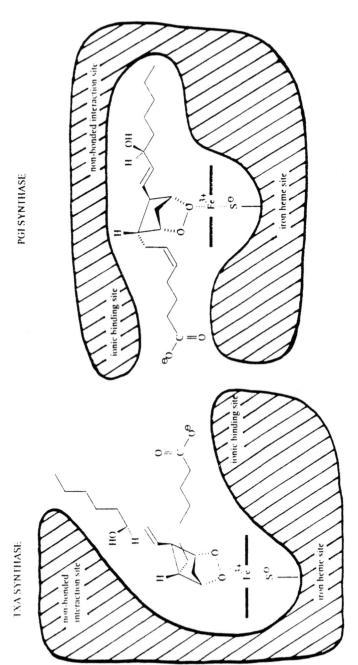

Scheme 2. Geometry of active sites of PGI_2- and TxA_2-synthase as proposed from inhibitor studies.

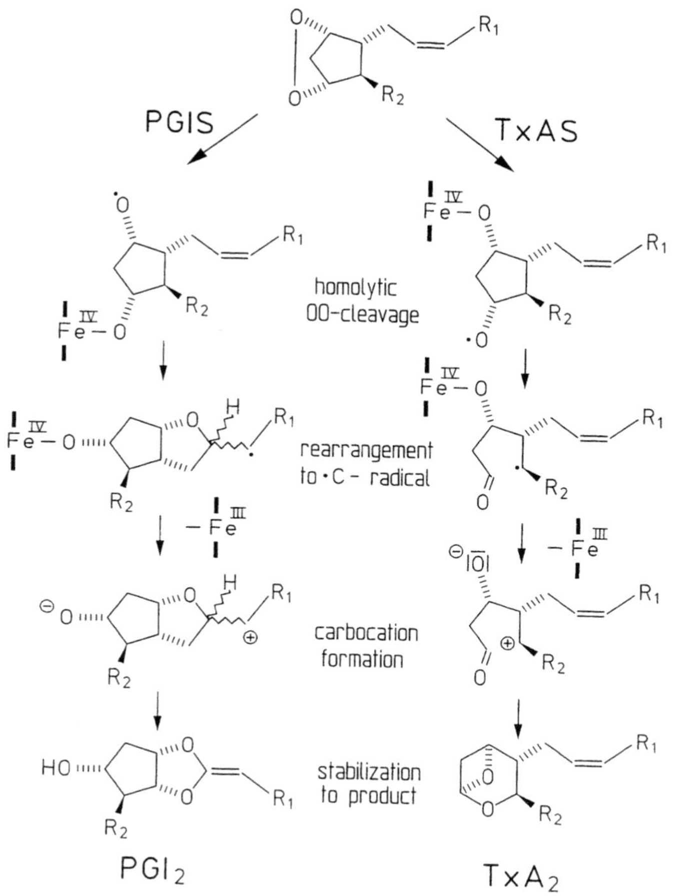

Scheme 3. Proposed isomerization mechanism of PGH$_2$ to PGI$_2$ and TxA$_2$.

gies to several P450 enzymes but not exceeding 36% as established with human nifedipine monooxygenase (P450 3A4).

Cysteine 479 could be identified by alignment as the ligand likely to coordinate to the heme group, and the 10 amino acids toward the N-terminus from this cysteine were found to be identical with those of human steroid 17α-monooxygenase from adrenals. Thus the P450 or hemethiolate nature could also be verified by the sequence data, and TxA$_2$-synthase can be considered as the first and so far only member of a new P450 family, called CYP5.

The sequence of PGI-synthase was expected to be highly interesting, since the use of the same substrate as TxA$_2$-synthase suggested a close relationship. The 500 amino acids exhibited a moderate homology of 31% with human cholesterol 7α-monooxygenase.[37] Since only one cysteine was present close to the C terminus, the thiol group that binds heme must belong to this amino acid. The sequence around the heme binding site had some homology with several other P450 monooxygenases and with TxA$_2$-synthase. Very surprisingly however, the total homology between PGI- and TxA$_2$-synthase was only 16%, which is close to a statistical value. Therefore PGI-synthase also forms a new family

Table 1. Effects of the PGH2 analog, U46619, on inhibition of prostacyclin synthase by peroxynitrite, and tetranitromethane

| | | % Inhibition | |
| | | With U46619 | |
Agent	Without U46619	10 μM	100 μM
Peroxynitrite			
1 μM	76.5 ±7.8	8.5 ± 10.7	1.7 ± 3.4
10 μM	87.4 ± 7.6	32.4 ± 11.1	16.4 ± 5.9
Tetranitromethane			
10 μM	22.7 ± 8.1	2.5 ± 1.5	1.1 ± 0.7
100 μM	67.3 ± 9.7	30.2 ± 9.3	15.4 ± 9.7
1000 μM	96.8 ± 2.1	54.5 ± 9.5	27.3 ± 6.7

Purified PGIS (113 pmol/ml) was incubated with the given concentrations of peroxynitrite, and tetranitromethane for 15 min in 0.1 ml of 100 mM KPi buffer, pH 7.4. U46619 was added 10 min before the incubation. The reaction was started by dilution to 1 ml with 100 mM KPi buffer, pH 7.4 and addition of 100 μM ^{14}C-PGH$_2$ and then incubated for 3 min with shaking. 6-keto-PGF$_{1\alpha}$ was extracted and analyzed after TLC-separation. PGIS activity was expressed as % inhibition compared with untreated enzyme. The data (means ± S.E.M.) represent 10 samples from 3 assays.

within the hemethiolate proteins, (CYP8) and no close phylogenetic relationship seems to exist between both synthases despite their functional relationship. This was puzzling, but since the enzymes generate two different products from the same substrate, the clue to the unexpected differences could reside in a different geometry of the peptide chains at the active site. Indeed, inhibitor studies indicate large differences in the substrate binding site, which tentatively allowed to depict the two active sites with their bound substrate in the following two conformations:

4. MECHANISTIC CONSIDERATIONS

The cartoon shows the substrate PGH$_2$ bound to the ferric heme iron coordinating to the endoperoxide group but differently to the two oxygen atoms. This could be concluded from spectral studies with PGH$_2$ analogues.[25] Starting out with the energetically most favored homolytic cleavage of the 0–0 bond as the subsequent step one can formulate the following sequences which are in line with earlier proposals and model studies.[38]

As the main characteristic features both pathways contain a conversion of a carbon-centered radical to a carbocation by electron transfer to the formally iron(IV)-heme, also called Compound II in heme biochemistry. We assume that the thiolate ligand plays a crucial role in this electron transfer but electronic calculations still have to prove this hypothesis.

5. REGULATION OF TxA$_2$- AND PGI$_2$-SYNTHASE ACTIVITY

Unlike the monooxygenases for drugs and xenobiotics no direct inducers for both enzymes are known. It rather appears that the constitutive expression of the enzyme is part of a differentiation program which comprises a series of enzymes required for the specific function of the cell. Likewise in human monocytes treatment with activin A and vitamin

D_3 causes an increase in TxA_2-synthase activity.[39] No corresponding studies with prostacyclin synthase have been reported but it seems that endothelial cells contain it constitutively. On the other hand posttranslational modifications of both, TxA_2-synthase and PGI-synthase, are likely and have been reported with hydroperoxides of unsaturated fatty acids. TxA_2-synthase can be inhibited by 12-hydroperoxy arachidonic acid, which is formed in platelets parallel to TxA_2 and therefore this reaction may have physiological significance.[40] A similar effect is known for 12- and 15-hydroperoxy arachidonic acid on the activity of PGI_2-synthase.[41] According to unpublished observations the underlying mechanism should be an oxidation of the thiolate ligand at the iron since the reduced CO-binding spectrum of the enzyme shows a shift to the P420 absorption. Since 5-hydroperoxy arachidonic acid does not cause this conversion one may anticipate a binding of these hydroperoxides to the active site and an oxygen transfer to the iron which will depend on a proper orientation of the peroxy group towards the iron. The resulting complex may be an equivalent to the oxenoid complex in P450 monooxygenases (also called Compound I in heme biochemistry) which is a potent oxidant and hydroxylating species but without a suitable acceptor may oxidize the adjacent thiolate ligand. An inhibition we also observed with hypochlorite which with an IC_{50} value of about 5 μM blocks PGI_2-synthase activity.[42] Under inflammatory conditions with activated leukocytes accumulating at the endothelial surface this inhibition may well be of physiological significance.

When screening for other oxidizing species we concentrated on possible inhibitions of PGI_2-synthase since this would affect endothelial function in a negative way whereas TxA_2-synthase inhibition would have no deleterious consequences. During these studies we found in peroxynitrite a highly potent inhibitor.[42]

6. PEROXYNITRITE AS A SELECTIVE ENDOGENOUS INHIBITOR FOR PGI$_2$-SYNTHASE

It was surprising that the reactive nitrogen monoxide (or nitric oxide, $\cdot N = O$) radical was identified as a mediator in many physiological processes. Produced in endothelial cells it exhibits similar actions as PGI_2 but unlike PGI_2 it does not increase cAMP in its target cells but cGMP through an activation of guanylyl cyclase.[43] The result is a profound relaxation of smooth muscle, an antiaggregatory action on platelets and a strong inhibition of leukocyte adhesion to the endothelium. It should be kept in mind that PGI_2 has the same spectrum of effects, so that through different mechanisms PGI_2 and NO work together synergistically to the benefit of a stable vascular tone.

NO is a radical and readily combines with oxygen radicals and dioxygen itself. Interestingly, the reaction with oxygen is termolecular and third order according to the equation

$$2\,NO + O_2 \rightarrow 2\,NO_2$$

and hence at the low physiological concentrations of NO it shows a rather long lifetime of 10–20 min. On the other hand superoxide combines with NO in an almost diffusion controlled reaction to give peroxynitrite[44]

$$\cdot NO + \cdot O_2^- \rightarrow \,^-OO\text{-}N = O$$

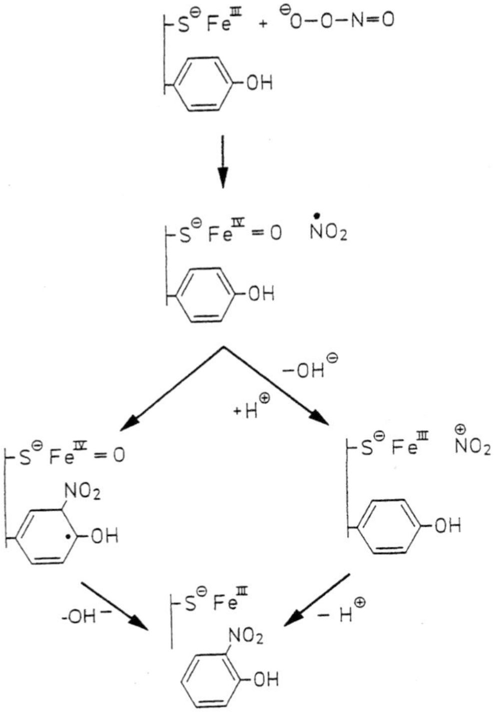

Scheme 4. Proposed nitration reactions for PGI$_2$-synthase.

Peroxynitrite can also be synthesized chemically from H$_2$O$_2$ and sodium nitrite and is stable under alkaline conditions. Upon addition of a proton, however, with a pK of 6.8 it is converted to the hydroperoxy compound which in a still undefined rearrangement forms a potent oxidizing and nitrating intermediate.[45] It is unavoidable that a small percentage of the cellular oxygen consumption forms superoxide so that by assuming a permanent endothelial production of NO peroxynitrite must be present in small concentrations. Pathological situations, like inflammation or ischemia-reperfusion, are known to increase superoxide generation dramatically and hence it was of interest whether peroxynitrite could influence the activity of PGI$_2$-synthase.

To our surprise we observed a strong inhibition already at 1 μM with an IC50-value as low as 60 nM.[42] This compares favorable with the inhibition by hypochlorite and also is completed already after a few seconds whereas hypochlorite inhibition develops over 1–2 min. When measuring a possible conversion to P420 we found that unlike hypochlorite peroxynitrite did not affect the heme-thiolate group. But also the oxidation of other sulfhydryl groups could not be the cause of inhibition since PGI$_2$-synthase only contains one cysteine, which provides the heme binding ligand. These results agree with the lack of inhibition of TxA$_2$-synthase indicating that at least at low concentrations up to 10 μM no conversion to P420 does occur.

What then is the molecular basis underlying the inactivation of PGI$_2$-synthase by peroxynitrite? It was easy to show by dilution experiments that this inhibition was irreversible indicating a chemical modification at an amino acid residue. This left nitration as a

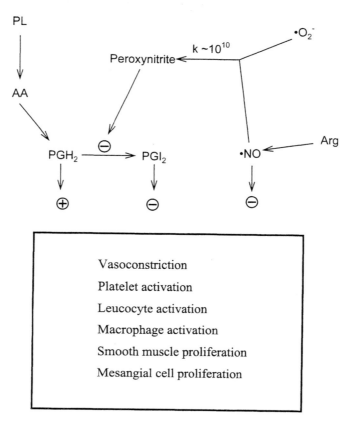

Scheme 5. Possible pathophysiological consequences of superoxide formation in the vascular system.

possible mechanism and indeed the wellknown tyrosine nitrating agent tetranitromethane could also block the activity.[46]

Additional information was obtained by the use of U46619, a substrate analog of PGH_2 known to bind to the active site without being metabolized. Peroxynitrite inhibition was now largely abolished, as was the inhibition by tetranitromethane. In contrast the oxidation by hypochlorite was not affected in agreement with the supposed action at the thiolate ligand. More evidence for a tyrosine nitration at the active site was provided by a positive reaction with an antibody against nitrotyrosine.[46] The Western blot showed a positive stain at the position of the purified enzyme and U46619 could decrease this staining. In addition HPLC analysis of the hydrolyzed enzyme revealed the presence of 3-nitrotyrosine and thus confirms nitration of a tyrosine as the most likely mechanism of inhibition.[46]

The question was still left open why this tyrosine residue showed such an exclusive reactivity. Since its nitration blocks enzyme activity it seemed to be essential in either the substrate binding or the catalytic mechanism. With regard to the latter one can envisage a role of a tyrosinate residue in abstracting a proton from the postulated cationic intermediate (see Scheme 3).

More important, however, and as a very likely explanation the ferric iron of the enzyme could react with the peroxide bond under homolytic cleavage to yield the nitrogen dioxide radical and a ferryl species (Compound II in heme chemistry). The nitration reaction could then proceed by electron transfer from the NO_2-radical to the ferryl species generating the chemically wellknown $NO_2^{\oplus}$-like intermediate as a strong electrophile. Instead, addition of the NO_2-radical to the tyrosine ring could lead to a radical adduct which then could be oxidized by the ferryl ion:

Irrespective of the pathway, both mechanisms require electron transfer from a radical intermediate to form a cationic intermediate, exactly as we have postulated for P450 or heme-thiolate enzymes.[38]

In order to support this general capability of heme-thiolate proteins to react with peroxynitrite in the proposed way we extended these investigations to other P450's and indeed could establish a very rapid reaction of $P450_{NOR}$[47] with peroxynitrite (M. Mehl, A. Daiber, V. Ullrich, unpublished). In this case no tyrosine nitration was observed but a nitration of phenol when added to the reaction mixture (A. Daiber, unpublished). Thus the same reaction scheme can be applied for this P450 model except that phenol as an exogenous substrate substitutes for the endogenous tyrosine residue.

7. PHYSIOLOGY OF THROMBOXANE A_2 AND PROSTACYCLIN

TxA_2 and PGI_2 act as a yin-yang system in the regulation of vascular tone.[48] PGI_2 is supported in its vasodilatory effects by nitric oxide ($^{\cdot}NO$) which is formed in the endothelium from arginine by way of a heme-thiolate dependent hydroxylation and oxidation process.[49,50] The effects of NO and PGI_2 are additive or even synergistic and independent since the receptor of PGI_2 transduces its signal to the smooth muscle by cAMP[51] whereas NO binds to the soluble guanylyl cyclase and generates cGMP.[52] Relaxation is a consequence through phosphorylation cascades or gating mechanisms affecting the contractile system of the smooth muscle.

Our findings with peroxynitrite provide a new and exciting new scenario for ischemic and inflammatory conditions. It is known that superoxide radicals are formed after an irreversible phase of ischemia by either xanthine oxidase[53] or still unknown mechanisms of mitochondrial autoxidations. Similarly, inflammation produces superoxide through activated leukocytes and the induction of the inducible NO-synthase in several cell types provides sufficient NO to generate peroxynitrite. A critical situation arises when superoxide production exceeds the rate of NO synthesis since the extremely rapid recombination of both radicals eliminates NO until very low levels and the relaxation by cGMP ceases. Not only NO as potent vasorelaxant would thus be eliminated but also PGI_2 starts to decrease progressively by the inhibitory action of peroxynitrite on PGI_2-synthase.

This allows vasoconstricting mediators like endothelins, thromboxane A_2, leukotriene C_4 or platelet activating factor to close the affected vessels which leads to a vicious cycle of less oxygen supply, more ischemia and cell death of first the endothelium and then the surrounding tissue. Although defense mechanisms even against this situation will be activated (like adenosine release) in severe cases the circulation will stop completely and cause organ dysfunction followed by death of the organism. Further work will be directed towards a proof for this hypothesis and pharmacological tools will be tested to interfere with the destructive action of peroxynitrite on PGI_2-synthesis.

REFERENCES

1. K.I. Williams, and G.A. Higgs, 1988, Eicosanoids and inflammation, *J. Pathol.* **156**: 101–110.
2. S. Moncada, and J.R. Vane, 1978, Pharmacology and endogenous roles of prostaglandin endoperoxides, thromboxane A_2, and prostacyclin, *Pharmacol. Rev.* **30**: 293–331.

3. S. Bergström, L.A. Carlson, and J.R. Weeks, 1968, The prostaglandins: a family of biological active lipids, *Pharmacol. Rev.* **20(1):** 1–48.

4. W.L. Smith, 1989, The eicosanoids and their biochemical mechanism of action, *Biochem. J.* **259(2):** 315–324.

5. O. Hayaishi, 1988, Sleep-wake regulation by prostaglandins D_2 and E_2, *J. Biol. Chem.* **263(29):** 14593–14596.

6. W.D. Watkins, M.B. Peterson, J.R. Fletscher (eds.), 1989, *Prostaglandins in clinical practice*, Raven Press, New York.

7. K. Schrör, 1984, *Prostaglandine und verwandte Verbindungen*, Thieme, Stuttgart, New York.

8. B. Samuelsson, 1983, From studies of biochemical mechanism to novel biological mediators: prostaglandin endoperoxides, thromboxanes, and leukotrienes. Nobel Lecture, 8 December 1982, *Biosci. Rep.* **3(9):** 791–813.

9. R. A. Johnson, 1985, Synthesis of thromboxanes, prostacyclin, and the endoperoxides, *Adv. Prostaglandin, Thromboxane and Leukotriene Res.* **14:** 131–154.

10. P. Needleman, J. Turk, B.A. Jakschik, A.R. Morrison, J.B. Lefkowith, 1986, Arachidonic acid metabolism, *Ann. Rev. Biochem.* **55:** 69–102.

11. W.L. Smith, L.J. Marnett, D.L. DeWitt, 1991, Prostaglandin and thromboxane biosynthesis, *Pharmac. Ther.* **49(3):** 153–179.

12. M. Hamberg, J. Svensson, B. Samuelsson, 1975, Thromboxanes: a new group of biologically active compounds derived from prostaglandin endoperoxides, *Proc. Nat. Acad. Sci. USA*, **72:** 2994–2998.

13. N. Witthaker, S. Bunting, J. Salmon, S. Moncada, J.R. Vane, R.A. Johnson, D.R. Morton, J.H. Kinner, R.R. Gorman, J.C. McGuire, and F.F. Sun, 1976, The chemical structure of prostaglandin X (prostacyclin), *Prostaglandins*, **12(6):** 915–928.

14. U. Diczfalusy, P. Falardeau, S. Hammarström, 1977, Conversion of prostaglandin endoperoxides to C17-hydroxy acids catalyzed by human platelet thromboxane synthase, *FEBS*, **84(2):** 271–274.

15. J. Fried, J. Barton, 1977, Synthesis of 13,14-dehydroprostacyclin methyl ester: a potent inhibitor of platelet aggregation, *Proc. Natl. Acad. Sci. USA*, **74(6):** 2199–2203.

16. A.D. Rahimtula, P.J. O'Brien, 1974, Hydroperoxide catalyzed liver microsomal aromatic hydroxylation reactions involving cytochrome P-450, *Biochem. Biophys. Res. Commun.* **60(1):** 440–447.

17. F. Lichtenberger, W. Nastainczyk, and V. Ullrich, 1976, Cytochrome P450 as an oxene transferase, *Biochem. Biophys. Res. Commun.* **70:** 939–946.

18. V. Ullrich, and H. Graf, 1984, Prostacyclin and thromboxane synthase as P-450 enzymes, *Trends Pharmacol. Sci.* **5(8):** 352–355.

19. S. Moncada, R. Gryglewski, S. Bunting, J. R. Vane, 1976, An enzyme isolated from arteries transforms prostaglandin endoperoxides to an unstable substance that inhibits platelet aggregation, *Nature* **263:** 663–665.

20. P. Wlodawer, S. Hammarström, 1979, Some properties of prostacyclin synthase from pig aorta, *FEBS* **97:** 32–36.

21. S. Bunting, R. Gryglewski, S. Moncada, J.R. Vane, 1976, Arterial walls are protected against deposition of platelet thrombi by a substance (prostaglandin X) which they make from prostaglandin endoperoxides, *Prostaglandins* **12 (5):** 685–713.

22. V. Ullrich, L. Castle, and P. Weber, 1981, Spectral evidence for the cytochrome P450 nature of prostacyclin synthase, *Biochem. Pharmacol.* **30(14):** 2033–2036.

23. H. Graf, and V. Ullrich, 1982, *Cytochrome P450, Biochemistry, Biophysics and Environmental Implications* (eds.: E. Hietanen, M. Laitinen, O. Hanninen), Elsevier Biomedical Press B.V., New York, 103–106.

24. D.L. DeWitt, and W.L. Smith, 1983, Purification of prostacyclin synthase from bovine aorta by immunoaffinity chromatography. Evidence that the enzyme is a hemoprotein, *J. Biol. Chem.* **258(5):** 3285–3293.

25. M. Hecker, and V. Ullrich, 1989, On the mechanism of prostacyclin and thromboxanes A_2 biosynthesis, *J. Biol. Chem.* **264(1):** 141–150.

26. L. Zhang, M.B. Chase, and R.-F. Shen, 1993, Moleclular cloning and expression of murine thromboxane synthase, *Biochem. Biophys. Res. Commun.* **194(2):** 741–748.

27. R.F. Shen, and H.H. Tai, 1986, Immunoafffinity purification and characterization of thromboxane synthase from porcine lung, *J. Biol. Chem.* **261(25):** 11592–11599.

28. S. Hammarström, and P. Falardeau, 1977, Resolution of prostaglandin endoperoxide synthase and thromboxane synthase of human platelets, *Proc. Nat. Acad. Sci. USA* **74(9)** 3691–3695.

29. M. Haurand, and V. Ullrich, 1985, Isolation and characterization of thromboxane synthase from human platelets as a cytochrome P-450 enzyme, *J. Biol. Chem.* **260(28):** 15059–15067.

30. S.C. Tang, S. Koch, G.C. Papaefthymiou, S. Foner, R.B. Frankel, J.A. Ibers, and R.H. Holm, 1976, Axial ligation modes in iron(III) porphyrins. Models for the oxidized reaction states of cytochrome P-450 en-

zymes and the molecular structure of iron(III) protoporphyrin IX dimethyl ester p-nitrobenzenethiolate, *J. Am. Chem. Soc.* **98(2):** 2414–2434.

31. T.L. Poulos, B.C. Finzel, and A.J. Howard, 1987, High resolution crystal structure of cytochrome P450$_{CAM}$, *J. Mol. Biol.* **195:** 697–700.

32. K.G. Ravichandran, S.S. Boddupalli, C.A. Haserman, J.A. Peterson, and J. Deisenhofer, 1993, Crystal structure of hemoprotein domain of P450$_{BM-3}$, a prototype for microsomal P450's, *Science* **261(5122):** 731–736.

33. P. Devaney, G.C. Wagner, P.G. Debrunner, and I.C. Gunsulus, 1980, Single crystal ESR of cytochrome P450$_{CAM}$ from pseudomonas putida, *Fed. Proc.* **39:** 1139–1142.

34. D.W. Nebert, D.R. Nelson, M.J. Coon, R.W. Estabrook, R. Feyereisen, Y. Fujii-Kuriyama, F.J. Gonzalez, F.P. Guengerich, I.C. Gunsalus, E.F. Johnson, J.C. Loper, R. Sato, M.R. Waterman, and D.J. Waxman, 1991, The P450 superfamily: Update on new sequences, gene mapping, and recommended nomenclature, *DNA Cell. Biol.* **10:** 1–14.

35. P.R. Ortiz de Montellano (ed.), 1995, *Mechanism, and Biochemistry*, Plenum Press, New York, London.

36. T. Tanabe, C. Yokoyama, A. Miyata, H. Ihara, T. Kosaka, K. Suzuki, Y. Nishikawa, T.Yoshimoto, S. Yamamoto, R. Nüsing, and V. Ullrich, 1993, Molecular cloning and expression of human thromboxane synthase, *J. Lip. Med.* **6:** 139–144.

37. S. Hara, A. Miyata, C. Yokoyama, H. Inoue, R. Brugger, F. Lottspeich, V. Ullrich, and T. Tanabe, 1994, Isolation and molecular cloning of prostacyclin synthase from bovine endothelial cells, *J. Biol. Chem.* **269(31):** 19897–19903.

38. M. Kishi, 1986, 11,12-Secoprostaglandin cyclization model in thromboxane A$_2$ biosynthesis, *J. Chem. Soc. Chem. Commun.*, 885–887.

39. R.M. Nüsing, S. Mohr, and V. Ullrich, 1995, Activin A and retinoic acid synergize in cyclooxygenase-1 and thromboxane synthase induction during differentiation of J774.1 macrophages, *Eur. J. Biochem.* **227:** 130–136.

40. D. Aharony, J.B. Smith, and M.J. Silver, 1982, Regulation of arachidonate-induced platelet aggregation by the liopxygenase, *Biochim. Biophys. Acta* **718(2):** 193–200.

41. B. Mayer, R. Moser, N. Gleispach, and H.R. Kukovetz, 1986, Possible inhibitory function of endogenous 15-hydroperoxy-eicosatetraenoic acid on prostacyclin formation in bovine aortic endothelial cells, *Biochim. Biophys. Acta* **875:** 641–653.

42. M.-H. Zou, and V. Ullrich, 1996, Peroxynitrite formed by simultaneous generation of nitric oxide and superoxide selectively inhibits bovine aortic prostacyclin synthase, *FEBS Lett.* **382:** 101–104.

43. S. Moncada, R.M.J. Palmer, and E.A. Higgs, 1991, Nitric oxide: physiology, pathophysiology, and pharmacology, *Pharmacol. Rev.* **43:** 109–142.

44. J.S. Beckman, 1996, Oxidative damage and tyrosine nitration from peroxynitrite, *Chem. Res. Toxicol.* **9:** 836–844.

45. K.N. Honk, K.R. Condroski, W.A. Pryor, 1996, Radical and concerted mechanisms in oxidations of amines, sulfides, and alkenes by peroxynitrite, peroxynitrous acid, and the peroxynitrite-CO$_2$ adduct: density functional theory transition structures and energetics.

46. M.-H. Zou, C. Martin, and V. Ullrich, 1997, Tyrosine nitration as a mechanism of selective inactivation of prostacyclin synthase by peroxynitrite, *Biol. Chem.* **378:** 707–713.

47. K. Nakahara, T. Tanimoto, K. Hatano, K. Usuda, and H. Shoun, 1993, Cytochrome P-450 55A1 (P-450 dNIR) acts as nitric oxide reductase employing NADH as the direct electron donor, *J. Biol. Chem.* **268:** 8350–8355.

48. S. Bunting, S. Moncada, and J.R. Vane, 1983, The prostacyclin-thromboxane A$_2$ balance:Pathophysiological and therapeutic implications, *Br. Med. Bull.* **39:** 271–276.

49. J.M. Hevel, K.A. White, and M.A. Marletta, 1991, Purification of the inducible murine macrophage nitric oxide synthase. Identification as a flavoprotein, *J. Biol. Chem.* **266:** 22789–22791.

50. K.A. White, and M.A. Marletta, 1992, Nitric oxide synthase is a cytochrome P450-type haemoprotein, *Biochemistry* **31:** 6627–6631.

51. S. Moncada, and V.R. Vane, 1979, Pharmacology and endogenous roles of prostaglandins, endoperoxides, thromboxane A$_2$ and prostacyclin, *Pharmacol. Rev.* **30:** 293.

52. S.A. Waldman, and F. Murad, 1987, Cyclic GMP synthesis and function, *Pharmacol. Rev.* **39:** 163–196.

53. S.A. Sanders, R. Eisenthal, and R. Harrison, 1997, NADH oxidase activity of human xanthine oxidoreductase. Generation of superoxide anion, *Eur. J. Biochem.* **245(3):** 541–548.

20

CELL ENGINEERING OF THE RAT AND HUMAN CYSTEINE CONJUGATE BETA LYASE GENES AND APPLICATIONS TO NEPHROTOXICITY ASSESSMENT

G. Gordon Gibson, Peter S. Goldfarb, Laurie J. King, Ian Kitchen,
Nick Plant, Claire Scholfield, and Helen Harries

Molecular Toxicology Group, School of Biological Sciences
University of Surrey, Guildford
Surrey GU2 5XH, England, United Kingdom

1. ABSTRACT

Halogenated alkenes such as hexachlorobutadiene (HCBD) are initially metabolised by glutathione conjugation, subsequent metabolic processing of which yields the cysteine conjugate. This cysteine conjugate subsequently serves as a substrate for the kidney enzyme cysteine conjugate beta lyase (CCBL), which is responsible for the production of nephrotoxic thiol metabolites.

We have used an antibody to CCBL to screen, isolate and sequence the full length rat cDNA, the latter being subesquenlty used to isolate the corresponding full length human cDNA. We have also produced cDNA riboprobes for CCBL and have developed an in situ hybridisation, video-based computer densitometry technique to demonstrate the specific localisation of CCBL mRNA to the P_3 segment of the proximal convoluted tubule, a location that is identical to that of HCBD-induced nephrotoxicity (necrosis).

The availabilty of both rat and human full length cDNAs has allowed us to isolate panels of transformed LLCPK-1 cells (pig kidney derived) that have stably integrated the cDNAs, thus allowing a comparison of the enzyme kinetics and substrate-induced cytotoxicity of the rat and human enzymes. We have therefore developed a platform for a mechanistically-relevant risk assessment of the nephrotoxicity of xenobiotic cysteine conjugates to man.

2. INTRODUCTION

There are many factors which predispose a particular tissue to the toxicity of xenobiotics or their metabolites and many compounds are targeted to the kidney.[1] Reasons for

Molecular and Applied Aspects of Oxidative Drug Metabolizing Enzymes,
edited by Arinç *et al.* Kluwer Academic / Plenum Publishers, New York, 1999.

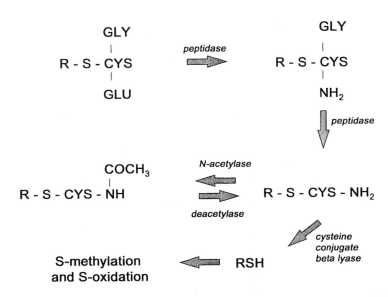

Figure 1. Metabolism of HCBD. R represents a halogenated alkene xenobiotic such as HCBD.

this selectivity may include the fact that the kidney receives 25% of the cardiac output, consists of multiple cell types, has the ability to concentrate blood-borne xenobiotics by active uptake mechanisms, the presence of enzymes of xenobiotic metabolism and its role as a major excretory organ. Several types of xenobiotic-induced kidney lesions are known including proximal tubule necrosis (HCBD), distal tubule necrosis (cisplatin), effects on the glomerular filtration rate/blood flow (cyclosporin A) and medullary toxins (paracetamol).[2] The kidney has several protection mechanisms against biologically reactive xenobiotics, a major one being conjugation with the tripeptide glutathione (Glu-Cys-Gly).[3] The glutathione conjugate may be excreted as such, or more usually be further processed (Figure 1) to the cysteine conjugate, which in turn, may be N-acetylated to the mercapturic acid and excreted.[4] Alternatively, cysteine conjugates may be metabolised by the kidney enzyme CCBL producing nephrotoxicity in the P₃ segment.[6]

    CCBL is a soluble enzyme found in the kidney, liver, GI tract, brain, bacteria and plants, has a monomeric molecular weight of approximately 48 kd and has pyridoxal phosphate as its prosthetic group.[7] In addition to beta lyase activity, the enzyme also exhibits transaminase activity,[8] i.e. two enzyme activities residing in the same protein molecule. The mechanism of nephrotoxicity of halogenated alkenes such as HCBD is summarised in Figure 2 and is the result of specific kidney uptake by the anionic transporter to the P₃ segment of the proximal convoluted tubule. In many cases, reactive thiols are the product of CCBL catalysis and these active metabolites are thought to modulate intracellular calcium homeostasis, impair mitochondrial respiration and the membrane potential, eventually leading to tissue necrosis, approximately 1 to 2 days after xenobiotic administration.[9]

    Although the above metabolic pathways and mechanism of toxicity have been reasonably well described in the rat and mouse,[10] the picture is not so clear for man. The perceived wisdom is that man is not susceptible to CCBL-dependent nephrotoxicty because of the low level of kidney enzyme expression, although the literature reports are somewhat contradictory. Accordingly, we decided to address the question of the human susceptibility to potential nephrotoxic metabolites generated by human CCBL, by cloning the

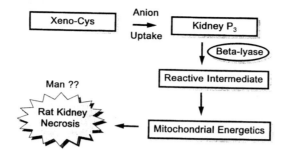

Figure 2. Molecular mechanism of beta lyase-mediated nephrotoxicity.

human cDNA and engineering cell lines that express CCBL activity, as a prelude to an in vitro toxicity evaluation.

3. CLONING OF RAT AND HUMAN cDNAs AND CELL TRANSFORMATION

When we first embarked on this programme of work, no CCBL gene had been isolated and hence we adopted the experimental approach outlined in Figure 3, which resulted in the isolation of full length cDNAs encoding the rat[11] and human[12] enzymes, both of which exhibited CCBL activity when transiently transfected into COS-1 cells (data not shown).

Although transient transfection of cDNAs in the appropriate expression plasmid is a useful way to examine the cognate enzyme activity, it somewhat time-consuming and we therefore decided to develop cell lines that had stably incorporated either the rat or human cDNAs into the genome of the LLCPK-1 cells. This cell line is pig kidney, epithelium derived and we chose it because it retains many of the phenotypic characteristics of the parent cell and additionally because it has been extensively used in the in vitro cytotoxicity testing of xenobiotics. Stable incorporation was achieved by inserting the cDNA, the human cytomegalovirus promoter and the E.Coli XPT gene in the expression plasmid pUS1000. This construct was transfected into LLCPK-1 cells and stable transformants were obtained by clonal selection in HAT medium/mycophenolic acid. As shown in Figure 4, several cell lines were isolated which expressed CCBL activity. Moreover, several cell lines of graded activity were isolated, presumably as a result of the genomic incorporation of multiple cDNA copies. These cell lines are very stable and retain enzyme activity up to at least 40 passages in cell culture, thereby providing a plentiful and enriched supply of transformed cells to compare the relative CCBL enzyme kinetics and cytotoxicity produced by these enzymes.

4. COMPARATIVE ENZYME KINETICS AND CYTOTOXICITY

As shown in Table 1, the enzyme kinetics determined in the stably transformed cells were very similar for the rat and human, indicating that there were no substantial intrinsic differences between the enzymes. This information is important because it demonstrates that the human enzyme has the *potential* to activate xenobiotic cysteine conjugates. Whether this potential is realised or not has to be tested by experimentation and we therefore compared the relative abilities of the rat and human cell lines to produce cytotoxicity in the presence of a xenobiotic cysteine conjugate, as assessed by cell survival in culture.

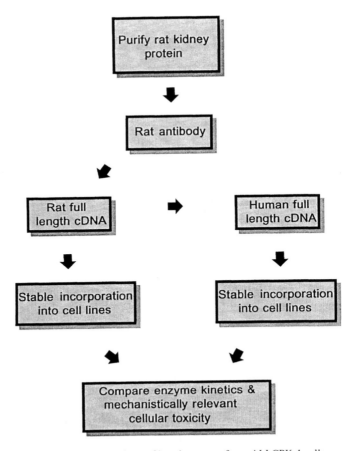

Figure 3. Isolation and use of beta lyase transformed LLCPK-1 cells.

The cytotoxicity of S-(1,2-dichlorovinyl)-L-cysteine (DCVC) in the transformed cell lines is shown in Figure 5 and it is clear that the cells transformed with the human enzyme are more susceptible to DCVC-induced cytotoxicity than those of the rat, with IC50 values of 0.1 and 0.2 mM for the human and rat respectively. The wild type cells were resistant to DCVC cytotoxicity, in keeping with the low level of CCBL activity in the parent, untransformed cell line. This observation is not inconsistent with the relative enzyme kinetic properties described above (Table 1).

5. IN SITU HYBRIDISATION STUDIES

Previous studies have shown that when HCBD is administered to the rat, localised necrosis occurs in the P_3 segment of the proximal convoluted tubule.[13] Notwithstanding the fact that this is the site of anionic uptake and hence would result in concentration of HCBD and its metabolites, we were interested to know if the P_3 segment was also the site of CCBL expression as an additional explanation for the site-specific nephrotoxicity. Accordingly, we developed an in situ hybridisation technique to localise CCBL mRNA in rat kidney sections as portrayed in Figures 6 and 7.

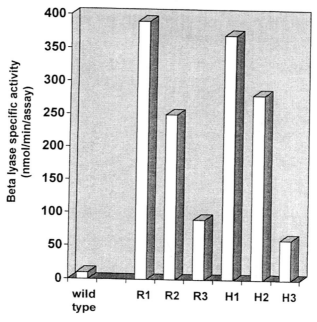

Figure 4. Cell lines stably expressing rat and human beta lyase cDNAs. Several cell lines were isolated by clonal selection and assayed for CCBL activity, derived from the rat [R] and human [H] cDNAs. Data are the average of 3 experiments, which did not deviate from each other by more than 10%.

Table 1. Kinetic properties of rat and human beta lyases in transfected LLCPK-1 cell lines

Substrate*	Km (mM)		Vmax (nmol/min/mg)	
	Rat	Human	Rat	Human
DCVC	0.5	0.7	6.4	11.6
TFEC	1.1	2.8	38.1	45.4
PCBC	0.8	0.9	8.3	8.6
CTFEC	1.6	2.6	30.4	46.6

*Beta lyase substrates used were DCVC (S-[1,2-dichlorovinyl]-L-cysteine), TFEC (S-[1,1,2,2-tetrafluoroethyl]-L-cysteine), PCBC (S-[1,2,3,4,4-pentachlorobutadi-enyl]-L-cysteine and CTFEC (S-[2-chloro-1,1,2-trifluoroethyl]-L-cysteine).
Data shown are the mean of 3 experiments, each in triplicate, with the replicates not differing by greater than 15 % from each other.

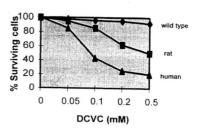

Figure 5. DCVC cytotoxicity in beta lyase engineered cell lines. Data shown are the mean of 3 experiments, each in triplicate, which did not differ from each other by more than 15%.

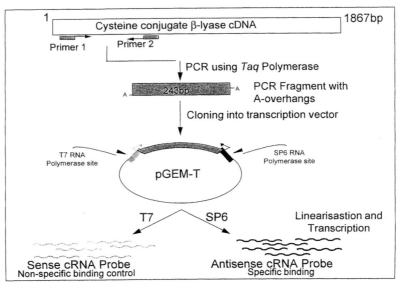

Figure 6. Preparation of cDNA probe for rat beta lyase mRNA.

As shown in Figure 8, the localisation of CCBL mRNA is exclusively sited in the P_3 segment (left panel, arrow) and after treatment with S-(1,1,2,2-tetrafluoroethyl)-L-cysteine (TFEC), the mRNA totally disappears, consistent with both the site of necrosis and the local destruction of CCBL mRNA generated by the cytotoxic TFEC beta lyase cleavage products. Thus the site-selective necrosis to the P_3 segment is due to a combination of both the specific kidney uptake and the presence of CCBL in this tubular region.

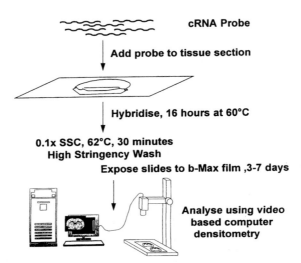

Figure 7. In situ hybridisation of rat kidney beta lyase mRNA.

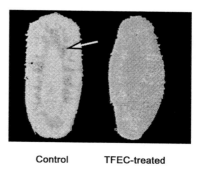

Figure 8. Localisation of beta lyase mRNA in control and TFEC-treated rat kidney by in situ hybridisation with a cRNA probe. The location of CCBL mRNA in untreated animals is indicated by the arrow (left section). Note the disappearance of the CCBL mRNA when rats were treated with TFEC (right section).

Control   TFEC-treated

6. REGULATION OF CCBL ACTIVITY

In light of the importance of CCBL in the nephrotoxicity of halogenated alkenes, an understanding of the regulation of the enzyme would be of considerable importance. Our initial attempts to induce CCBL activity in the rat with the classical inducers of mixed function oxidase activity (phenobarbitone, beta-naphthoflavone, ethanol, dexamethasone or clofibrate) resulted in no induction of the enzyme (data not shown). However, treatment of rats with a metabolite of HCBD (N-acetyl-S-{1,2,3,4,4-pentachloro-1,3-butadienyl}-L-cysteine) resulted in a small induction(approximately 1.5 to 2.0-fold) of CCBL activity.[13] Using different dosing protocols (altering the dose or repeat administration) did not increase the extent of induction and may best be explained by the induction per se and the concurrent necrotic effect of the xenobiotic.

In addition to its beta lyase activity, CCBL also exhibits transaminase activity in the kidney, specifically glutamine transaminase K[8], and therefore it would be instructive to examine the influence of a transaminase substrate on the enzyme catalysed beta lyase reaction. As shown in Figure 9, the presence of alpha-keto-gamma-methiolbutyrate (KMB) substantially increased beta lyase activity towards TFEC in LLCPK-1 cells transformed with the rat enzyme, which may be rationalised as demonstrated in Figure 10.

During the concurrent turnover of the beta lyase enzyme, the pyridoxal phosphate cofactor is converted to the pyridoxamine derivative, as a result of the transaminase reaction. The pyridoxamine form does not support the beta lyase reaction,[8] but the alpha-keto acids present in the cellular cytosol used would regenerate the active pyridoxal form of the

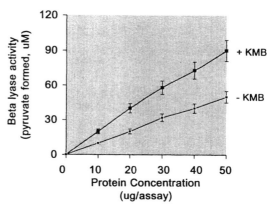

Figure 9. Influence of the alpha-keto acid KMB on cysteine conjugate beta lyase activity in LLCPK-1 cells transformed with rat beta lyase cDNA. Varying amounts of cytosolic protein from the R1 cell line (Figure 4) were cultured with TFEC (0.5mM), either in the absence or presence of KMB (0.2mM).

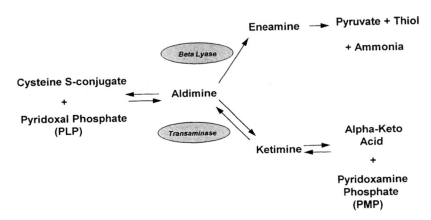

Figure 10. Concurrent transamination and beta lyase activities.

enzyme by transamination. The cellular supply of the appropriate alpha-keto acids are obiously limited and at relatively low concentrations and the transaminase reaction would therefore gradually switch off. However, the effect of adding a large excess of KMB (0.2 mM) would also regenerate the active pyridoxal form, as evidenced by the data in Figure 10. Therefore it would appear that the in vivo regulation of beta lyase activity is relatively complex with an on-going competition between cysteine conjugate and alpha-keto acid substrates, the outcome of which would be determined by the relative affinities of both substrates for CCBL. In addition, the regeneration of the active pyridoxal form of the enzyme would be dependent on the cellular pool of appropriate alpha-keto acids and hence the prevailing metabolic status of the proximal tubular cell.

7. SUMMARY

The transformed cell lines described herein provide a useful addition to our understanding of the risk associated with human exposure to halogenated alkenes. Whether this risk is perceived or real must await a more extensive examination of a wider range of cysteine conjugates, development of the in vitro system to encompass a toxicologically more specific endpoint rather than cell survival and a realistic assessment of the level of beta lyase expression in the human kidney.

REFERENCES

1. T.J.Monks, M.I.Rivera, J.J.W.M.Mertens, M.M.C.G.Peters and S.S.Lau, 1996, The kidney as a target for biological reactive intermediates, in : *Biological Reactive Intermediates V: Basic Mechanistic Research in Toxicology and Human Risk Assessment* (R.Snyder, J.J.Kocsis, I.G.Sipes, D.J.Jollow, H.Greim, T.J.Monks and C.M.Witmer, eds.), pp 203–212, Plenum Press, New York.
2. P.H.Bach, F.W.Bonner, J.W.Bridges and E.A.Lock (eds.), 1982, *Nephrotoxicity : Assessment and Pathogenesis*, Wiley, Chichester.
3. N.P.E.Vermeulen, G.J.Mulder, W.H.M.Peters and P.J.van Bladern (eds.), 1996, *Glutathione S-transferases : Structure, Function and Clinical Implications*, Taylor and Francis, London.
4. W.Dekant, 1996, Biosynthesis and cellular effects of toxic glutathione substrates, in : *Biological Reactive Intermediates V : Basic Mechanistic Research in Toxicology and Human Risk Assessment* (R.Snyder,

J.J.Kocsis, I.G.Sipes, D.J.Jollow, H.Greim, T.J.Monks and C.M.Witmer, eds.), pp 297–312, Plenum Press, New York.

5. W.Dekant, S.Vamvakas and M.W.Anders, 1994, Formation and fate of nephrotoxic and cytotoxic glutathione S-conjugates : cysteine conjugate beta lyase pathway, in : *Conjugation-dependent Carcinogenicity and Toxicity of Foreign Compounds*, (M.W.Anders and W.Dekant, eds.), pp 115–162, Academic Press, San Diego.

6. J.Ishmael and E.A.Lock, 1982, Necrosis of the pars recta (S3 segment) of the rat kidney produced by hexachlorobutadiene, *J.Pathol*, **138** : 99–113.

7. A.J.L.Cooper, Enzymology of cysteine conjugate beta lyase, in : *Conjugation-dependent Carcinogenicity and Toxicity of Foreign Compounds*, (M.W.Anders and W.Dekant, eds.), pp 71–114, Academic Press, San Diego.

8. L.H.Lash, R.M.Nelson, R.A.van Dyke and M.A.Anders, 1990, Purification and characterisation of human kidney cytosolic cysteine conjugate beta lyase activity, *Drug Metab.Dispn*, **18** : 50–54

9. A.Wallin, T.W.Jones, A.E.Vercesi, I.Cotgreave, K.Ormstad and S.Orrenius, 1987, Toxicity of S-pentachlorobutadienyl-L-cysteine studied with isolated rat renal cortical mitochondria, *Arch.Biochem.Biophys.*, **258** : 365–372.

10. E.A.Lock, 1988, Studies on the mechanism of nephrotoxicity and carcinogenicity of halogenated alkenes, *CRC Crit.Rev.Toxicol.*, **19**, 23–41.

11. S.Perry, C.Scholfield, M.MacFarlane, E.A.Lock, L.J.King, G.G.Gibson and P.S.Goldfarb, 1993, The isolation and expression of a cDNA coding for rat kidney cysteine conjugate beta lyase, *Mol.Pharmacol.*,**43** : 660–665.

12. S.Perry, H.Harries, C.Scholfield, E.A.Lock, L.J.King, G.G.Gibson and P.S.Goldfarb, 1995, Molecular cloning and expression of a cDNA for human kidney cysteine conjugate beta lyase, *FEBS Lett.*, **360** : 277–280.

13. M.MacFarlane, M.Scholfield, N.Parker, L.Roelandt, M.David, E.A.Lock, L.J.King, P.S.Goldfarb and G.G.Gibson, 1993, Dose-dependent induction or depression of cysteine conjugate beta lyase in rat kidney by N-acetyl-(1,2,3,4,4-pentachloro-1,3-butadienyl)-L-cysteine, *Toxicol.*, **77** : 133–144.

INDEX